R. Ferrario

Reliability, Availability, Maintainability and Safety Assessment

Volume 2

Reliability, Availability, Maintainability and Safety Assessment

Volume 2
Assessment, Hardware, Software and Human Factors

Alain Villemeur
Electricité de France
France

Main translators
Anne Cartier
and
Marie-Christine Lartisien

JOHN WILEY & SONS
Chichester • New York • Brisbane • Toronto • Singapore

Translated from Alain Villemeur's book 'Sûreté de fonctionnement des systèmes industriels' (Collection des Etudes et Recherches d'Electricité de France) published by Editions Eyrolles – Paris)
Revised and updated 1991

Main translators:-
Anne Cartier
and
Marie-Christine Lartisien

Copyright © 1992 by John Wiley & Sons Ltd
                    Baffins Lane, Chichester
                    West Sussex PO19 1UD, England

All rights reserved.

No part of this book may be reproduced by any means, or transmitted, or translated into a machine language without the written permission of the publisher.

Other Wiley Editorial Offices

John Wiley & Sons, Inc., 605 Third Avenue,
New York, NT 10158-0012, USA

Jacaranda Wiley Ltd., G.P.O. Box 859, Brisbane,
Queensland 4001, Australia

John Wiley & Sons (Canada) Ltd, 22 Worcester Road,
Rexdale, Ontario M9W IL1, Canada

John Wiley & Sons (SEA) Pte Ltd, 37 Jalan Pemimpin 05-04,
Block B, Union Industrial Building, Singapore 2057

Library of Congress Cataloguing-in-Publication Data:
Villemeur, Alain.
　　[Sûreté de fonctionnnement des systèmes industriels.　English]
　　Reliability, availability, maintainability & safety assessment : methods and techniques / Alain Villemeur.
　　　p.　cm.
　　Translation of: Sûreté de fonctionnement des sytèmes industriels.
　　Includes bibliographical references and index.
　　ISBN 0 471 93048 2 (v. 1) — ISBN 0 471 93049 0 (v. 2)
　　1. Reliability (Engineering)　I. Title.　II. Title: Reliability, availability, maintainability, and safety assessment.
TA169.V5513　1991
620'.00452—dc20　　　　　　　　　　　　　　　　91-14329
　　　　　　　　　　　　　　　　　　　　　　　　　CIP

A catalogue record for this book is available from the British Library

Typeset by Keytec Typesetting Ltd, Bridport, Dorset
Printed in Great Britain by Courier International, East Kilbride, Lanarkshire

This book is dedicated to:

	Marie-Dominique
	Julien
Odile	**Benjamin**
André	**Laurent**
	Corentin

CONTENTS

Foreword by Paul Caseau — xv
Foreword by Arnould d'Harcourt — xvii
Acknowledgements — ixx
Preface — xxi
Main Notations — xxv

PART 3 SPECIFIC METHODS — 365

15 Dependent and common-cause failures — 367
15.1 Introduction — 367
15.2 Interdependences between failures — 368
15.3 Examples of critical dependences among failures in the aeronautical, oil and nuclear industries — 370
15.4 Common-cause and cascade failures — 373
    15.4.1 History of these concepts and their introduction — 373
    15.4.2 Definitions — 374
    15.4.3 Dependent failures and primary, secondary and command failures — 375
15.5 Classification of common-cause failures according to their nature — 376
    15.5.1 Environmental hazards — 378
    15.5.2 Design errors — 379
    15.5.3 Manufacturing errors — 382
    15.5.4 Assembly errors — 383
    15.5.5 Operating errors — 383
15.6 Using event reports to study common-cause failures — 384
15.7 Dependent failure assessment methods — 387
    15.7.1 Main assessment methods — 388
    15.7.2 Other methods — 391
15.8 Dependent and common-cause failure probabilities — 394
    15.8.1 Explicit methods — 394
    15.8.2 Parametric methods — 395
15.9 Importance of these failures and means to prevent them — 400
References — 402

16 Human factors — 405

- 16.1 Introduction — 405
- 16.2 Historical overview — 405
- 16.3 Behaviour of the human operator — 407
    - 16.3.1 Functioning of the human operator — 407
    - 16.3.2 Some outstanding characteristics — 411
    - 16.3.3 Classifying tasks — 413
    - 16.3.4 Human errors — 413
- 16.4 Major concepts — 418
- 16.5 Phases of a human reliability assessment — 421
    - 16.5.1 Identification of potential human errors — 422
    - 16.5.2 Selecting significant errors — 423
    - 16.5.3 Detailed analysis of significant errors — 424
    - 16.5.4 Integration with system modelling — 425
    - 16.5.5 Quantification — 426
    - 16.5.6 Presentation of the approach used and its results — 426
    - 16.5.7 Some remarks — 426
- 16.6 Quantification models — 427
    - 16.6.1 TESEO — 427
    - 16.6.2 THERP — 428
    - 16.6.3 HCR — 432
    - 16.6.4 Simulation models — 432
    - 16.6.5 Conclusions — 433
- 16.7 Data — 433
    - 16.7.1 Data collection — 434
    - 16.7.2 Data banks — 437
    - 16.7.3 Range of values — 439
- References — 441

17 Mechanics — 446

- 17.1 Introduction — 446
- 17.2 General considerations — 447
- 17.3 Stress and strength — 448
- 17.4 Fatigue — 449
    - 17.4.1 Statistical approach — 449
    - 17.4.2 Probabilistic fracture mechanics — 453
- 17.5 Dependability of mechanical systems — 457
- References — 458

18 Software domain — 461

- 18.1 Introduction — 461
- 18.2 Concepts — 462
- 18.3 Main features of the software system — 466
    - 18.3.1 Life-cycle stages — 466
    - 18.3.2 Software testing — 467
    - 18.3.3 Quality assurance — 469
    - 18.3.4 Reliability — 470
- 18.4 First indications on software reliability — 471
    - 18.4.1 Complexity measurements — 471
    - 18.4.2 Test-related measurements — 473

18.5	Software reliability prediction	474
	18.5.1 Introduction to models	474
	18.5.2 'Perfect debugging' model	475
	18.5.3 'Imperfect debugging' model	480
	18.5.4 'Random debugging' model	482
	18.5.5 'Bugs with different occurrence rates' model	483
	18.5.6 Parametric models	484
	18.5.7 Model validation	484
	18.5.8 Conclusions	487
18.6	Reliability of a software system	487
18.7	Availability and maintainability of a software system	491
18.8	Fault-tolerant software systems	494
	18.8.1 Architecture	494
	18.8.2 Reliability	495
	18.8.3 Examples	496
18.9	Computer system dependability	497
References		498

19 Assessing safety 504

19.1	Introduction	504
19.2	The risk concept	505
19.3	The actual risks	506
	19.3.1 Risks from human and natural sources	506
	19.3.2 Risk acceptance	508
	19.3.3 Risk perception	511
19.4	Probabilistic risk assessment (PRA)	514
	19.4.1 Nuclear industry	514
	19.4.2 Petrochemistry	523
19.5	Quantified safety goals	529
	19.5.1 Principles of the approach	529
	19.5.2 Aircraft	531
	19.5.3 Chemical plants	534
	19.5.4 Off-shore oil rigs	534
	19.5.5 Nuclear power plants	536
References		541

PART 4   COMPUTERIZED METHODS 547

20 Computer codes for dependability assessment 549

20.1	Introduction	549
20.2	Cause tree analysis codes	550
	20.2.1 Qualitative analysis computer codes	551
	20.2.2 Quantitative analysis computer codes	553
	20.2.3 Computer codes for qualitative and quantitative analysis	554
	20.2.4 Computer codes for direct analysis	559
20.3	Codes for the consequence tree method	560
20.4	Codes for the state-space method	562
20.5	Codes for common-cause failure analysis	564
20.6	Codes for uncertainty evaluation	565

20.7	Monte Carlo simulation codes	566
	20.7.1 Principle	567
	20.7.2 Distribution generation	568
	20.7.3 Simulation characteristics	569
	20.7.4 Application	569
20.8	Miscellaneous programs	569
	References	572

21 Automatic assessment of dependability — 579

21.1	Introduction	579
21.2	Modelling by logical operators and inductive analysis: the GO program	580
	21.2.1 Introduction	580
	21.2.2 Modelling by logic operators	580
	21.2.3 The analysis principles	582
	21.2.4 The GO program	583
21.3	Modelling by decision tables and cause tree construction: the CAT program	583
	21.3.1 Introduction	583
	21.3.2 Modelling with decision tables	584
	21.3.3 Cause tree construction	587
	21.3.4 The CAT program	588
21.4	Electronic gate-based modelling and failure simulation: ESCAF and S.ESCAF systems	589
	21.4.1 Introduction	589
	21.4.2 Modelling and analysis principles	590
	21.4.3 ESCAF	591
	21.4.4 S.ESCAF	592
21.5	Using expert systems for modelling: the EXPRESS program	593
	21.5.1 Introduction	593
	21.5.2 How does an expert system work?	594
	21.5.3 Example of a system reliability analysis	596
	21.5.4 Prospects	600
21.6	Conclusion	600
	References	600

PART 5   CONCLUSIONS — 603

22 Dependability assessment approach — 605

22.1	Introduction	605
22.2	Advantages and drawbacks of the various analysis methods	605
22.3	Comparison of the various analysis methods	609
	22.3.1 Inherent features	609
	22.3.2 System-dependent features	611
22.4	Criteria for choosing the methods	612
22.5	Dependability assessment: approach definition	613
22.6	Limitations of dependability assessment	615
	22.6.1 Limitations of the qualitative assessment	615
	22.6.2 Limitations of the quantitative assessment	616
22.7	Validation of the dependability assessment	617
22.8	Organization and management of dependability assessment	619
	References	620

23 Dependability assessment application — 621

- 23.1 Introduction — 621
- 23.2 System design — 622
  - 23.2.1 Dependability assessment in design — 622
  - 23.2.2 Deterministic design and probabilistic design — 623
- 23.3 System operation — 625
- References — 626

PART 6   CASE STUDIES — 627

24 Analysis of a set of elementary systems by different methods — 629

- 24.1 Goals — 629
- 24.2 Presentation of the system — 629
  - 24.2.1 General description — 630
  - 24.2.2 Detailed description of the elementary systems — 631
- 24.3 Failure modes and effects analysis — 633
- 24.4 Cause tree method — 645
  - 24.4.1 Beginning of the construction — 645
  - 24.4.2 Cause tree — 648
- 24.5 Gathered fault combination method — 653
  - 24.5.1 Identification of internal gathered faults — 653
  - 24.5.2 Identification of external and global gathered faults — 653
  - 24.5.3 Undesirable event — 662
  - 24.5.4 Discussion — 663
- 24.6 Consequence tree method — 664
  - 24.6.1 Initiating events — 664
  - 24.6.2 Consequence tree — 665
  - 24.6.3 Links with the gathered fault combination method — 666
- 24.7 Cause–consequence diagram method — 667
- 24.8 Quantitative analysis — 670
  - 24.8.1 Occurrence probability of the undesirable event — 670
  - 24.8.2 Conclusions; lessons — 672
  - 24.8.3 Quantitative analysis method: discussion — 673
- References — 674

25 Human reliability assessment — 676

- 25.1 Presentation of the example — 676
- 25.2 Identification of potential human errors — 676
- 25.3 Selecting significant errors — 678
- 25.4 Detailed analysis of significant errors — 678
- 25.5 Integration with system modelling — 679
- 25.6 Quantification — 680
- 25.7 Results — 682

APPENDICES 685

Appendix 1 Main definitions 687

Appendix 2 Availability of a component and a system on standby and periodically tested 715

A2.1 Introduction 715
A2.2 Availability of a component on standby and periodically tested 715
A2.3 Mean availability 721
A2.4 Optimizing intervals between tests 723
A2.5 Availability of a system on standby and periodically tested 724
References 724

Appendix 3 Importance factors 725

A3.1 Introduction 725
A3.2 Importance factors 725
References 727

Appendix 4 Assessment of uncertainties 728

A4.1 Introduction 728
A4.2 Processing of uncertainties 729
A4.3 Propagation of uncertainties 730
    A4.3.1 Monte Carlo-type simulation 730
    A4.3.2 Moments method 731
References 732

Appendix 5 Operating rules for cases where a safety system is observed to be unavailable 733

A5.1 Introduction 733
A5.2 Method 734
A5.3 Examples of application 735
    A5.3.1 Aeronautical applications 735
    A5.3.2 Nuclear applications 736
References 737

Appendix 6 Semi-Markovian process: MUT, MTTR and MTTF 739

A6.1 MUT, MTTR and MTTF calculation 739
A6.2 Asymptotic frequency method 741
A6.3 Asymptotic duration method 742
A6.4 Examples 743
Reference 744

Appendix 7 International standards on dependability   745

Index   I-1

FOREWORD

First of all I would like to say I am very grateful to the author for the importance he gave to the history of reliability engineering: not only does he dedicate the whole of Chapter 1 to this topic, but, throughout the book, he clearly explains why a new method had to be developed. I am sure this will be very much appreciated by readers who love significant anecdotes and examples.

On reading this history, two main points attract our attention. First, we cannot help but be struck by the importance reliability engineering has achieved within a few decades. From a virtually unknown discipline, reliability engineering has grown by leaps and bounds since World War II. The rich diversity of methods, practical or theoretical progress described in this book obviously testify to the ever-evolving nature of this discipline. Indeed, each breakthrough is a step closer to perfection.

Even though the evolution of reliability engineering is characterized by the multiplicity of developments, it has always formed a coherent whole. Unlike other disciplines which are stubbornly locked up in their world, there are no industrial barriers in reliability engineering. Thus the reader is constantly captivated as the author guides him from Three Mile Island to the Ekofisk platform, from the accident at Ermenonville to the large fires in oil refineries.

Electricité de France, the French utility, contributed to the progress of reliability engineering. Quite naturally so since, right from the inception of the nuclear program, exceptional safety requirements had to be met. However, the responsibility of the utility extends far beyond nuclear power plants. The complexity and multiple interactions characteristic of the electric power system seem expressly designed as a challenge to reliability engineers.

The team of researchers the author has contributed to set up at the Direction des Etudes et Recherches (Research and Development Division) has taken a significant share in the advancement of the discipline. As the bibliography shows, this team has developed many methods. Certain characteristics peculiar to the working methods of the Division, such as the importance placed on computer science and the role envisaged for artificial intelligence, can be found in the book.

One of the major merits of this work is to provide a description opening onto an extensive computerization project. We seem to have reached an age where man's reasoning can be formalized by artificial intelligence and computers can be used for a more efficient, homogeneous and exhaustive processing of problems.

Since the book succeeds in attaining two aims, it is reasonable to hope that it can reach two types of public.

On the one hand, it presents a panorama of the history of a discipline, stresses its diversity without glossing over the problems or the shortcomings and finally depicts the present developments, not hesitating to take sides. In fact it is this aspect of the book which is sure to appeal to laymen seeking to understand the 'whys' without a detailed analysis of the 'hows'.

On the other hand, the book is definitely intended for specialists and can be used as a training manual by beginners and as a reference book by experts. The rich bibliography will certainly be invaluable in this respect.

Therefore it deserves to be among the books published by the Direction des Etudes et Recherches. It is a testimony to the quality of the teams who have been working for several years on the study of dependability, a testimony to the author's mastery of his subject and finally a testimony to his deep attachment to one of the key disciplines of contemporary industry.

Paul Caseau
Directeur des Etudes et Recherches
Electricité de France

FOREWORD

A. Villemeur's book fills a gap in the field of dependability. The first existing publications are no longer up to date due to the continual evolution of the concept. Villemeur's book is the fruit of several years of practice and reflection, personal and team research work. The author offers long awaited solutions to concrete problems. All these contribute to make the book interesting reading for it is neither a compilation nor a rearrangement of knowledge already written elsewhere.

At the same time, the author adopts a good pedagogical method. He refuses to choose the easy way which consists in offering the reader the illusion of knowledge by plunging him straightaway into the comfortable world of mathematics. He rather chooses to embrace the more difficult process of explaining the methods proposed, thus allowing each reader the chance of finding the solution to the concrete problem which might face him.

The reader is thus made to understand that a knowledge of the correct way of framing and analysing a question is of prior importance before adopting any of the numerous methods proposed by the author. Experience has shown that in this branch it is quite rare that two systems belonging to the same field can be analysed in the same way. Therefore, adepts in this discipline are always called upon to be inventive, and this is one of the particular virtues of dependability engineering.

The interesting historical account, a reflection of the importance the Latin tradition places on general knowledge, is a smooth way of introducing the reader to the main subject. It also allows readers to realize the considerable time it takes for a proposal to be finally adopted in a large sector of the industry. An average lead time of ten years is a not over pessimistic picture of the actual situation.

The publication, though limited to a certain number of important fields of dependability, treats each field in an exhaustive way. This choice, deliberately made by the author, contributes to the scholarly character of his book, also confirmed by the importance he places on presentation as evidenced by the excellent lexicon and the index, which the readers, I am sure, will be delighted with.

In addition to the remarkable qualities of this publication, I personally owe thanks to Mr Villemeur for the interesting and frank debate with which he introduces the difficult subject of software reliability. I would like to thank him as well for his effective contribution to international standardization as Convenor of the Working Group 'Analysis Techniques for System Reliability' in the framework of the International Electrotechnical Commission (IEC) program.

Arnould d'Harcourt
Chairman of the Technical Committee 'Dependability' of the
International Electrotechnical Commission

ACKNOWLEDGEMENTS

I would like to thank Mr Paul Caseau, Director of the Direction des Etudes et Recherches (Research and Development Division) of Electricité de France (EDF) and Mr Arnould d'Harcourt, Chairman of the Technical Committee 'Dependability' of the International Electrotechnical Commission (IEC) for doing me the honour of writing the forewords to this book.

I wish to express my deep gratitude to the Direction des Etudes et Recherches (DER), and its successive directors, Maurice Magnien, Claude Bienvenu and Paul Caseau, who already in the early 1970s sensed the importance of this yet fledgling science and gave constant support to work in this field. My thanks also go to the Service Réacteurs Nucléaires et Echangeurs (Nuclear Reactors and Heat Exchangers Branch), and its successive heads, Guy Malhouitre and François Boulot, for actively supporting the publication of this work and for their ever-innovating policy in the field of dependability.

Many directors and engineers at Direction des Etudes et Rescherches made an invaluable contribution for helping EDF to achieve excellence in the field of dependability. This is evident from numerous references in this book to their work. I would like to take this opportunity to thank them all warmly here and ask their forgiveness for not naming each of them, as they are so numerous.

May I nonetheless be allowed to recall two personal memories: the beginning of my career as an engineer and the period during which I wrote this book.

My career as a young engineer was marked by particularly propitious circumstances. The first probabilistic risk assessment of a nuclear power plant, which stands out as a landmark in the 1970s, was carried out under the direction of Professor Rasmussen and had just been published in the United States. In France, at the Commissariat à l'Energie Atomique (French Atomic Energy Commission), Mr Pierre Tanguy, then Director of the Institut de Protection et de Sûreté nucléaire (Institute of Nuclear Safety and Protection), had set up a Bureau of probabilistic studies as early as 1972 led by Mrs Annick Carnino. At EDF Michel Hug, then Director of the Direction de l'Equipement (Construction Division), launched the first probabilistic safety assessment program for the Fessenheim nuclear power plant under the leadership of Mr Bernard Gachot. With skill, enthusiasm and courage, Mrs

Annick Carnino and Mr Bernard Gachot were able, in those pioneering days, to direct work on an international level, and they strongly influenced my career as a young engineer. I would also like to express my gratitude to Mr Mathieu Israel, who was then head of the Départment Physique des Réacteurs (Reactor Physics Branch), and who first accepted and nurtured this new activity in his branch, and later entrusted me with the care of leading these studies.

This book was written mainly between 1983 and 1990; I wish to address my sincere thanks to all those who contributed their innovative work, their creative thoughts and fruitful discussions, and who also participated directly in its writing. This is true of all the engineers in the Division Etudes Probabilistes (Probabilistic Studies Group) whose competence, energy and spirit of innovation greatly encouraged me to carry out this work.

Within the IEC and the Technical Committee 'Dependability', Messrs Frey (Switzerland), Holm (Sweden), Monshaw (United States), Onodera (Japan), Pullum (United Kingdom), and Wild (Canada), the members of the working group on analysis techniques I am in charge of, were of invaluable help to me, with their wealth of experience and their reflections on the concepts and methods.

I would like to extend special thanks to Mr Jean-Pierre Berger from the Direction de l'Equipement, and Messrs Bernard Magnon and Thierry Meslin from the Direction de la Production et du Transport (Generation and Transmission Division) whose energy and competence considerably broadened the scope of this discipline—and of the supporting methods—in the field of nuclear power plant design and operation.

Mr Fréderic Mosneron-Dupin deserves particular thanks for his active participation in drafting Chapter 16 and developing the example discussed in Chapter 25. I also wish to thank Mrs Claudie Ancelin for her useful cooperation in the writing of Chapter 24. Messrs Jacques Boisseau, Marc Bouissou, Alain Dubreuil-Chambardel and Jean-Michel Moroni were also of great help to me with their comments and constructive criticism.

Jeanne and Michel Le Roux's affectionate attention was a constant encouragement, and they efficiently and courageously reread all the proofs, thereby greatly improving the form of the book.

Last but not least, I wish to thank Mrs Sylvette Israel for having aided me with her expertise and stimulating advice first as the head of the Group I worked in upon my arrival at EDF and second as the Director of the Book Collection of the Direction des Etudes et Recherches.

For the English version of the book, I am particularly grateful to the Direction des Etudes et Recherches for its support and specifically to Mr François Boulot and Mrs Sylvette Israel.

Special acknowledgement is gratefully given to Anne Cartier and Marie Christine Lartisien who translated most of the book with patience and diligence and Claire Mays, psychologist and human factors consultant, who translated Chapters 16 and 25.

PREFACE

The need to have safe and reliable tools has been part of man's worries since time immemorial, probably since he designed and fashioned the first tools. As he gained experience, man gradually improved upon his tools. For thousands of years, faults and accidents were the only schools for learning to make safer and more reliable equipment.

World War II marked a radical epistemological change. Because of the pressing need to master the reliability of increasingly complex systems such as the first missiles, composed of several hundreds of components, new questions and new concepts emerged: can reliability be predicted as early as at the design phase and how can it be measured? Thus the first reliability prediction model was born and a measure was worked out from the concept of probability. From a model of past systems and systems yet to be, engineers hoped to predict dependability—a genuine but utopic ambition at a time when knowledge of components reliability was sadly insufficient and did not permit accurate assessment.

Since this period, concepts have multiplied and the associated methods and measures have considerably developed in numerous industrial sectors, the aeronautical, space and nuclear industries playing a major part. To the concern for reliability and safety was added that for availability and maintainability, thus giving rise to the development of a formal engineering science called dependability.

Dependability is a science undoubtedly as are physical sciences: it is so because it involves reliability models, the corresponding equations for calculating the probability of an undesirable event from its causes or combinations of elementary causes, numerous measures (probability, MTBF, fault number, etc) and the validation of these measures against operating experience with systems and components.

The increasing complexity of industrial systems, the prominent part they now play in economic and social life, the need for design and operation cost reduction in a highly competitive market are as many reasons for carefully attending to dependability. Now mere experience, as reflected by the rule book and standards, is no longer adequate for mastering economic, human or environmental risks. Prevention of these risks through their prediction is a true challenge to our world.

Electricité de France, and particularly the Research and Development Division, have long been conscious of this challenge to knowledge and technology. That is why the Division has been working on the research and development of industrial system predictive analysis since the early 1970s.

In 1973, the extensive program of nuclear power plant construction launched by Electricité de France was a prime mover of this research and development work. As a result, the plant design and operation benefitted from the lessons of the probabilistic approach to dependability. At the same time, the requirements of power system planning led to the refinement of probabilistic assessment methods.

Against this background, in 1980, Alain Pages and Michel Gondran published a book *System Reliability*. The work offers a remarkable synthesis of methods used in the 1970s. The detailed mathematical analyses and the original ways of solving Markovian processes it contains make it a valuable reference book. Since 1980, the methods have significantly developed and their field of application considerably extended so that they have now reached a good level of maturity.

Quantitative assessments multiplied, often becoming major decision aids for system design or operation. Currently, Electricité de France uses these methods to improve the dependability of its nuclear power plants and electric power system, not to mention robots and electrically powered equipment.

In my opinion, this explosion of concepts, methods and applications demanded synthetic thinking. The result is the present publication. The book offers a description of the concepts and methods used for assessing and measuring the dependability of industrial systems. It synthesizes all the different techniques existing in various industrial sectors. In order to work out a genuine theory of assessment and measurement, I have chosen to classify and break down the different methods—which too often are developed separately—and to analyse their interactions.

Since a scientific publication is supposed to give precise definitions of the concepts used, a lot of time and energy was devoted to this task. The publication can thus boast of roughly 250 technical terms, the meaning of which often differs from one industrial sector to another. It is my earnest hope that these terms will contribute in the long run to the development of a common language between specialists, in accordance with the recommended methods.

I am sure it will come as a surprise to no one that man and software are the two most difficult systems to assess! The present state of methodological art indicates clearly the need for improvements in the coming years. Since man and software are important components of every complex industrial system, the need for progress in this field cannot be overemphasized. Since there is no such thing as 100% safety for the workers, the public and the environment, quantitative approaches to safety are essential. These approaches are used by a growing number of engineers, decision-makers and people responsible for controlling industrial risks.

Models for assessing the dependability of large industrial systems now take into account thousands of events (faults, repairs, ...) in the life of components and systems. Assessment software is becoming essential; that is why the major software programs in the world have been listed. Since modelling is complex, engineers are tempted to have them automatically built by computers; hence the importance of research work on study computerization and especially of investigations into artificial intelligence with the very promising future they hold.

Every industrial system needs or will need one day that dependability concepts and techniques be used to some degree from the first stages of the design up to the completion and operation of the project. Dependability is thus becoming a fundamental aspect of engineering. It is my fervent wish that this book will be useful to engineers and that it will contribute to the development of dependability, thereby upholding the honour of technology.

Alain Villemeur

MAIN NOTATION

Methods and studies

CCDM	Cause–consequence diagram method
CQTM	Consequence tree method
CTM	Cause tree method
DTM	Decision table method
FMEA	Failure modes and effects analysis
FMECA	Failure modes, effects and criticality analysis
GFCM	Gathered fault combination method
HRA	Human reliability assessment
PHA	Preliminary hazard analysis
PHRA	Preliminary hazard and risk analysis
PRA	Probabilistic risk assessment
PSA	Probabilistic safety assessment
SDM	Success diagram method
SSM	State-space method
TTM	Truth table method

Probabilities and probability laws

$P[E]$	Probability of the event E
$P[A\|B]$	Probability of the event A given that the event B has occurred
X	Random variable
$f(x)$	Probability density function of the random variable X
$F(x) = P[X \leq x]$	Distribution function of the random variable X
$E[X]$	Expectation or mean of the random variable X
$V[X]$	Variance of the random variable X
$\sigma[X]$	Standard deviation of the random variable X

Events (and states)

E_i	Event i or state i
E_i	Intermediate event

IE	Initiating event
C_i	Minimal cut set i
B_i^j	Event j of the minimal cut set C_i (or of the minimal tie set T_i)
AC	Acceptable consequences
UC	Unacceptable consequences
E_i^O	Operating state i
E_i^F	Failed state i
ε	Any set of states
ε^O	Set of operating states
ε^F	Set of failed states
ε_M^O	Set of minimal operating states
ε_M^F	Set of minimal failed states
F	Undesirable final event
fr or $fr[E]$	Frequency of one event (E)
T_i	Minimal tie set i
m	Number of minimal cut sets for one undesirable final event
EGF or E	External gathered fault
GGF or G	Global gathered fault
IGF or I	Internal gathered fault
S_{qi}	Events (or states) sequences i

System

S	System
S_i	Elementary system i
c_i	Component i of the system
n	Number of system components
p	Number of system states
l	Number of operating states of a system

Dependability measures

$R(t)$	Reliability of an entity or of a system
$r(t)$	Reliability of a system component
$\bar{R}(t)$	Unreliability of an entity or of a system
$\bar{r}(t)$	Unreliability of a system component
$A(t)$	(Instantaneous) availability of an entity or a system
$A(\infty)$	Asymptotic availability of an entity or of a system
$a(t)$	(Instantaneous) availability of a system component
$a(\infty)$	Asymptotic availability of a system component
$\bar{A}(t)$	(Instantaneous) unavailability of an entity or of a system
$\bar{A}(\infty)$	Asymptotic unavailability of an entity or of a system
$\bar{a}(t)$	(Instantaneous) unavailability of a system component

MAIN NOTATION

$\bar{a}(\infty)$	Asymptotic unavailability of a system component
$M(t)$	Maintainability of an entity or of a system
$m(t)$	Maintainability of a system component
$\bar{M}(t)$	Unmaintainability of an entity or of a system
$\bar{m}(t)$	Unmaintainability of a system component
MDT	Mean down time
MTBF	Mean time between failures
MTTR	Mean time to repair
MTTF	Mean time to failure
MUT	Mean up time
T	Random variable measuring the operating time of the entity
$U(t)$	Failure density (with random variable T) of an entity or of a system
$u(t)$	Failure density (with random variable T) of a system component
$G(t)$	Repair density (with random variable measuring the repair time) of an entity or of a system
$g(t)$	Repair density (with random variable measuring the repair time) of a system component
$\Lambda(t)$	Instantaneous failure rate of an entity or of a system
$\lambda(t)$	Instantaneous failure rate of a system component
Λ_a	Standby failure rate of an entity or of a system
λ_a	Standby failure rate of a system component
Γ	Failure rate upon demand of an entity or of a system
γ	Failure rate upon demand of a system component
γ_d	Failure rate (upon demand) ascribable to the starting of a system component
$M(t)$	Instantaneous repair rate of an entity or of a system
$\mu(t)$	Instantaneous repair rate of a system component
$W(t)$	Failure intensity of an entity or of a system
$w(t)$	Failure intensity of a system component
$V(t)$	Repair intensity of an entity or of a system
$v(t)$	Repair intensity of a system component
$N_F(t_1, t_2)$	Expected number of failures during time interval $[t_1, t_2]$
$N_R(t_1, t_2)$	Expected number of repairs during time interval $[t_1, t_2]$
$P_i(t)$	Probability of being (at time t) in (operating or failed) state i when assessing the reliability of a system
$Q_i(t)$	Probability of being (at time t) in (operating or failed) state i when assessing the availability of a system
A	Transition rate matrix for assessing the availability
a_{ij}	Transition rate matrix from state i to state j of a system
A'	Transition rate matrix for assessing the reliability
A'_l	Reduced transition rate matrix (for assessing the reliability)
F	Transition probability matrix for assessing the availability
F_{ij}	Transition probability

F'	Transition probability matrix for assessing the reliability
F'_l	Reduced transition probability matrix (for assessing the reliability)
B	Asymptotic transition probability matrix for assessing the availability
b_{ij}	Asymptotic transition probability
B'_l	Asymptotic reduced transition probability matrix (for assessing the reliability)

Part 3
SPECIFIC METHODS

15

DEPENDENT AND COMMON-CAUSE FAILURES

15.1 INTRODUCTION

In predictive system dependability assessments, there are two categories of failures likely to affect the components:

- random, individual failures, which are independent from one another;

- dependent failures characterizing dependences between failures of multiple components.

This second category poses problems that are often difficult: hence we decided to devote a chapter to this category with special emphasis on common-cause failures.

First we will illustrate the importance of these failures by examples of accidents having occurred in various industries and, second, we will recall the history of these concepts and suggest some definitions. Then, this chapter will address:

- a classification of common-cause failures according to their nature;

- the characteristics of these failures derived from the analysis of the failures having occurred in the aeronautic and nuclear industries;

- an inventory of the methods available to predict these failures together with an examination of their respective merits;

- the probabilistic aspects of these failures: on the one hand, the statistical lessons drawn from their occurrence and, on the other, models chosen to compute their probabilities;

- an inventory of the various preventive means available, since these failures are very important both because of their consequences and because of their occurrence rate.

Such are the questions treated in this chapter.

15.2 INTERDEPENDENCES BETWEEN FAILURES

With the growing complexity of industrial systems and the increasingly stringent safety and reliability requirements, attention focused on interdependences between failures. Thus a safe design of nuclear power plants is inseparable from that of safety systems which must be highly reliable so as to guarantee the safety of the plant in the event of an accident. This reliability is achieved, among others, thanks to the component redundancy, the high quality of the equipment used, the precautions taken when putting the system into service and frequent periodic tests.

Redundancy is effected by multiplying the components the system needs to fulfil its functions. Thus it is possible to prevent one or several failures from having too serious consequences. However, failures with the same origin (for instance, inadequate design of the components) can affect redundant components simultaneously, thus making the redundancy ineffective. Such failures are called common-cause failures and form a subcategory of dependent failures. These concepts will be defined more precisely below.

Obviously, dependability assessment for such a system is only valid if, at the same time, all the dependent and common-cause failures which may affect the system are studied as thoroughly as possible. The probability of these failures is generally difficult to quantify.

To facilitate the dependability assessment, classifications of such failures were proposed as well as means for eliminating them or reducing their probability. However, it is almost impossible to predict them all since they often reflect the limitations of our know-how and of our expertise in a given technique. We could at least help to define these failures better so as to be able to assess them systematically and to discover the means to minimize their influence.

Let us recall that two failures E_1 and E_2 are dependent (see Section 4.2) if and only if

$$P[E_1.E_2] = P[E_1] \times P[E_2|E_1] \neq P[E_1] \times P[E_2] \qquad (15.1)$$

In other words, the probability of event $E_1.E_2$ cannot be expressed simply as the product of the probabilities of events E_1 and E_2. On a more general level, let us consider a set of failures $E_1, E_2, \ldots, E_n$. If these failures are dependent:

$$P[E_1.E_2.\cdots.E_i.\cdots.E_n] = P[E_1] \times P[E_2|E_1] \times \cdots \times P[E_i|E_1.E_2. \\ \cdots .E_{i-1}] \times \cdots \times P[E_n|E_1.E_2. \cdots \\ .E_{n-1}]$$

$$(15.2)$$

Thus the probability of each successive failure depends on the preceding failures in the sequence. The probability of dependent failures generally exceeds the product of the probability of these supposedly independent failures.

Let us give a more physical and more operational definition of dependent failures. *These are failures affecting multiple entities simultaneously or concurrently and united by relations of dependence.*

This last definition calls for the following remarks:

- *Failure dependence level*. The failures affect numerous entities (components, elementary systems, ...); the components may belong to the same elementary system (S_1, for example) or to different elementary systems (S_1 and S_2, for example). In the first instance, they are called dependent failures of system S_1 and, in the second, dependent failures of systems S_1 and S_2.

Of course, the dependent failures of a component can also be defined; the failures then affect numerous parts of the component.

Often it is interesting to consider dependent failures of physical systems or of the 'human operator' system: some of the latter are related to human errors.

- *Simultaneity and concurrence*. These failures occur simultaneously, i.e. at the same time, or concurrently. In the latter case, they occur within a given time interval, to be defined, which will be called 'critical time interval'. Generally, this definition may be based on:
  - operating experience and, in particular, the analysis of definitely observed failures,
  - the functional analysis when critical dependent failures are being predicted for a given function.

  Let us take, for example, two redundant pumps in an aircraft hydraulic system. The loss of a first pump during a flight and the loss of the second ten flights later can be dependent failures. The time lag between their occurrences should be underlined as the consequences of these failures and their criticality could have been quite different, had they occurred during the same flight.

Thus this definition may induce the analyst to connect these failures to their criticality with respect to one or several functions.

- *Interdependence*. It reflects the physical, functional, human or other relations (for instance, statistical relations) between failures. As an illustration, let us mention a cause which produces an increase in the failure rate of other components. There is no deterministic link between the cause and the failures.

To clarify this concept of dependent failures, we shall examine the *classification of these failures* proposed in the *Guide to the Performance of Probabilistic Risk Assessments for Nuclear Power Plants* [1]. There are three different categories of dependent failures:

- *Common-cause initiators*. These include the facility internal and external events that may initiate an accident while simultaneously producing multiple elementary system failures.

  These events usually have serious consequences on the facility, its components and structures. Examples include fires, floods, earthquakes, aircraft crashes, for external events, and the loss of a busbar or the loss of an essential elementary system, for internal events.

- *Intersystem dependences*. Events or failure causes that create interdependences among the undesirable events in numerous elementary systems. There are several types of such dependencies:

  - *functional dependences*: dependences among the functions of elementary systems resulting from the facility design. One example is an elementary system that is only required when another repairable elementary system has failed.
  - *shared-equipment dependences*: dependences resulting from the existence of shared equipment. For example, two systems pumping water from the same tank.
  - *physical interactions*: events that increase the probability of multiple system failures though they are not common-cause initiators.
  - *human-interaction dependences*: dependences resulting from human action at all the stages of the elementary system life-cycle, i.e. its design, manufacture, assembly and operation.

- *Intercomponent dependences*. Events or failure causes resulting in a dependence among undesirable events of numerous components. In this category also, four types of dependences are defined, but here dependences concern components rather than elementary systems.

15.3 EXAMPLES OF CRITICAL DEPENDENCES AMONG FAILURES IN THE AERONAUTICAL, OIL AND NUCLEAR INDUSTRIES

Three accidents will be examined to illustrate these failures and their great diversity.

Accident of the DC 10 large transport aircraft in Ermenonville (France)

On 3 March, 1974, a large transport aircraft (DC 10; 346 passengers) took off from Orly airport (France) with its maximum load and its cargo door badly locked. At 4000 m, there was an explosive depressurization, causing the cabin floor to cave in and wrecking vital control mechanisms located below this floor. Deprived of its control mechanisms, the plane crashed near Ermenonville [2].

Thus the malfunction of the fastening mechanism of the cargo door had ruined all the efforts made to devise safe control systems with a level 3 redundancy.

The malfunction of the fastening mechanism with no warning of the alarm system provided for that purpose, the rear cabin floor being unable to withstand a depressurization and control circuits all being located in the same part of the plane can all be regarded as design errors.

Uncontrolled blow-out on the Bravo Production Platform (Ekofisk, Norway)

On 22 April 1977 an oil and gas blow-out accident occurred on the 'Bravo' production platform drilling petroleum from a North Sea oil field. This accident occurred while the casing was being mended [3]. For this type of operation, the well is sealed with mud, the 'Christmas tree' used during normal operation is withdrawn and a special system, the blow-out preventor, is installed. The blow-out preventor is designed to preserve safety whenever an operating tube is removed or when the safety valves mounted in the lower parts of the casing are pulled out.

The inquiry revealed that while the Christmas tree was being dismantled, the blow-out preventor was stripped down in two pieces. The additional work necessary to mount this device was incompletely performed within 36 hours. Indeed this work was performed by overworked men who did not take the rest they should have before undertaking such a vital operation. Moreover the men ignored the signs showing that the first mud plug was gradually giving way.

The inquiry commission concluded that the oil and gas blow-out could have been avoided if the blow-out preventor had been hanging, ready to be used, on the winch and that the accident essentially originated in the failure of overworked men [3].

Accident at the Three Mile Island Power Plant (United States).

On 28 March, 1979 [4–6], a failure in the feedwater system caused the turbine and the nuclear reactor to trip and resulted in transient conditions in the primary system cooling the reactor (sharp pressure rise). In Babcock and Wilcox reactors, this causes the pressurizer relief valve to open (Figure 15.1). As the valve remained stuck open, operators were confronted with 'conventional' accident conditions provided for in the design. The loss of water through the relief valves resulted in a further depressurization of the system, thus triggering the safety injection system which is used to maintain the water level inside the reactor core.

372 DEPENDENT AND COMMON-CAUSE FAILURES

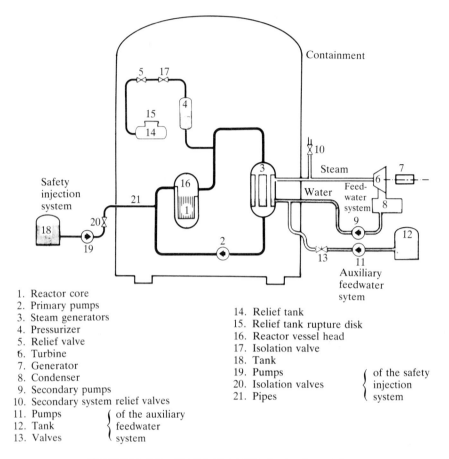

1. Reactor core
2. Primary pumps
3. Steam generators
4. Pressurizer
5. Relief valve
6. Turbine
7. Generator
8. Condenser
9. Secondary pumps
10. Secondary system relief valves
11. Pumps ⎫
12. Tank   ⎬ of the auxiliary feedwater system
13. Valves ⎭
14. Relief tank
15. Relief tank rupture disk
16. Reactor vessel head
17. Isolation valve
18. Tank ⎫
19. Pumps ⎬ of the safety injection system
20. Isolation valves ⎭
21. Pipes

Figure 15.1 Flow diagram of a pressurized water reactor.

Note that the general operation of this type of nuclear power plant is described in Section 19.4.

When the safety injection system operates properly, serious damage to the core is avoided. However, at the Three Mile Island plant, the operators did not grasp the true nature of the accident and stopped the safety injection system. This deliberate shutdown was the real starting point of the accident which resulted in heavy damage to the nuclear reactor core.

The safety injection system is redundant: at TMI its failure was due to the operators' inadequate action induced by a diagnosis error itself permitted by a design error. Let us clarify these errors closely.

Among the plethora of incoming data, the operator focused on the 'pressurizer level' and was worried by the level rise; he feared that the primary system might 'go solid' (i.e. contain only water) and that he might be unable to control the pressure inside the system. However, in this case (opening of the relief valve), the pressurizer level indication was erroneous, and the actual level decreased although the sensors indicated a level increase.

That is why the operators decided to stop the safety injection. The indication 'relief valve closed'—whereas in fact, the valve was open—displayed in the control room confirmed the operators in their diagnosis.

Numerous data were in fact available to the operators which could have helped them to detect the valve failure. It took the operators 2 hours and 20 minutes and a phone call to the nuclear boiler constructor to isolate the relief valve by closing the isolation valve.

Note also that that auxiliary feedwater system (redundant system) did not operate at the beginning as redundant valves were closed, probably due to errors made during power plant operation. Eight minutes into the accident, the valves were reopened. Most likely the operators' engrossment in the problem was detrimental to the diagnosis.

Let us also mention that there were warning events: twice already in other nuclear power plants, the pressurizer relief valves had stuck open, resulting in diagnosis problems. Here again, all the lessons from these events had failed to be drawn!

The consequences of these accidents were out of all proportion to their causes: these causes (design error, operators' fatigue, operators' inadequate action) produced interdependences among failures causing the safety devices to be ineffective.

15.4 COMMON-CAUSE AND CASCADE FAILURES

15.4.1 History of these concepts and their introduction

The importance of certain interdependences among failures was rapidly recognized by the specialists designing and operating experimental nuclear reactors [7]. As early as 1957, Siddall [8] proposed to use the expression 'common failure' to refer to possible failures in redundant reactor core instrumentation channels which could have limited the effect of a higher redundancy level whose adoption was then contemplated. In 1960, Laurence [9], in his article on reactor safety in Canada, called certain failures common-cause failures. Other terms were also used (systematic failure [10], crossed linked faults [11]) until the generic term common-mode failure was coined at a meeting of the Atomic Energy Commission in 1968. In 1969, discussions started on whether to use common-mode or common-cause failures [12] to characterize the multiple aspects of these failures.

From 1970 to 1980, both expressions were indiscriminately used although in the first probabilistic risk assessment of nuclear power plants [13], the term 'common mode failure' is employed. We must stress, however, that the term definition is quite broad since it covers 'multiple, dependent failures resulting in higher failure probability'. Likewise, in 1975, the Institute of Electrical and Electronics Engineers (IEEE) defined common-mode failures as 'multiple failures ascribable to a common cause'[14]. Nevertheless the various terms were not always given the same meaning. During the 1970–80 decade, many studies and publications were devoted to the subject and a bibliography was written in 1979. Recently, a number of publications tried to clarify these concepts [1, 15, 16]. Our own effort will try to be in line with these publications. The proposed definitions are a result of these efforts and we shall endeavour to clarify what precisely these definitions cover.

15.4.2 Definitions

- **Common-cause failures** Dependent failures having the same direct cause. Thus these failures make up a subclass of dependent failures. This definition calls for the following comments:

  - *Same cause*. The following open list of examples of 'same causes' can be given:

    —environment-related events;
    —human error during the different phases of the component development;
    —human error during operation;
    —failure of another component.

    This cause is not necessarily unique in so far as it may result from a combination of causes, but it will always be referred to as 'the cause'.

  - *Direct cause*. This is a very important concept since it is used to differentiate common-cause failures from cascade failures defined further below.

    Let us consider event E producing the simultaneous failures of components A and B, given that failure A causes the failure of component C; failures A, B and C are dependent failures. Are they common-cause failures?

    — E is a common failure cause for A, B and C;
    — E is a direct failure cause for A and B; E is an indirect failure cause for C;
    — A is a direct failure cause for C.

    Thus, only A and B are common-cause failures. Of course, this concept depends on the preliminary definition of the various elementary systems (and their components) that will have to be analysed.

- **Common-mode failures** Common-cause failures characterize the same failure mode of the components. These failures therefore make up a subclass of common-cause failures. The concerned components should normally be very similar to have the same failure mode.

- **Cascade failures of the nth order** Dependent failures that can be arranged in a chronological sequence $A_1, \ldots, A_i, \ldots, A_n$ such that each failure A_i ($i \neq n$) is the direct cause of the next failure A_{i+1}.

When such failures affect n components, they are called cascade failures of the nth order. Common-cause failures and cascade failures are mutually exclusive.

Most specialists agree on the essential points of these definitions. Note, however, that the broadest definitions have been deliberately chosen to allow the user to tailor them to specific studies or industrial fields.

15.4.3 Dependent failures and primary, secondary and command failures

We will examine in this section how the concepts of primary, secondary and command failures (see Chapter 2) are related to the concepts of common-cause and cascade failures.

Let us consider common-cause or cascade failures affecting n components. Can we identify, for each component, its primary, secondary and command failures?

- **Primary failure** There are broadly two categories of primary failures:

  - *Independent primary failure*. Primary failure whose direct cause affects only one component;

  - *Parallel primary failures of the nth order*. Primary failures having a same direct cause and affecting n components. Example: a design error affecting n identical components. These failures are common-cause failures.

- **Secondary failure** There are broadly two categories of such failures:

  - *Cascade secondary failures of the nth order* Cascade failures of the nth order consisting of secondary failures given that the first failure can be a primary or a command failure. These failures are cascade failures.

  - *Parallel secondary failures of the nth order* Secondary failures having the same direct cause and affecting n components. Example: a human error during operation resulting in the deterioration of n identical components. These failures are common-cause failures.

- **Command failures** There are broadly two categories of command failures.

  - *Cascade command failures of the nth order*. Cascade failures of the nth order consisting of command failures given that the first failure can be a primary or a secondary. These failures are cascade failures.

  - *Parallel command failures of the nth order*. Command failures having the same direct cause and affecting n components. Example: an error during operation resulting in the shutdown of n identical components. These failures are common-cause failures.

Obviously, combinations of the above-defined failures are dependent failures.

By way of illustration, let us consider the following dependent failures:

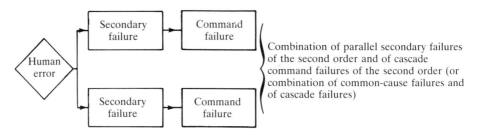

Figure 15.2 Example of dependent failures.

15.5 CLASSIFICATION OF COMMON-CAUSE FAILURES ACCORDING TO THEIR NATURE

Numerous common-cause failure classifications have been proposed; they generally result from an analysis of incidents having occurred in industrial facilities, in particular in the aeronautical and nuclear fields. They are meant to:

- facilitate the analysis of actual failures by helping to identify some of their characteristics;

- serve as a guide for the prediction of such failures by drawing the analysts' attention to failure categories.

Up to now, classifications have been restricted to common-cause failures: as a matter of fact, they are also valid, to a certain extent, for dependent failures.

A classification based on generic causes is presented here. It is used by Electricité de France (EDF) to analyse dependent (mainly common-cause) failures of safety-related systems in nuclear power plants [17, 18].

Failures affecting components or elementary systems are considered; they generally result either from environmental hazards or from (human) errors committed during design, manufacture, assembly or operation.

Therefore five failure categories were defined according to the nature of the failure causes. Note that they are described with reference to an elementary system belonging to a set of interacting elementary systems (herein called 'the facility'). Of course, this classification can be easily

Table 15.1 Classification of common-cause failures according to their generic causes

1. *Environmental hazards*
   - Internally or externally generated normal environment (dust, dirt, moisture, temperature, vibrations, corrosive atmosphere, ionizing radiation ...)
   - Extreme natural environment
     — extreme meteorological conditions (frost, wind, ...),
     — earthquakes,
     — floods
   - Internally generated accident environment
     — environmental conditions resulting from an accident,
     — pipe whip,
     — missile,
     — local flood,
     — fire,
     — explosion, ...
   - Externally generated accident environment
     — plane crash,
     — dam collapse induced flood,
     — explosion,
     — fire, ...

2. *Design errors*
   - system component not adapted to its mission,
   - system diagram revealing potential common-cause failures,
   - inadequate or damageable periodical tests,
   - system (or component) difficult to operate,
   - system (or component) difficult to maintain,
   - inadequate design optimization for common cause,
   - inadequate design optimization for common-cause failures,
   - omission or negligence in the design studies

3. *Manufacturing errors*
   - non-conformity to manufacturing technical specifications
   - technology errors

4. *Assembly errors*

5. *Operating errors*
   - during (normal, incident, accident) operating conditions,
   - during inspection and tests,
   - during maintenance operations.

extended to a facility or to another group of systems. The failure categories are as follows:

- *Environmental hazards.* Events related to the environment outside the facility or inside the facility but, in any case, outside the studied elementary system.

- *Design errors.* Errors made during the component and elementary system design studies (operating principles, components used, operating and test procedure definition, etc) which impair the elementary system functions.

- *Manufacturing errors.* Errors made in manufacturing the elementary system components.

- *Assembly errors.* Errors committed in the course of the component assembly (at the factory, on the site, etc) and during preoperational tests performed on the components and elementary system.

- *Operating errors.* Errors committed when operating the components and elementary system previously found fit for service.

We will examine each of these categories below and give examples of such (actual) failures.

15.5.1 Environmental hazards

Environmental hazards represent the most common potential sources of dependent failures. Protection means against these types of failures are implemented during the design studies of a component or system.

The level of protection a system (or component) should be provided with is defined according to the conditions the system should perform its mission in and according to the importance of this mission for the safety of the unit it belongs to.

Environmental hazards are so numerous and diversified that they should also be subdivided. Thus, in a nuclear power plant, there are normal internal or external conditions, extreme natural conditions and internally or externally generated accident conditions. Table 15.1 contains a list of these hazards, which serves as a guide for nuclear power plant components.

The systematic analysis of the impact of environment-related factors on the component dependability is usually performed in two phases:

- selection of the factors likely to have an impact on the components and threshold values above which this impact begins to tell;

- assessment of the variation range of the above-mentioned factors under all the facility operating conditions. The architect–engineer is the one who can best define these parameters in the technical specifications. If the above-mentioned thresholds can be exceeded, measures must be taken to strengthen the protection against the said factor.

To illustrate these failures, let us consider an elementary system (Figure 15.3) consisting of two parallel, redundant lines pumping water from a tank upon a safety signal and injecting water into another cicuit. The tank is located outside the industrial facility.

When the vent pipe is plugged by ice and snow due to harsh meteorological conditions, a vacuum is created in the tank when the pump starts sucking, resulting in the tank break and in the feedwater loss. This common-cause failure is due to environmental hazard. Of course, this failure could also be regarded as a design failure, thus demonstrating how difficult it is to classify these failures!

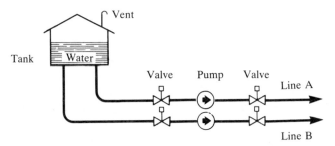

Figure 15.3 Example of an elementary system.

15.5.2 Design errors

These are the most difficult failure causes to foresee since they are closely related to our know-how limitations. They concern both components and elementary systems and can assume various aspects which will be gone into for illustration's sake:

- **Component or elementary system ill-adapted to its mission** Because the exact conditions of a specific mission were insufficiently known, a component or an elementary system may prove incapable of fulfilling its function satisfactorily. Generally, this type of error is detected during a system's preoperational tests or periodic tests. Unfortunately, there are instances where the exact mission conditions cannot be simulated during the tests, often for lack of money or of time or because real-life conditions cannot be recreated.

  Systems designed to limit the consequences of accidents likely to be extremely damaging to the facility or its environment are the most typical examples. The loss of coolant accident in a nuclear power plant belongs to this category of accidents with an extremely low probability given the means implemented to prevent it from occurring. The full-scale simulation of this type of accident is of course out of the question, and we may hope we shall never experience such an accident.

  The only available method to check whether the systems designed to meet this type of situation are well adapted to their mission is to simulate the accident conditions with computer programs which are improved by feeding into the program data collected during tests on certain phases of the accident sequence. Some inadequacies may however be detected during these studies.

- **Configuration of elementary systems with potential common-cause failures** The configuration of an elementary system or of its connections with other systems (vital auxiliary systems, for instance) may be such that the failure of a component may induce the failure of the whole system if not of several systems. This risk may be concealed by an apparently satisfactory redundancy level and only be identified after a detailed reliability assessment. Thus an elementary system fulfilling several functions is often at the origin of such failures.

- **Inadequate or harmful periodic tests** Periodic tests are essentially aimed at maintaining the availability of a standby elementary system or component at a satisfactory level. These tests are only performed because equipment may deteriorate while not working. However their design may lead to common-cause failures.

  For instance, the periodic test of a fluid system—during which the position of the valves inside the system must be changed—for the purpose of the test may be harmful (human error: failure to place the valve back in the proper position).

A periodic test can also be harmful because it may lead to the wear of several identical components simultaneouly.

- **Not readily operable elementary system (or component)** When the design of a system is such that it cannot be easily operated or monitored by the operator, the probability of operating errors increases. Therefore the simplest design of the system configuration (with not too many components) and operation is a requisite.

    Designing a system whose function is to control an accident, that is to face a rare event, is a real problem. The designer may choose either to automate the system operation entirely, leaving only a totally passive part to the operator (but will the operator remain passive in the event of an accident?), or to automate the system partially with the operator performing some actions within prescribed times.

- **Elementary system (or component) difficult to maintain** If the design of an elementary system is such that access to the system is difficult, and tagging, disassembly and reassembly operations are problematic, the probability of operational errors will obviously increase. Therefore, corrective or preventive maintenance conditions as well as operating conditions should be carefully studied with a view to minimizing the probability of human errors as much as possible.

- **Inadequate design optimization concerning common-cause failures** Protection measures against a common-cause failure may directly result in or prepare the ground for another such failure. Two examples can be given here:

    - Whip restraints to limit pipe movements in a fluid system during accident conditions may go against the required level of freedom in the pipe movements during normal operating conditions.

    - Placing the redundant lines of a safety-related system into separate bunkers in nuclear power plants has been regarded as the best guarantee against environmental hazards. However, this type of protection has increased the system vulnerability to ventilation malfunctions, the ventilation system even becoming a vital auxiliary system of the safety-related system whereas it was not so in the past. Note that bunkers are also in conflict with the easiness of access mentioned above.

    The critical analysis of the design should help to optimize the protection against common-cause failures and to avoid a protection being given more importance than another and thus being detrimental to it or a protection creating favourable conditions for a common-cause failure without the designer's knowledge.

- **Omission or blunder in the design studies** The study quality control is normally meant, among others, to limit the occurrence rate of this type of errors.

  Common-cause failures which are not taken sufficiently into account or simply forgotten about at the design phase deserve special attention. These failures may be related to the system mission conditions in the event of accidents. For instance the seismic resistance of a nuclear power plant safety-related system or of one of its vital auxiliary systems must be provided for in the design of the system and of its supporting elements. A possible omission can only be detected by checking the calculations since not a single test can reveal it after.

  There are several ways of controlling the quality of the studies performed to detect protection failures:

  - by checking the design as regards protection devices whose omission cannot be detected at a later stage;

  - by looking for visable omitted protections in the facility (preferably on the facility drawing);

  - by checking the design adequacy during preoperational tests as regards all the other protections.

Let us now quote some design errors for the elementary system mentioned earlier as an example (Figure 15.3). Suppose there is a high point on each line with no vent pipe and that, following maintenance operations, gases have accumulated in this area. Entrainment of these gases results in the failure of both pumps. This common-cause failure is due to a design error.

Another example will illustrate how difficult it is to pinpoint design errors. A somewhat more complex configuration of the circuit (system S) represented in Figure 15.3 is taken here (Figure 15.4). This system operates only when pressure P_B inside system B connected to system S falls below a critical threshold value P_0. During the first operating phase of system S, $P_B > P_0$ and water is circulated in a closed loop via the minimum flow line through the open check valves C_i and C'_i and isolation valves V'_i of the minimum flow line. Check valves C''_i prevent the water from flowing back from B to S.

During the second operating phase, when $P_B < P_0$, check valves C''_i open, and water is injected from S to B. The pressure drop in the minimum flow lines is such that the water flow rate inside the injection lines, Q_i, exceeds the flow rate inside the minimum flow line. When flow rate Q_i in line i, recorded by an electromagnetic flow meter D_i, exceeds a specified maximum, a closing signal is sent to valve V'_i. All the flow circulated by the pumps is injected from S to B.

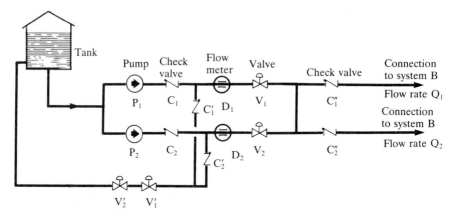

Figure 15.4 Diagram of a thermal-hydraulic system with a cascade failure.

Suppose pump P_1 is lost during the first operating phase ($P_B > P_0$), while P_2 is still operating. Check valves C_1 and C'_1 apparently prevent any malfunction of this circuit which has only one line operating. In fact, water may be circulated via V_1 and C'_1 through flow meters D_2 and D_1. This flow rate is sufficient to cause the minimum flow line to close and to damage pump P_2 as a consequence.

The design error lies in the fact that the minimum flow lines have been connected *after* check valves C_i provided for the downstream isolation of the pumps, instead of being connected *before* these check valves between P_i and C_i. This design error results in a cascade failure.

15.5.3 Manufacturing errors

Manufacturing errors often lead to common-cause failures affecting several components of the same type and make, distributed in the same elementary system and its vital auxiliary systems.

Two types of manufacturing errors should be examined in greater detail:

- *Errors consisting in the failure to comply with technical manufacturing specifications.* Quality controls of the manufactured products should be organized so as to minimize the frequency of such errors. They should include not only controls to check that the specifications regarding the materials and dimensional tolerances for the components are complied with but also tests of the required functional performance levels on test benches or at the factory.

- *Technological errors.* These errors can create a weak point in the design of a given type of components. Undetected by conformity tests, they can only be spotted during tests at the factory, on a test bench or on the site.

Note, however, that a large number of such errors can be eliminated, before production is launched, if the detailed component diagrams are inspected in time by specialists of these components. Such an inspection helps to optimize tests at the factory, on test benches or on the site in order to facilitate the discovery of possible weak points in a component.

Note finally that, regarding proven components that are bought because of their good operating records, attention should focus on those seemingly minor changes which can remain unnoticed or are claimed to be improvements over a previous series, though there is no evidence to prove it. As a matter of fact, this type of improvement is a usual common-cause failure source.

By way of illustration, let us go back to the above-mentioned elementary system (Figure 15.3): suppose that, because of a manufacturing defect, the pumps inject a water flow below the required flow rate. This common-cause failure—a manufacturing defect—will most certainly be identified during preoperational tests.

15.5.4 Assembly errors

Assembly errors often lead to common-cause failures affecting several identical components. These errors may be due to:

- non-compliance with good engineering practice;
- non-compliance with cleaning up rules;
- non-compliance with technical specifications;
- defective welds on several components;
- inadequate or botched preoperational tests.

In the elementary system already studied, suppose that, after the system has started, the pump flow rate proves to be largely insufficient. An inspection of the two system lines shows that filters placed during the system assembly for cleaning purposes remained in the system and are seriously clogged. Indeed the filters were not removed as initially prescribed by the system commissioning procedures. This common-cause failure originates in an assembly error.

15.5.5 Operating errors

A list derived from nuclear power plant operating experience is given here by way of example:

- *Errors committed during operation (normal, incident and accident conditions)*

- Erroneous actions due to failure to apply, or improperly applied, procedure, misinterpreted data, misunderstood processes, . . .;
- disregarded alarm or alarm misinterpreted as a failure of the alarm system itself:
- inadequate transmission of orders among operators;
- wrong diagnosis during incidents or accidents, etc.

- *Errors committed during inspections and tests*
  - Omission of periodic tests or incorrect performance of periodic tests;
  - calibration or adjustment errors:
  - use of obsolete or incomplete periodic test procedures;
  - negligence or omission resulting in the system being left in a state different from the state specified for the end of the tests, etc.

- *Maintenance errors*
  - Failure to comply with maintenance procedures;
  - use of obsolete or incomplete maintenance procedures;
  - negligence or omission resulting in the system being left in a state different from the state it should be in at the end of maintenance operations.

We will now illustrate this point with our elementary system (Figure 15.3). Suppose now that a maintenance operation is performed on both pumps successively (for instance, retightening the packings). The operator responsible for this task makes a mistake when working on the first pump (excessive tightening of the packing) and then makes the same error on the second pump. This error causes the pumps to trip when the system is started.

Generally speaking, other classifications have been proposed for these failures [7, 14, 19–23]; they are very similar to the classifications used by Electricité de France. Table 15.2 thus summarizes another classification proposed by Watson and Edwards [20, 21]. It is based on the distinction between 'technical' failures, relating to design and manufacture, and 'operational' failures covering procedures, maintenance, tests and environment.

15.6 USING EVENT REPORTS TO STUDY COMMON-CAUSE FAILURES

Knowledge on common-cause failures may be derived from the operating experience of industrial systems. In the aeronautical and nuclear industries, for instance, event reports are written to this end so that these failures may be concretely analysed and methodically categorized. Distinctive properties and characteristics can thus be identified.

The aeronautical industry has systematically studied common-cause failures [20, 21]. Interesting results are also obtained with the nuclear power plant operating experience which has now become quite extensive [18, 21–24]. We shall merely present the results of two different studies.

Table 15.2 Common-cause failures classification proposed by Watson and Edwards

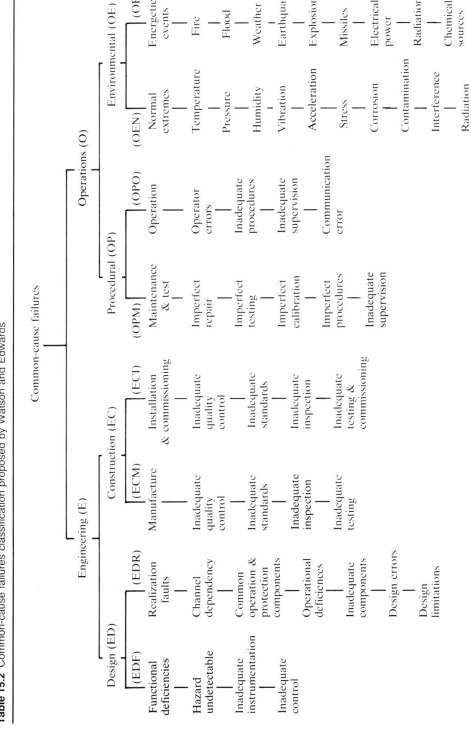

The first study concerned the operating experience of a number of nuclear power plants connected to the grid before the end of 1975. The safety injection system and the auxiliary feedwater system are the two elementary systems analysed. Data were derived from two complementary American data bases: the Nuclear Power Experience data base and the Nuclear Safety Information Center data base [18, 24]. The statistical data obtained by analysing event reports are presented in Table 15.3.

The percentage of such failures according to their generic causes, as listed in Section 15.5, can be found in Table 15.4.

Rather similar percentages of failures of each category are found for each elementary system. The following main conclusions can be drawn from the figures in Table 15.4.

- The largest failure number corresponds to design and operating errors.

- The other failures (environmental hazards, manufacturing errors and assembly errors) are very small in number. This demonstrates the remarkable efficiency of the component preoperational tests performed at the factory and on the site, especially as regards manufacture and assembly.

It is quite impossible here to describe all the failures observed. However, we should note that most failures discussed in Section 15.5 have actually been recorded on these systems. Examples quoted in that section have been chosen to illustrate the nature of these failures. The incidents are remarkably simple. Their causes are clearly identified. Unfortunately, this is not always the case. Too often the failure causes remain partially unaccounted for, in particular where operators' errors are concerned. Moreover the circumstances of the incident are often unclear and the consequences of the said incident on the

Table 15.3

Elementary system	Safety injection system	Auxiliary feedwater system
Cumulative period of reactor-years	54	69
Total number of incidents recorded	220	75
Number of common-cause failures	21	14

Table 15.4 Classification of common-cause failures (%)

Environmental hazards	10
Design errors	50
Manufacturing errors	0
Assembly errors	10
Operating errors	30

system availability are not well defined. All this makes it generally difficult to assess the failure probabilities (see Section 15.8).

The second study [20] specifies how common-cause failures are distributed among the previously defined categories, first, for the safety injection system of a nuclear power plant and, second, in the aeronautical field. For the safety injection system, the incidents having occurred in pressurized water power plants and in boiling water reactors have been added up. As a matter of fact, the safety injection systems of both types of power plant have a relatively similar design. A total of 933 incidents has thus been recorded, corresponding to 84 common-cause failures, over a period of 218 reactor-years. In the aeronautical field, the data recorded cover the 1959–1975 period and all the plane types. 123 common-cause failures have been spotted among the 3118 incidents of all sorts recorded.

If these failures are classified according to their generic causes, the results in Table 15.5 are obtained.

The results derived from incident statistics in the aeronautical industry confirm the conclusions obtained previously except for the impact of the environment, which is obviously much more important in the aeronautical field. Moreover the distribution of these failures according to their generic causes is more or less the same as that derived from the analyses mentioned in the previous section.

These studies show how important design and operating errors are and, in consequence, the present limitations of our know-how in these fields.

15.7 DEPENDENT FAILURE ASSESSMENT METHODS

The main methods (see Part 2) for predicting the dependability of industrial systems reveal interdependencies between failures. However, since it is extremely difficult to predict all these dependencies, more numerous and specific approaches are used.

We shall first discuss the essential merits of the main assessment methods described in Chapters 6–14. Then we shall describe the methods more specifically applied to the study of these failures.

The various advantages of these methods classified according to their impact on a set of interacting systems (see Section 15.7.2) are summarized in

Table 15.5

	Safety injection system (nuclear) (%)	Incidents on various elementary systems (aircrafts) (%)
Environmental hazards	5	32
Design errors	40	30
Manufacturing and assembly errors	10	0
Operating errors	45	38

Table 15.6 where the sign '+' gives an approximate qualitative indication of the merits of each of these methods.

15.7.1 Main assessment methods

We shall now see how the main methods used to predict system dependability take dependent failures into account. We shall only discuss the FMEA, CTM, the gathered fault combination method and the consequence tree method because of their importance for dependent failure identification.

15.7.1.1 The failure modes and effects analysis (FMEA)

When applied to the components of a system, this highlights:

- cascade failures originating in a failure mode of a system component;

- common-cause failures affecting the system components, and possibly the equipment of other systems, and originating in the component failure mode.

Moreover, by looking for the causes of the failure modes systematically, it is possible to identify all the causes common to component failures, whether internal or external to the system. Among external causes we can mention environmental effects, initiating events and human errors. However, looking for internal causes is not easy in the rigid framework of an FMEA and other methods should also be used to complete this search. It is also particularly complicated to study human-interaction dependences with this method. Nevertheless, the same kind of method can also be applied to analyse operating procedures. It is then a task analysis.

Generally speaking, the FMEA is a first fruitful way of approaching these failures. This is confirmed by the prediction of the scenario of the DC 10 1974 accident (see Section 15.3) by means of such an analysis.

15.7.1.2 The cause tree method (CTM)

To construct the cause tree of an undesirable event (affecting a system), the failure causes must be systematically identified at all the tree levels. Intercomponent dependences are modelled in the tree, but vital dependences for the undesirable event are essentially characterized during the tree reduction and the construction of minimal cut sets. This method has the same advantages when it is applied to a group of systems.

In this respect, the computer programs developed in the United States from this method to help taking these failures into account should be mentioned [25]. Indeed, when minimal cut sets are obtained for the undesirable event,

Table 15.6 Merits of dependent failure assessment methods

Dependent failure assessment method	Initiating events		Dependences between elementary systems				Intercomponent dependences			
	external	internal	Functional dependences	Shared-equipment dependences	physical interactions	Human-interaction dependences	Functional dependences	Shared-equipment dependences	Physical interactions	Human-interaction dependences
MAIN METHODS										
• FMEA (of an elementary system)							+	+	+	
• Cause Tree Method for an undesirable event		+	+	+			+	+	+	+
• Gathered fault combination method for a set of interacting elementary systems		+	+	+	+	+	+	+	+	+
• Consequence tree method (used in connection with another method)		+	+	+	+	+	+	+	+	+
SPECIFIC METHODS										
• Specific studies of initiating events	+	+								
• Studies of generic causes	+	+								
—prediction by generic cause classification			+	+	+	+	+	+	+	+
—zone analysis					+				+	
—analysis of human factors in operating conditions						++				++
• analysis of operating experience	+	+	+	+	+	+	+	+	+	+

the susceptibility of these cut sets to certain common and potential failure causes is not known. Hence, parallel primary failures are difficult to detect. This is why the following method was proposed:

- Categories of potential common-cause failures are defined. For instance, component manufacturer, component location, temperature, moisture, radiation, dust, preventive maintenance, etc.

- The various items of the above-defined categories are inventoried. Thus the items of the 'manufacturer' category will be the codes of the various manufacturers of the studied system components.

- The conventionally obtained minimal cut sets are analysed by the computer program to identify those which are susceptible to a given category or item. Example: those among all the minimal cut sets which involve failures of the same make are listed.

The COMCAN computer code [26, 27] is one of the first to perform this type of analysis. Other similar computer programs exist (see Section 20.5).

15.7.1.3 The gathered fault combination method

This method, preceded by an FMEA, has primarily and obviously all the advantages of this technique. Internal, external and global gathered faults once identified, all the interactions—especially physical interactions—among components and systems can be inventoried. The method is particularly valuable when applied to a set of elementary systems. It then reveals:

- functional dependences;

- shared-equipment dependences: these generally induce failures inherent in this equipment;

- intersystem physical interactions: these are modelled using external gathered faults;

- human-interaction dependences: to identify these, it is enough to consider an elementary system consisting of the sole operator. The effects of his errors are modelled with external gathered faults that are subsequently integrated into the analysis to identify global gathered faults.

15.7.1.4 The consequence tree method

The identification and analysis of event sequences by means of the consequence tree method generally reveal:

- initiating events inherent in the facility;

- dependences among elementary systems and especially functional dependences and interactions between elementary systems.

The consequence trees can be further reduced by using additional methods such as the CTM or the gathered fault combination method; relevant intersystem and intercomponent dependences, especially shared equipment dependences, then appear in an event sequence.

15.7.2 Other methods

Numerous other methods exist: they generally belong to one of the following three main categories:

- specific studies of events initiating dependent failures;

- studies of generic causes;

- operating experience analysis.

15.7.2.1 Specific studies of events initiating dependent failures

These studies are aimed at analysing the effects of external events (aircraft crash, earthquake, hurricane, etc) or of internal events (design-basis accident, loss of electric power supplies, pipe whip, etc) in order to bring down their consequences to an acceptable level. They are used to design the concerned elementary systems and components.

15.7.2.2 Studies of generic causes

These studies are aimed at predicting and analysing common-cause failures due to one or several generic causes (see Section 15.5). Let us mention, for instance:

• **Prediction by generic cause classification** This method has been used by Electricité de France (EDF) [28] to predict dependent failures affecting nuclear power plant safety systems. On the basis of the classification discussed in Section 5.1, the analyst tries to picture what happens when such failures occur in the studied system. For each of these potential failure sources, the analyst makes sure that, thanks to all the constructive provisions, they can be considered unimportant.

Moreover, the analyst examines those previous incidents that revealed such dependent failures in similar systems. Whenever possible, the probabilities of these failures are assessed. The results of this analysis are presented in tables. Table 15.7 is an example. It comprises the following columns:

- *Generic causes of common-cause failures*. The generic cause is specified as defined in the above-mentioned classification.

- *Specific causes of failures*. The specific causes identified by the analyst and belonging to the generic cause are entered. For instance, humidity is a specific cause belonging to the generic cause category 'Internally or externally generated normal environmental hazard'.

- *Potential consequences on the studied system and on the other systems*. The actual or potential consequences of the specific cause on the components and vital functions of the elementary system (or neighbouring systems) are listed. Redundant components are given special attention.

- *Detection means*. The various means available to detect such failures (their causes and/or their consequences) are stated.

- *Protection means—operator's actions*. The means available (if any) to protect against this specific cause as early as the design stage as well as the actions the operator may have to take to mitigate the consequences are recorded.

- *Incidents having occurred in similar systems*. It is very useful to list the incidents, if any, due to such causes in identical or similar systems.

- *Failure probabilities*. The probability of these failures is assessed, if possible, using the various means available (see Section 15.9).

- *Comments*. This column is for comments on the type of analysis performed.

- **Zone analysis** This method is widely used in the aeronautical field [29]. It is aimed at analysing failure dependences due to the location of some components or subsystems. It is based on

  - the definition of installation rules;
  - the component inventory in each zone;
  - the recording of possible faults;
  - the identification of human errors during maintenance.

Faults and human errors identified for the zone and having serious consequences on the elementary system or group of elementary systems are subsequently identified; they usually are further defined during inspections of system mock ups.

Table 15.7 Assessment table of common-cause failure

POWER PLANT: SYSTEM STUDIED:	COMMON-CAUSE FAILURE PREDICTION BY GENERIC CAUSE CLASSIFICATION:					PAGE:
Generic causes of common-cause failures	Specific failure causes	Possible consequences on system studied or on other systems	Detection means	Protection means, operator's actions	Incidents identified on similar systems	Failure probabilities	Comments

- **Analysis of human factors during operation** Because of the vital importance of these factors, this type of predictive analyses has considerably developed in numerous industrial fields. They are discussed in Chapter 16.

15.7.2.3 Analysis of operating experience

The detailed analysis of the operating experience of industrial systems is an inexhaustible source of dependent failures. It is based on the use of event reports in which the incidents affecting components or systems are systematically recorded. The richness of this analysis largely depends on the quality of the event data base, on its exhaustivity as well as on the care taken in recording and investigating these incidents. Let us mention, for instance, the works undertaken in the nuclear field to prepare a common-cause failure classification which will help to analyse the failures having occurred during operation and recorded in a file [30].

15.8 DEPENDENT AND COMMON-CAUSE FAILURE PROBABILITIES

Two types of methods are available to compute the probabilities of dependent and common-cause failures.

- *Explicit methods*. These are based on very precise analysis and knowledge of the causes of these failures.

- *Parametric methods*. These are based on the modelling of the failure effects, with no identification or inventory of the failure causes.

15.8.1 Explicit methods

Suppose all the dependent failures have been inventoried and their mechanisms and causes are thoroughly known. The conditional probability theorem discussed in Section 4.2 is then applied. Once identified, these failures can generally be used in the dependability assessments performed by the main available methods (see Part 2).

As regards cascade failures or common cause failures, their probability is reduced respectively to the first failure or to the probability of the common-cause when the connections between failures are purely physical and deterministic. Otherwise the nature of the dependences should be taken into account.

The probabilistic assessment of operational errors often requires specific methods and models. They will be discussed in Chapter 16.

15.8.2 Parametric methods

These methods essentially apply to common-cause failures.

15.8.2.1 The beta-factor method

The beta-factor method worked out by Fleming [31] is the most widely known and also the easiest to use. Similar approaches have also been developed [32, 33]. The model assumes that each component has two types of failures:

- independent failures (independent primary failures): the failure rate is then λ_i
- dependent failures (of the common-cause type): the failure rate is λ_c.

The total failure rate for a component is then

$$\lambda = \lambda_i + \lambda_c \tag{15.3}$$

A parameter β is thus defined:

$$\beta = \frac{\lambda_c}{\lambda} = \frac{\lambda_c}{\lambda_i + \lambda_c}, \quad 0 < \beta < 1 \tag{15.4}$$

so that

$$\lambda_i = (1 - \beta)\lambda, \quad \lambda_c = \beta\lambda \tag{15.5}$$

Hence, the parameter β is the fraction of the total failure rate attributable to dependent failures. Of course, failures during operation or upon demand can be modelled in this way.

The method has been primarily used to quantify common-cause failures of systems having identical redundant components. It assesses the availability or reliability of such systems. Consider, for example, a redundant system composed of two pumps in standby. The pump failure rate upon demand is

$$\Gamma \simeq (1 - \beta)^2 \gamma^2 + \beta\gamma \tag{15.6}$$

where γ = failure rate of a pump upon demand.

Note that the state-space method yields an exact formula. The first term of this approximate formula (15.6) defines the contribution of multiple independent failures whereas the second gives the contribution of common-cause failures.

Given the conditions $\beta < 0.1$ and $\gamma < 0.1$, the first term becomes negligible compared to the second:

$$\Gamma \simeq \beta\gamma \tag{15.7}$$

In a first approximation, $1/\beta$ corresponds to the unavailability reduction achieved with a level-2 component redundancy. Moreover, a maximum estimator of β is

$$\beta = \frac{N_c}{N/2} = \frac{2N_c}{N} \qquad (15.8)$$

where:

N = total number of observed failures on a set of systems with a level-2 redundancy,
N_c = number of recorded common-cause failures of this set.

What values do parameter β assume? Values have been suggested for nuclear power plant components. They were derived from an analysis of the incidents having affected redundant components in safety systems. Indeed, these components are closely monitored, and any incident affecting them must be reported. The first values of β were published by Fleming and Raabe [34]; they were drawn from a small failure sample (operating experience prior to 1978) and computed by the above-mentioned method. Consequently, they are in all likelihood maximum values.

These values must be used with caution [16]. As a matter of fact:

- Because of the small number of observed failures, parameter β is fraught with a high level of uncertainty.

- Event reports do not always have the level of detail necessary to allow dependent failures to be analysed and categorized easily. Thus, it seems essential to differentiate between 'actual' and 'potential' common-cause failures. As a matter of fact, a component may fail and the same cause-inducing mechanism exist for the redundant component although without resulting in its actual failure (or in its recorded failure): this common-cause failure is then regarded as 'potential'. Therefore, the introduction of this type of failures in statistics may significantly modify the values of β [16].

Accordingly, the values used in dependability assessments of nuclear power plant safety systems are generally below these figures. Two important factors are then taken into account:

- These values were derived from analyses of the operating experience acquired from 1970 to 1980 with safety systems designed in the 1960's or in the early 1970's at most. Because the design of these systems was improved and provisions were made to minimize the importance of these failures, certain failures can no longer occur. However, we should also remain conscious that some of these failures mark the limits of our know-how and are therefore unpredictable: indeed, despite the designers' efforts, design-induced failures, for instance, will continue to occur although probably less often. This is why it is so difficult to predict what this parameter β will be in the future.

Table 15.8

Components	Value of β
High-pressure pump:	
• failure to start	0.14
• during operation	0.06
Motor-operated valve	0.23
Diesel generators	
• failure to start	0.13
• during operation	0.14

- These values cover an entire range of failure causes. Explicit methods attempt to list and quantify these failures. However, one must be careful not to take certain common-cause failures into account twice when a parametric method is used as a complement to other methods.

Therefore, these values should be adapted to the systems analysed by taking account of improvements in the system design as well as of the characteristics of the prediction method. These values will be further discussed in Section 15.8.2.3 on the shock method.

15.8.2.2 Multiple greek letters method

The method of multiple greek letters is the generalized version of the preceding method used when more than two components are considered [35]. It has been used in the framework of a Probabilistic Safety Analysis performed in a nuclear power plant in the United States [36]. The method is based on the following parameters:

- component failure rates taking the contributions of independent and common causes into account;

- conditional probabilities.

As an example, consider a set of three standby components. We have the following parameters:

λ_s = the rate of failure upon demand of each component due to independent and common-cause failures;

β_s = conditional probability that a failure cause upon demand be shared by one or more other components;

δ_s = conditional probability that a failure cause upon demand shared by one or more other components be also shared by the three components considered.

The parameters of the method once known, the various probabilities of common-cause failures are deduced for all the components.

15.8.2.3 The shock method

Other methods are proposed [37–40]. They can be grouped under the name 'shock method'.

A particular model has been developed by Atwood [41–43]. Based on the binomial law, it is called the 'binomial failure rate common-cause model'. Take a system consisting of m components likely to undergo common-cause failures. The following parameters are defined:

λ_i = independent failure rate of each component;
μ_{NL} = occurrence rate of a non-lethal shock;
p = conditional failure probability of a component after a non-lethal shock;
ω_L = occurrence rate of a lethal shock.

Two types of shocks are thus differentiated:

- the lethal shock: all the components are then failed;

- the non-lethal shock: it does not necessarily result in the failure of all the components. Once this shock has occurred, the failure probability of a given component is p.

Let us take an example to illustrate these concepts. Suppose a system consisting of three pumps with a common suction line. Air has been sucked into the pipe. Two of these pumps cavitate, while the third is still operating. This is considered a non-lethal shock. Indeed, the shock, although it increases significantly the pump failure probability, does not necessarily lead to this failure.

Let us now assess the failure rate of a component resulting from all possible causes:

$$\lambda = \lambda_i + \mu_{NL}p + \omega_L \tag{15.9}$$

The failure rate of a given group k among the m components is computed using the following approximate formula:

$$\lambda_k = \mu_{NL}p^k + \omega_L \tag{15.10}$$

Let us now consider how the estimators of parameters λ_i, μ_{NL}, p, ω_L can be computed from the analysis of the operating record of a system incorporating m components. The following parameters are defined as follows:

T = cumulative operation time of the system,

N = number of independent failures,
N_{+i} = number of non-lethal shocks affecting i components,
$$N_+ = \sum_{i=1}^{m} N_i,$$
N_L = number of lethal shocks.

The estimators of λ_i and ω_L are the following:

$$\lambda_i = \frac{N}{mT} \qquad \omega_L = \frac{N_L}{T} \qquad (15.11)$$

The calculation of p is written as follows:

$$p = v[1 - (1-p)^m] \qquad (15.12)$$

where $1 - (1-p)^m$ is the probability—when a non-lethal shock has occurred—that at least one component be affected, and where v is the proportion of components affected by a non-lethal shock having caused at least one failure:

$$v = \frac{\sum_{i=1}^{m} iN_{+i}}{mN_+} \qquad (15.13)$$

p is therefore the solution to the following equation:

$$p = \frac{\sum_{i=1}^{m} iN_{+i}}{mN_+}[1 - (1-p)^m] \qquad (15.14)$$

To compute μ_{NL} we write

$$\lambda_+ = \mu_{NL}[1 - (1-p)^m] \qquad (15.15)$$

where λ_+ is the occurrence rate of non-lethal shocks resulting in the failure of at least one component. Hence

$$\mu_{NL} = \frac{N_+}{T[1 - (1-p)^m]} \qquad (15.16)$$

A significant part of the operating experience recorded for safety system components of American nuclear power plants was analysed with this method [44–47]: parameter values were computed for components such as standby generators, pumps, valves and various sensors. The values must be used with extreme caution since these data are known to be fraught with a great level of uncertainty. Indeed the failure sample was selected using a very extensive definition of common-cause failures and of non-lethal shocks. Moreover there are important differences in the design of the safety systems and equipment considered, and this does not contribute to the consistency of the data collected.

The operating experience for the safety system components of French nuclear power plants has also been analysed with this method [48]. The analysis covered an operating experience of 140 reactor-years; there are 104 common-cause failures among the 1235 failures (of all sorts) recorded on such safety system components as sensors, valves, pumps, etc. An equivalent of parameter β (denoted β_2^2) is given for a group of two components in Table 15.9. This parameter is deduced from the shock method parameters:

$$\beta_2^2 = \frac{\lambda_2}{\lambda}$$

This method is also used to assess failure rates for groups of 3 or 4 components, that is parameters of the β_3^3 and β_4^4 types. Despite existing uncertainties, the reduction of β_i^i-type values seems to become ever smaller [44–48]. When the group grows from 2 to 3 or 4 components, parameter β decreases only by a maximum factor of 2 to 10. Thus, for the components mentioned above, the failure rate gain from a 2 to 4 redundancy increase rapidly reaches a limit. This obviously has important consequences on the design of such systems. There is no point in excessively increasing the redundancy of these systems. Their reliability or availability should be improved by other means (for instance by diversifying the components).

15.9 IMPORTANCE OF THESE FAILURES AND MEANS TO PREVENT THEM

The analysis of the Ermenonville DC 10 crash and of the accidents at the Ekofisk oil rig and Three Mile Island nuclear power plant has shown how important these failures and their effects were: indeed, they constitute a real challenge to the designers and operators of highly dependable redundant systems. The study of their causes—design errors or human errors during operation or both—highlights the present limitations of our know-how. It reveals the specific points the designers and operators should strive to predict and take into account if the dependability of those redundant systems that are often used for the safety of industrial plants is to be improved.

Indeed, an analysis of the operating record of safety systems in American and French nuclear power plants and of dependent failures in these systems highlights the fairly large share of these failures in incidents. It also

Table 15.9

Components	β_2^2 values
(Temperature, flow rate, pressure, level) sensors	0.05
Pumps (upon starting and during operation)	0.05
Motor-operated valves (upon demand)	0.06
Air-operated valves (upon demand)	0.07

demonstrates that availability or reliability gains from a redundancy of two (or more) are lower than initially thought. Thus, for the most important and most frequently used components of safety systems (e.g. pumps, valves, diesel generator sets), the gain could be of the order of 5–20 (for a level-2 redundancy), with the parameters β roughly varying from 2.2 to 5×10^{-2}. A redundancy increase—up to a level of 3 or 4—results in only low additional gains—at most 2–10—for the components mentioned above. Above these values, the redundancy increase no longer produces significant gains.

What prevention means are generally used to lessen the importance of these failures? The main prevention means, from the most conventional to those most recently introduced in the nuclear industry, are listed below. This list is given as an example and should in no way be regarded as exhaustive.

At the design level

1. Means to prevent (incident or accident) initiating events generating common-cause failures: design provisions to control the impact of:
   (a) the environment and its hazards (see Section 15.5.1);
   (b) an internally generated accident environment: the main conceivable or hypothetical accidents will be conservatively taken into account.

2. Means to prevent dependences between elementary systems:
   (a) physical and geographical separation of redundant systems;
   (b) separation of safety functions provided by different systems;
   (c) functional diversity and system diversity;
   (d) different auxiliary systems;
   (e) provision for periodic tests;
   (f) optimization of man–machine interface: for instance:
      (i) automation of those human actions which cannot be performed in time with a good reliability,
      (ii) understandable, precise system operating procedures tested on simulators,
      (iii) allowance for forseeable human errors;
   (g) systematic identification of these dependences using predictive analyses (see Section 15.8).

3. Means to prevent intercomponent dependences:
   (a) physical and geographical separation of redundant components;
   (b) diversity of the redundant components: different designers and manufacturers;
   (c) safe component failure modes;
   (d) provision for periodic tests;
   (e) systematic identification of these dependences by predictive analyses (see Section 15.8).

During the operating phase

1. Means to prevent dependent failures:

(a) systematic and detailed analysis of all the incidents and accidents having occurred in the facility.

2. Means to prevent human errors:
   (a) during normal operating, incident and accident conditions: importance of training operators and of maintaining their motivation, presence of several operators, incident diagnoses made by two independent crews using different means;
   (b) during maintenance operations: no simultaneous maintenance on redundant components, maintenance operations on important components checked by other teams.

REFERENCES

[1] PRA Procedures Guide (1983) *A Guide to the Performance of Probabilistic Risk Assessments for Nuclear Power Plants*, NUREG/CR 2300, US NRC.
[2] P. Eddy, E. Potter and B. Page (1977) *Destination désastre*, Grasset.
[3] *Uncontrolled Blow-out on Bravo*, Report from the Commission of Inquiry appointed by the Royal Decree of 26 April 1977.
[4] F. Pouliquen (1979) L'analyse de l'accident de la centrale de Three Mile Island. Compte rendu de la conférence, débat organisé par la SFEN le 14 juin 1979, Note EDF-DER, HT 13/38/79.
[5] A. L. Ferguson (1979/1980) *A Reliable Plant is a Safe Plant, Lessons learned From TMI*. Proceedings Annual Reliability and Maintainability Symposium, San Francisco, 22, 24 janvier 1980. *Staff Report on the Generic Assessment of Feedwater Transients in Pressurised Water Reactors Designed by the Babcock and Wilcox Company*, Office of Nuclear Reactor Regulation, US Nuclear Regulatory Commission, Rapport NUREG-0560.
[6] TMI-2 (1979) *Lessons Learned Task Force Status Report and Short-term Recommendations*, US Nuclear Regulatory Commission, NUREG-0578.
[7] E. W. Hagen (1979) *Common-Mode/Common-Cause Failure: a Review and a Bibliography*, Report NUREG/CR 0566, ORNIL/NUREG/NSIC-148. Oak Ridge National Laboratory, Nuclear Safety Information Center, USA.
[8] F. Siddall (1957) Reliable reactor protection, Atomic Energy of Canada, Chalk River, Ontario, *Nucleonics*, **15**(6).
[9] G. C. Laurence (1960) Reactor safety in Canada, Atomic Energy of Canada, Chalk River, Ontario, *Nucleonics*, **18**(10).
[10] E. P. Epler (1961) Dangers in safety systems, Oak Ridge National Laboratory, *Nuclear Science*, (4).
[11] C. G. Lennox, A. Pearson and P. R. Tunnicliffe (1962) Regulation and protective system design for nuclear reactors, Atomic Energy of Canada, Chalk River, Ontario, *AECL*, 1495.
[12] E. P. Epler (1969) Common mode failure considerations in the design of systems for protection and control, Oak Ridge National Laboratory, *Nuclear Safety*, **10**(1).
[13] US Nuclear Regulatory Commission (1975) *Reactor Safety Study. An Assessment of Accident Risks in US Commercial Nuclear Power Plants*, Report WASH, 1400/NUREG-75-01.
[14] IEEE (1975) *Guide for General Principles of Reliability Analysis of Nuclear Power Generating Station Protection Systems* (an American National Standard) ANSI N41.4, 1976. IEEE Std 352, 1975 (revision of IEEE Std, 1972).

[15] A. M. Smith and I. A. Watson (1980) Common cause failure. A dilemma in perspective, *Reliability Engineering*, **1**(2).
[16] K. N. Fleming, A. Mosleh and A. P. Kelley (1983) On the analysis of dependent failures in risk assessment and reliability evaluation, *Nuclear Safety*, **24**(5).
[17] B. Gachot (1977) *Les défauts de cause commune*. Cycles de conférences CEA-EDF sur 'la fiabilité et disponibilité des systèmes élémentaires et de leurs composants', Jouy-en-Josas.
[18] M. Llory, A. Villemeur and R. Portal (1979) *Les défaillances de mode commun. Identification et prévention. Enseignements tirés de dépouillements de fichiers d'incidents*, EDF-DER-HT/13/28/79.
[19] W. C. Gangloff (1975) Common mode failure analysis, *IEEE Transactions on Power Apparatus and Systems*. **PAS-94**(1) 27–30.
[20] I. A. Watson and G. T. Edwards (1978) *Common Mode Failures in Redundancy Systems*, E.N.S./A.N.S. International Topical Meeting on Nuclear Power Reactor Safety, Brussells.
[21] G. T. Edwards and I. A. Watson (1979) *A Study of Common-Mode Failures*, Report SRD.R.146, Safety and Reliability Directorate, UKAEA..
[22] *Task Force on Problems of Rare Events in the Reliability Analysis of Nuclear Power Plants* (1978) Committee on the safety of nuclear installations (ESNI)-OECD, Nuclear Energy Agency, Report No.51.
[23] CSNI (1977) *Task Force on Rare Events*, Research Sub-Group on Common-Mode Failure, Interim Report, System Reliability Directorate, UKAEA.
[24] M. Llory and A. Villemeur (1980) *Les méthodes d'analyse de la fiabilité des systèmes de sûreté des centrales nucléaires*, Séminaire européen sur la sécurité des systèmes, Bordeaux.
[25] N. H. Roberts, W. E. Vesely, D. F. Haasl and F. F. Goldberg (1981) *Fault tree Handbook*, NUREG-0492, U.S. Nuclear Regulatory Commission, January 1981.
[26] G. R. Burdick, N. H. Marshall and J. R. Wilson (1976) *COMCAN. A Computer Program for Common Cause Analysis Aerojet*, Nuclear Company, ANCR-1314.
[27] D. M. Rasmuson et al. (1979) *COMCAN II.A. A Computer Program for Automated Common Cause Failure Analysis*, TREE, 1361, EGG, Inc. INEL.
[28] C. Ancelin, J. Boisseau, E. Bourgade and A. Villemeur (1984) *Les méthodes d'analyse prévisionnelle de la fiabilité des systèmes de sûreté des centrales nucléaires*, Quatrième colloque International de Fiabilité et de Maintenabilité, Perros-Guirec, France.
[29] C. Lievens (1976) *Sécurité des systèmes*, Cepadues Editions, Toulouse.
[30] G. L. Crellin, I. M. Jacobs and A. M. Smith (1984) *A study of Common Cause Failures*, Phase 1: A Classification system, EPRI NP-3383.
[31] K. N. Fleming et al. (1975) AIPA Risk Assessment Methodology, Vol. II, GA-A13617, pp. 4.13, 24.38.
[32] J. Hartung (1981) *Common Cause Failure Theory*, Atomics International Division, Rockwell International, NOOOIT1000 145.
[33] J. R. Taylor (1978) *A Study of Failure Causes Based on U.S. Power Reactor Abnormal Occurrence Reports*, IAEA-SM-195/16.
[34] K. N. Fleming and P. H. Raabe (1978) *A Comparison of Three Methods for the Quantitative Analysis of Common Cause Failures*, USDOE Report GA-A-14568, General atomic Company, NTIS.
[35] K. N. Fleming, A. Mosleh and R. K. Deremer (1985) *A Systematic Procedure for the Incorporation of Common Cause Events into Risk and Reliability Models*, Nuclear Engineering and Design, International Post-conference Seminar, SMIRT 8, Brussels, Belgium.
[36] Seabrook Station Probabilistic Safety Assessment (1983) *Prepared for Public Service Company of New Hampshire and Yankee Atomic Electric Company*, Pickard, Lowe and Garrick, PLG-0300.

[37] G. E. Apostolakis (1976) The Effect of a Certain Class of Potential Common Mode Failures on the Reliability of Redundant Systems. *Nuclear Engineering and Design*, 36(1), pp. 12-3-133.

[38] B. B. Chu, D. P. Gaver (1977) *Stochastic Models for Repairable Redundant Systems Susceptible to Common Mode Failures*. International Conference on Nuclear Systems Reliability Engineering and Risk Assessment, Gathinburg, TN, 20–24 June.

[39] W. E. Vesely (1977) *Estimating Common Cause Failure Probabilities in Reliability and Risk Analysis*. International Conference on Nuclear Systems Reliability Engineering and Risk Assessment. Gathinburg, Tennessee, 20–24 June.

[40] T. Mankamo. Common Cause Failure of Reactor Pressure Components. IAEA-SM, 218/5.

[41] C. L. Atwood (1980) *Estimators for the Binomial Failure Rate Common Cause Model*. NUREG/CR-1401, EGG, Idaho, Inc.

[42] C. L. Atwood (1983) *Data Analysis using the Binomial Failure Rate Common Cause Model*. N.R.C.–NUREG/CR-3437.

[43] C. L. Atwood *User's Guide to Binomial Failure Rate*. N.R.C.–NUREG/CR-2729.

[44] C. L. Atwood and J. A. Steverson (1976/8) *Common Cause Fault Rates for Diesel Generators* Estimates Based on Licensee Event Reports at U.S. Commercial Nuclear Power Plants.

[45] C. L. Atwood (1983) *Common Cause Fault Rates for Pumps*. N.R.C.–NUREG/CR-2098.

[46] C. L. Atwood (1983) *Common Cause Fault Rates for Instrumentation and Control Assemblies*, N.R.C.–NUREG/CR-3289.

[47] C. L. Atwood (1983) *Common Cause Fault Rates for Valves*. N.R.C.–NUREG/CR-2770.

[48] T. Meslin and E. Bourgade (1987) *Common Cause Failure Analysis and Quantification on the Basis of Operating Experience*. International Topical Conference on Probabilistic Safety Assessment and Risk Management, Zurich, Switzerland, August 31–September 4.

16

HUMAN FACTORS *

16.1 INTRODUCTION

Man has always been present in industrial installations, whether it be in their design or in their operation. Thus it is correct to speak of the man–machine system, in which the human being is a system component. Undeniably, human errors can be identified at all stages of the system life cycle: errors in design, in calculation, in wiring, in assembly, and in running the installation, etc, whence the importance of Human Reliability Assessment (HRA). Methods and data in this branch of assessment have grown progressively since the 1960s.

Following a short history of how human factors have been taken into account in system safety, we will present methods and data which are current notably in the field of nuclear energy. We will examine with particular attention human errors committed by operators in large industrial systems (aircraft, nuclear power plants, ships, etc.). At the outset, let us define a human operator (or operator) as any individual, team or human organization whose function is to accomplish an action or carry out a mission in the course of the different phases of a system life cycle.

16.2 HISTORICAL OVERVIEW

Following World War II, the field of reliability developed considerably, concentrating on reliability of hardware and leaving somewhat to the side human reliability. The first focusing on human reliability came about in aeronautics, in regard to piloting [1, 2].

At the end of the 1950s, research carried out in various industries demonstrated the importance of human factors in system failure; for example, it was noted that 20–50% of missile malfunctions resulted from human error [3]. Analogous conclusions were drawn in the fields of nuclear weapons construction [4, 5], in aeronautics [6], and later in the electronuclear industry

---

* This chapter was written with the collaboration of M. Mosneron-Dupin (Electricité de France).

[7]. The first reliability assessments taking into account human errors and their quantification were carried out from 1957 [8].

At the same time as this research developed, the demands made by the growing complexity of industrial systems became less and less compatible with operators' capabilities and limits; bad design characteristics were revealed to be important sources of human error in operations. A new discipline, called 'ergonomics' in Europe and 'human factors' in the United States, began to develop in the early 1950s and took off in the 1960s, as evidenced by a number of important references [9–12]. The area covered by human factors ranges from the adaptation to operators of control rooms or workstations, through to training, task or work load analysis, etc. Development is most active in the field of aeronautics, given the necessity of reducing human errors seen to contribute to almost two thirds of air disasters [13].

From the start of the 1960s work was carried out in the development of human reliability predictive methods, especially by Meister [14, 15] and Swain [16–18]; both are American Army psychologists. An outcome is THERP (Technique for Human Error Rate Prediction) [4, 16, 17]. The first human reliability data bank was created in the 1960s as well, by the American Institute for Research (AIR). The bank provided human error rates for so-called basic tasks, based on statistical analyses of electronics industry failures [14, 19, 20]. By the end of the 1960s another data bank existed at the Sandia National Laboratories [21, 22].

During the early 1970s, in the context of a Probabilistic Risk Assessment carried out in two nuclear power plants in the United States (the 'Rasmussen Report' [23]), the first large-scale human reliability assessment was performed. Published in 1975, the study took into account human errors from hardware maintenance to post-accident operation. Led by Swain, these assessments used the methods already developed and contained human reliability data: they represented thus the 'state of the art' for the mid-70s. This HRA inspired numerous studies in all parts of the world.

A combination of design faults, equipment failures and human errors, the accident at the Three Mile Island nuclear power plant then played an important role in bringing forward the importance of the human factor. As stated by the investigating commission [24], 'what the Nuclear Regulatory Commission and the nuclear industry underestimated, is that the people who run the plants are in themselves an important safety back-up system.'

Since that time, probabilistic risk assessments in nuclear plants have multiplied and generally include an HRA. These studies go hand in hand with methodological development aiming at:

- rendering HRA methods more reliable and more exhaustive;

- modelling operator behaviour, particularly for the prediction of diagnostic and strategic errors;

- improving the qualitative and quantitative data on which these models might be based.

At the request of the American authorities, Swain and Guttmann published in 1983 [25] an important treatise on methods and data in HRA applied to nuclear power plants. The first international conferences and workshops were organized by the IEEE (Institute for Electrical and Electronics Engineers) on the topic of human factors and nuclear power safety [26, 27]. A recent general work on human reliability analysis [28] reflecting the American approach (with nuclear power plant applications) deserves mention.

In Europe as well, numerous studies were carried out in the nuclear domain. In Denmark, Rasmussen [29–36] carried out studies on the variability of human performance and on the causes and mechanisms of human errors. Carnino led groups of experts in research carried out under the aegis of the Commission of the European Communities [37].

In France, in the field of aeronautics, predictive accounting for human factors and errors (including for example the use of simulators) has been a constant concern in the evaluation of safety system dependability for the Concorde and Airbus aircraft [38–40]. In the nuclear domain, Electricité de France (EDF) introduced the concept of human redundancy [41] in terms of both personnel and equipment for the diagnosis of incidents or accidents. EDF has also developed the use of simulators for training and study of operator behaviour in incident or accident situations [42–44]. Among the outcomes have been the definition and widespread use of methods for the analysis of incidents due to human error [45]. HRAs have also been carried out at EDF in the context of safety evaluations for nuclear power plants; the experience acquired in this domain forms the basis for this chapter. Finally, two works published merit citations: one on the subject of human reliability by Leplat [46], the other on the role of human factors in accidents by Nicolet *et al.* [47].

An emergent current in the field of human factors focuses attention on reintegrating the information obtained through assessment methods into real-time operation. This focus calls upon concepts from the fields of social and cognitive psychology. Traditionally, human factors studies have been the domain of experts; however, involving personnel from all parts of the organization in decentralized risk analysis [48] has shown promising worker safety results in several industrial contexts.

16.3 BEHAVIOUR OF THE HUMAN OPERATOR

Following a sketch of how the human operator functions, we will enumerate the principal aspects of human behaviour which can aid in explaining errors. The tasks carried out by operators in large industrial systems are then classified. Finally, we define human errors and their principal characteristics.

16.3.1 Functioning of the human operator

How does the human operator function? The answer to this question remains a subject of research. Nonetheless, human functioning while a task is performed can be schematically described according to several phases:

- *Acquisition of information*. Calling upon the human sensory apparatus, of which the receptors include the eye, ear, inner ear, limbs, the entire body, etc.

  Perception is not a simple copy of outside information in the mind of the operator. Rather, subtle mechanisms, often unconscious, of selection and inhibition are called into play. One and the same stimulus may be perceived differently according to the context in which it is embedded, the experience and the needs of the operator; it may even not be perceived at all in certain cases.

  The resulting conclusion can be of prime importance in understanding human errors: putting an instrument in front of the operator does not guarantee that the information it provides will be received!

- *Information processing*. The information gathered is processed in the mind. Various types of processing can be identified, among which are:
  - logic processes (e.g. diagnosis of an accident);
  - calculation of parameters (calculation of foreseeable evolution in a parameter, calculation of a new parameter by applying mathematical rules, etc.). In general, this type of processing requires access to thought programs stored in long-term memory, while the current information is held in short-term memory. It is generally agreed that long-term memory is the seat of the representation or models constructed by the operator of the operating system's reality. This active construction evolves, of course, as a function of training received and accumulated experience.

- *Decision making*. Processing information leads the operator to decision choices. For example:
  - undertake no new action but allow the system to evolve;
  - undertake corrective actions to reduce the gap between actual values and desired values of the parameters observed.

  The decision is made according to the predefined cognitive model as well as goals selected by the operator such as:
  - goal of availability: the operator gives priority to system availability;
  - goal of safety: the operator gives priority to safety;
  - personal goals: for instance, the operator gives priority to finishing his task as quickly as possible....

- *Physical responses*. The operator's 'output', by which he accomplishes (whether appropriately or not!) his task. He will use, for instance, hands, feet or voice.

The functioning of the human operator is represented schematically by Figure 16.1. In summary, every task involves sensory activity, mental activity and physical activity.

Contrary to a long-held belief, mental activity is present in all tasks, including repetitive work, even that limited to a few simple movements. The mental activity of the person is simply more or less demanding and complex according to the task.

Needless to say, human functioning is much richer and more complex than Figure 16.1 can convey, in that there is constant interplay among the sensory, mental and physical levels.

Two aspects of this representation deserve to be taken up in detail: one is the sequential arrangement of sensory, mental and physical activity; the other concerns the role played by reflex or conscious mental activity in shaping behaviour. Wanner, a human factors specialist in French aeronautics, attempts the following demonstration [39]:

> 'Operations such as reading an instrument, interpreting the reading, elaborating strategies and tactics, acting upon the controls, etc., are accomplished in succession and not in parallel, contrary to widespread belief. To attack this statement, one might point out for example that a person is capable of taking notes at the same time as she listens to a lecture. In reality, though, speech is very redundant: it is not necessary to hear every word in a phrase nor every syllable of a word in order to understand. Thus the brain can function on a "time sharing" system, alternating between listening (data gathering) and interpreting and then transcribing on paper. It should be noted as well that once the order to write a word has been conveyed by the brain, part of this operation can be carried out by what we would call "reflex", leaving the brain available for data gathering operations. A certain proportion of operations thus are carried out in a reflex mode, leaving the mind of the operator free for other

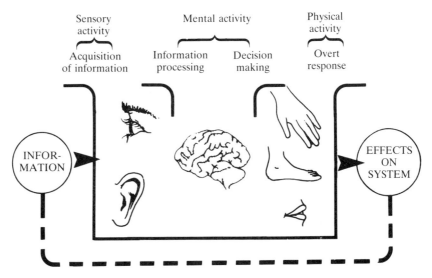

Figure 16.1 Schematic representation of human operator functioning.

operations such as reading, interpretation, elaboration, handling controls, ... Indeed, it is during these "reflex" periods that the brain can rest or be available for new tasks. This is particularly striking in the case of driving an automobile, the greater part of which is accomplished in reflex mode. Every driver has noticed that often one doesn't remember having passed some familiar part of the route. It is thanks to this kind of "automatic piloting" that one can drive six or eight hours (whereas new drivers who have not yet had the practice to acquire this reflex skill are often incapable of driving more than one hundred kilometers without becoming exhausted). Thanks to this mode too, one can listen to the news on the radio while driving; however, it should be noted that such reflex piloting will give way to vigilant piloting as soon as an incident arises. What driver has never missed some interesting and awaited information on the radio because he has returned to vigilant driving, which has forced his brain to give itself over entirely to the task of controling the car, abandoning meanwhile the auditory information?.'

After analysing a number of human errors encountered in the running of nuclear power plants, Rasmussen [36] proposed three major classes of human functioning or behaviour. They are distinguished mainly by the relative importance of reflex or conscious mental activity. These classes are:

- *Skill-based behaviour*. Behaviour which is practically automatic, where the operator 'acts mechanically'. Conscious mental activity is hardly activated, while the operator limits himself to reproducing in almost reflex manner a series of perfectly mastered behavioural schemas, acquired through practice or intensive training. The operator displays this behaviour when he must react very quickly and almost automatically to an important alarm.

- *Rule-based behaviour*. Behaviour calling on conscious mental activity and consisting of a coordinated execution of tasks according to learned or written procedures or rules. The rules in question are not well enough integrated by the operator to be applied automatically.

    The tasks which can be accomplished by rule-based behaviour do not demand complex choices among several options. This type of behaviour is displayed for example when the operator carries out routine checks or calibrates a sensor.

    This type of behaviour is targeted by HRA through a process of breakdown into basic tasks and application of a method such as THERP (see Section 16.6.2).

- *Knowledge-based behaviour*. This is the behaviour of an operator in less familiar situations, based upon a complex conscious mental activity seeking to resolve problems and plan tasks. This behaviour is displayed by the operator when he proceeds to diagnose a complex and unaccustomed situation, drawing on numerous elements of information; or when he elaborates a new strategy to master an accident after observing that the procedure provided is, in the given case, inadequate.

Note that knowledge-based behaviour can also take the form of discrimination in the choice of future behaviour, as when the operator hesitates between rule-based behaviour (a well-practised routine, for example) and knowledge-based behaviour (less practised, but which could allow him to gain time, for example).

16.3.2. Some outstanding characteristics

The operator's behaviour is of course very different from that of the equipment which he has designed or operates [49]. It is characterized notably by intelligence and by the capacity for innovation and invention, by the faculty of adaptability and by the extreme sensitivity of some of his sensory receptors. Some characteristics linked to these special qualities should be looked at more closely to gain understanding of the mechanisms of human error.

- **Variability** We can never carry out the same action twice in an absolutely identical fashion. Human performance is greatly variable. This variability is linked to the complexity of the sensory, mental and physical processes called into play for task resolution. Variability implies a certain percentage of actions that fall 'outside of tolerance', which can be the basis of human errors. In fact, some degree of variation can be tolerated for many tasks; human error appears when the performance variations exceed the tolerance limit.

  This type of error is the counterpart of man's qualities of adaptability, which allow him to face changing circumstances but which play against him when the task at hand calls for rigorous performance of a series of actions.

- **Need for information and predictive capabilities** In general, an absence of information is felt as unpleasant by the human mind; our response is to seek information by all our available sensory channels. Wanner describes well this phenomenon [39]:

  > 'Everyone has noticed that he is incapable of staring for more than five seconds at the red light that has stopped him at a corner: the red light provides, in fact, nul information! Thus one's gaze tends to seek out information all around, useless information as it happens, but information all the same. It is only once we have picked up the green light turning yellow on the cross street that we are ready again to watch the red light, because now we have the means to make a prediction: we know that in less that ten seconds the light will turn green again. Anticipating the predictable event renders the absence of new information tolerable; but should the wait be longer than predicted, the eye wanders off again in search of new information.

This example, chosen among many other possible ones, sheds light on the concept of prediction. This faculty of the human operator for extrapolating from the current situation has two consequences:

- a positive consequence in that the operator thus can give part of his attention to watching other parameters once he has observed that a given parameter, after checking, is evolving favorably;

- an unfavorable consequence consisting of making a bold extrapolation of a stationary situation and believing, in consequence, that nothing new will happen in the next hours....

Should this situation be accompanied by an absence of information (if the situation is stationary, the parameters are constant and thus the information provided by instruments or the outside world is nul), the operator concludes that no incident is possible and in order to remedy the lack of information, creates some artificially by thinking of something else. He projects his own private inner film, separating him entirely from the process he is supposed to be tending; this is what we generally call loss of vigilance.'

- **The capacity to compensate for variations in task difficulty by increasing workload without introducing performance variations** An increase in the difficulty of a task does not always produce diminished operator performance. Thus,

    'The precision with which a task is accomplished is maintained independently of its difficulty, for the operator increases the frequency of his checking operations, hence augmenting his workload as the difficulty increases. Of course, the workload furnished by the operator cannot rise indefinitely. A maximum value exists, dependent upon what we will call the physical and mental state of the operator; beyond this maximum the operator can no longer compensate for increases in difficulty. At this point performance worsens sharply, one at least of the parameters necessary to function no longer being held at its reference value [39].'

- **Stress** Although some authors [50] equate 'stress' and 'stressful situation', the following distinction [51] seems clearer:

    - *Stressful situation*. A situation in which the subject perceives a gap between the requirements of the situation and his readiness to meet them.

    - *Stress*. An ensemble of physiological reactions to the stressful situation. Selye [51] makes a classic description of the 'general adaptation syndrome'; alarm reaction (notably, secretion of hormones such as adrenaline); resistance, then exhaustion if the situation goes on. This syndrome has the remarkable character of being independent of the type of stressful situation. Repeated stressful situations can lead to sickness.

Stress has psychological repercussions: these take the form of a feeling of tension and have significant impact on performances. In extreme situations, stress can go so far as to 'paralyse' the operator.

16.3.3 Classifying tasks

The tasks carried out by an operator in large industrial systems can be classified in the following manner:

- simple tasks;
- complex tasks;
- tasks requiring vigilance;
- verification tasks;
- post-incident or post-accident tasks.

Let us spell out the principal characteristics of these tasks:

Simple tasks: these correspond to actions making up part of sequential operations, requiring slight decision making (example: go and open a manual valve) [52];

Complex tasks: corresponding to sequential operations which are well enough defined but which require decision making; some latitude is left to the operator (example: carrying out of an accident diagnostic).

Tasks requiring vigilance: corresponding to signal or alarm detection; the probability of error here is affected by the length of the wait, the degree of motivation, the type of signal, its frequency and the type of action which must be engaged when it is recognized [52].

Verification tasks: these correspond to the surveillance and verification of multivariant processes where action decisions must be made. The operator must prevent disturbances from evolving into serious failures.

Post-incident or post-accident tasks (or emergency tasks [52]): corresponding to an operator's activity after an incident or accident; these can vary considerably, ranging from an automatic learned response to the seeking out of a new strategy. The probability of errors occurring in these tasks rises sharply when the incident is sufficiently serious that the operator can find himself in a state of elevated tension.

This classification of tasks is useful for a first conceptualization of the associated human errors; further on we will look at the probabilities that can be attributed to these errors.

16.3.4 Human errors

Intuitively, error is defined as a deviation from an action, an action sequence or a strategy which is thought to be optimum and which serves as a reference.

Error results from a 'malfunction' on the level of the operator's sensory, mental or physical activity. From the system's point of view, error need not be taken into account until it produces an inappropriate action upon the system; the effects of such an action are then called error mode. These concepts will be examined in Section 16.4.

Human errors are often dependent: one human error may bring about another. Predictive analyses must imperatively take these dependencies into account.

Ergonomists in particular have put the accent upon the role of the work situation in human error. Swain has defined a list of 'performance shaping factors' [25] affecting human activity and forming possible sources of error. He distinguishes external factors (work context, equipment), internal factors (training, physical condition) and stress factors stemming from physiological or psychological sources (see Table 16.1).

The influence of these factors explains the great diversity of causes of human error in man–machine interface, as numerous studies have demonstrated [53].

Accidents observed in various industries reveal some 'surprising', but genuine, behaviours:

- Numerous flight staffs, devoting their entire attention to holding difficult parameters during landing, have been unable to see that the 'landing gear up' signal was lit! Even more striking, when all members of one piloting team were busy locking in emergency landing gear, not one noticed the rapid descent of the aircraft nor heard the ground proximity warning system sound off; the aircraft crashed [38].

- In a refinery control room, warning lights started flashing and instruments gave abnormal readings. A fire was at the source. The first reaction of the operator present was to leave the control room to find the instruments technician and ask him 'what's going on with the instruments?'; in this way, he lost precious time in limiting the effects of the fire [25].

A certain number of human malfunctions can explain such errors. The following list is not exhaustive:

- *Saturation.* The human operator is unable to absorb an overload of information coming at him concerning parameters and their variation. In such cases he cannot proceed to make a global analysis.

- *Loss of vigilance.* The human operator no longer seeks to gather pertinent information, but starts thinking of other things, or drops off, or gets stuck on a relatively insignificant problem. He may not notice then that a parameter is evolving dangerously, or he may react too late.

- *Incredulity.* The human operator does not believe the information transmitted to him. In case of accident, the operator who is overly confident in the design of his hardware ('something like that can't happen') will first condemn the alarms which go off 'for no reason'! This can lead him to construct a reassuring explanatory model, even if it implies ignoring contradictory pieces of information or making up incorrect explanations for them. Such a 'model' may not not be called into question by its contradictions until very late in the game; this was the case in the Three Mile Island accident (see Chapter 15).

- *Stress.* The human operator recognizes immediately how serious the accident is and the risks it implies—including the risk to his own life—and undergoes weighty psychological tension leading him to react too late or to take inappropriate actions.

What sort of mechanisms can explain such malfunctions? At the current state of knowledge, the answer appears difficult to find. Rasmussen did, though, suggest certain mechanisms of human operator malfunction in large industrial systems:

- Conscious mental activity is activated but consists of validating an intuitive hypothesis based on an incomplete subset of information; the operator believes he perceives the event he is expecting as soon as a certain minimum of information is obtained.

- Conscious mental activity is activated and adequate information is present but the processing of this information is incorrect owing to difficulty in taking into account available information, especially in case of tension.

- Incorrect task execution can be due to insufficient development or defects in the model used by the operator.

Understanding and predicting such error mechanisms obviously require, on the one hand, a thorough analysis of the mental activity implied by a given task and, on the other hand, knowledge of the models used by the operator to represent reality.

Furthermore, and this fact is fundamental, the human operator is capable of realizing his errors and of correcting them. When he observes that his own errors are difficult to correct, he may sometimes invent a new strategy to void or minimize their effects. Such factors which may bring the operator to discover his errors to prevent them or to limit their effects, will be called recovery factors. As examples of recovery factors, we can cite alarms tied to the effects of human error, checks performed by an independent operator, etc.

An HRA must take into account all such recovery factors. Indeed, an error does not always have immediate effects. Recovery factors in the system can enable the operator to become aware of error and take steps to block its potential before serious effects appear. Hence the importance of recovery factors for both system design and reliability assessment.

Table 16.1 Some performance-shaping factors (PSFs) in man–machine systems [25]

External PSFs		Internal PSFs
Situational characteristics:	Task and equipment characteristics:	Organismic factors:
Those PSFs general to one or more jobs in a work situation	Those PSFs specific to tasks in a job	Characteristics of people resulting from internal & external influences
Architectural features	Perceptual requirements	Previous training/experience
Quality of environment	Motor requirements (speed, strength, precision)	State of current practice or skill
• temperature, humidity, air quality and radiation	Control-display relationships	Personality and intelligence variables
• Lighting	Anticipatory requirements	Motivation and attitudes
• Noise and vibration	Interpretation	Emotional state
• Degree of general cleanliness	Decision-making	Stress (mental or bodily tension)
Work hours/work breaks	Complexity (information load)	Knowledge of required performance standards
Shift rotation	Narrowness of task	Sex differences
Availability/adequacy of special equipment, tools, and supplies	Frequency and repetitiveness	Physical condition
Manning parameters	Task criticality	Attitudes based on influence of family and other outside persons or agencies
Organizational structure (e.g. authority, responsibility, communication channels)	Long- and short-term memory	Group identifications
Actions by supervisors, co-workers, union representatives, and regulatory personnel	Calculational requirements	
Rewards, recognition, benefits	Feedback (knowledge of results)	
	Dynamic vs. step-by-step activities	
	Team structure and communication	
	Man–machine interface factors:	
	• Design of prime equipment	
	• Test equipment, manufacturing	
	• Equipment, job aids, tools	
	• Fixtures	

	Stressor PSFs	
	Psychological stressors:	
	PSFs which directly affect mental stress	
	Suddenness of onset	
	Duration of stress	
	Task speed	
	Task load	
	High jeopardy risk	
	Threats (of failure, loss of job)	
	Monotonous, degrading, or meaningless work	
	Long, uneventful vigilance periods	
	Conflicts of motives about job performance	
	Reinforcement absent or negative	
	Sensory deprivation	
	Distractions (noise, glare, movement, flicker, colour)	
	Inconsistent cueing	

Job and task instructions: Single most important tool for most tasks	Physiological stressors: PSFs which directly affect physical stress
Procedures required (written or not written) Cautions and warnings Work methods Plant policies (shop practices)	Duration of stress Fatigue Pain or discomfort Hunger or thirst Temperature extremes Radiation G-force extremes Atmospheric pressure extremes Oxygen insufficiency Vibration Movement constriction Lack of physical exercise Disruption of circadian rhythm

16.4 MAJOR CONCEPTS

The concepts linked to human reliability form a subject of controversy. For this reason, we will put forward and concentrate upon definitions which are coherent with those used in system safety analysis. This should encourage the integration of HRA in predictive analyses relating to the safety of engineered systems.

- **Human error (or error)** Gap between an operator's actual behaviour and behaviour defined as appropriate, when this gap exceeds acceptable limits under given conditions.

'Given conditions' are here defined as excluding malevolent intent. The definition of human error thus excludes malevolent behaviour.

This definition contains no notion of blame and thus no moral judgement is implied.

'Acceptable limits' may be defined according to:

- the characteristics of the prescribed action (as defined by professional judgement or by formal procedures);
- the consequences of the human error, especially on the system.

These limits may be explicit or implicit, or become explicit only after error is manifested.

As an example, imagine the case of an operator who reads 220 °C on a temperature gauge rather than 200 °C; a 5 °C error at most in reading is admissible given the scale graduation. Moreover, a reading error of less than 10 °C has no consequences for the installation.

According to his manner of defining acceptable limits, the analyst will or will not consider the reading to be a human error!

A priori, we can advise the analyst to define acceptable limits in function of the prescribed action and independently of consequences. His judgement thus can be brought to bear on the consequences as a function of specific circumstances. This allows us to classify under human error those actions which have no consequences in one set of circumstances, but which can have serious consequences in another.

By 'given conditions', we mean for example:

- operational conditions under which the operator must accomplish his tasks;
- the aptitudes required of the operator: he must be seen to be apt (physically and intellectually) to carry out the tasks.

Let us take an example: the operator must go and close a valve in a given place, but the presence of a fire prevents him from carrying out his task. Was there human error? No, because the operational conditions were not fulfilled and the operator is thus incapable of completing his task; the term 'human incapability' should be used instead.

Human errors can be classified according to the different phases of a system life-cycle during which they may be committed: design errors, manufacturing errors, assembly errors, operating errors, etc.

As far as operating errors are concerned, we can distinguish, for example: errors in operation under normal, incident or accident conditions, maintenance errors, test errors, etc.

Human errors committed in specific tasks generally earn names used in everyday language: reading error, diagnostic error, error in calculation, typo, etc.

Three other types of human error can also be distinguished:

- omission: non-performance of a required action (or task);

- commission: incorrect accomplishment of a required task, or accomplishment of a task at other than the required moment;

- non-required action: accomplishment of a task not required.

- *Human incapability* (or incapability). Gap between the operator's actual behaviour and behaviour defined as appropriate, when the required conditions for completing the actions are not present. For example, a fire prevents the operator from carrying out his task. 'Required conditions' for task accomplishment are not found, thus there is no human error but rather incapability.

- *Human failure*. Halt to the human operator's own ability to accomplish a required mission.

What are the causes of human failure? Generally, three categories of cause can be distinguished:

— human error;
— human incapability of internal origin (sickness, heart attack);
— human incapability of external origin: disturbance in operator's working conditions (fire, flood), insufficient training, food poisoning, etc.

Note that in large industrial systems, precautions are taken to reduce the likelihood of human incapability or to limit its effects; whence, for example:

— the 'dead man' system on trains bringing the train to an automatic stop if the operator does not manifest his presence by pressing a button regularly;

— serving meal trays from different sources in order to prevent simultaneous food poisoning of both the pilot and copilot of an aircraft.

- *Human failure mode.* Effect by which a human failure is observed.

- *Error mode.* Effect by which a human error is observed. In practice, error and error mode are often confused. In order to clarify these concepts, let us take an example. The operator forgets to close a valve, although he was personally in condition to do it and the required operating conditions were present.

    – The human error here is: the operator forgets to close the valve;
    – The error mode is: omission of valve closing (effect seen in operator's responses); the valve stays open (effect on the system).

- *Incapability mode.* Effect by which a human incapability is observed.

- *Human fault.* Prolonged state of a human operator characterized by the inability to accomplish a required mission.

This concept deserves some explanation; indeed, contrary to hardware failures, human failure does not necessarily lead to human fault. 'Failure' is an event, as distinguished from 'fault', which is a state. Let us imagine that an operator has a series of tasks to execute; he commits an error in performing one of these tasks. Has he become inept to continue? No, he can go on to correctly execute all the rest of his tasks, and even correct his previous error. The term 'human fault' is thus reserved for characteristic, prolonged inaptitude: some examples might be the inaptitude resulting from taking medicines producing drowsiness or sleep in the operator, or the inaptitude resulting from a profound error in accident diagnosis by an operator.

Note that everyday vocabulary makes a distinction between 'make an error' (human failure) and 'to be in error' or 'be mistaken' (human fault).

- *Human reliability.* Outside the general meaning of 'science of human failures', this term has the following meaning: 'aptitude of a human operator to accomplish a required mission, in given conditions, within a given time period.'

It is supposed that, at the beginning of the mission, the operator has been recognized as apt to carry it out.

By 'given conditions', we mean the conditions under which the mission is or will be completed. Examples: outside environmental conditions (temperature, luminosity, etc.); task execution conditions (time constraints, etc.). These conditions are not always easy to identify and list, inasmuch as the human operator is sensitive to such a wide range of physiological, psychological and sociological factors.

Human reliability is measured by the probability that a human operator will accomplish a required mission, in given conditions, within a given time period.

- *Human availability*. The ability of a human operator to be in a state permitting the accomplishment of a required mission under given conditions and at a given time.

  Availability is generally measured by the probability that a human operator will be in a state permitting the accomplishment of a required mission at time t under given conditions.

- *Human maintainability*. Aptitude of a human operator to maintain himself in or to restore himself to a state in which he can accomplish his required mission under given conditions.

  In order to maintain this aptitude at a constant level over time, the operator will for example participate periodically in update training to refresh his knowledge. In order to facilitate the return of an operator to a functional state, nuclear power plant managers for example will provide procedures after an incident or accident.

  Two different measures are associated with human maintainability: automaintainability and recoverability.

  Automaintainability is measured by the probability that the human operator will return to appropriate behaviour (end of human fault) before time t, when he has left it at time $t = 0$.

  We have seen that certain human errors bring about only momentary inaptitude; in these cases, we define the recoverability of the error.

  Recoverability is measured by the probability that the human error will have been corrected before time t, given that it was committed at $t = 0$. Irrecoverability is measured by the probability that the human error will not have been corrected before time t, given that it was committed at $t = 0$.

16.5 PHASES OF A HUMAN RELIABILITY ASSESSMENT

The most frequently employed approach used in past years in the nuclear industry is that developed by Swain. The methods and models used are described in detail in [25]. Their sequencing is presented and illustrated by examples of applications in [54]. Other methods do exist, generally less complete, but sometimes better adapted to specific problems.

In order to guide the user, the Electric Power Research Institute (EPRI) has developed SHARP (Systematic Human Action Reliability Procedure) [55]; it aids in choosing among different methods and in conducting the study in a systematic and verifiable manner. SHARP defines the principal phases of an HRA, and gives advice on how to carry them out. At each step, SHARP indicates without exclusion the principal methods that may be used.

The approach presented here is an adaptation of SHARP, based on working experience gained at Electricité de France. It is illustrated by a simple example presented in Chapter 25.

In parallel to the HRA, the system with which the persons interact will also be studied. The two studies, closely interwoven, constitute the study of the man–machine ensemble; we will refer to it as the 'overall study'.

Although it was developed for probabilistic risk or safety assessments in nuclear power plants, the approach presented here can easily be adapted to other industrial contexts.

16.5.1 Identification of potential human errors

The first phase in the study is the search for 'unforeseen' actions (unforeseen by procedures, operating rules or practices) that could be effected by the operator.

These actions can be beneficial (spontaneous positive actions) or harmful (errors or operator incapabilities). However, to simplify, we will use the term 'errors' in what follows.

To guide the search, it is helpful to use error classifications. One can thus systematically seek out errors from each category. Hannaman [55] and Swain [25] give examples of typologies. One can also use the classification already described:

- non-accomplishment of a required action;
- incorrect accomplishment of a required action;
- extraneous action.

To find errors of the first two types, we must identify the required actions. We should thus consult procedures for operation, periodic testing and maintenance. The most important procedures will become the object of detailed examination, but we will not neglect the analysis of actions which appear secondary. Indeed, serious errors can be committed in carrying out relatively unimportant actions.

Example In an accident procedure, it is specified that the minimal flow line of a pump must be isolated. This action is of secondary importance. However, the minimal flow line valve control is located on the control panel just next to the pump command itself. There is thus a risk that the operator may confuse the two controls and stop the pump rather than isolating its minimal flow line. This action will have serious consequences.

Seeking out errors of the third kind (extraneous actions) is more delicate. What kind of actions may operators come up with in the course of complex scenarios, in circumstances which are hard to foresee (for example, situations already disturbed by previous errors)? The great adaptability of human behaviour makes such prediction difficult.

In order to alleviate this difficulty somewhat, we can look at the way operators have behaved in situations analogous to those studied:

- on similar real installations; this is the study of operating feedback (see Section 16.7.1.1);

- on simulators (see Section 16.7.1.2).

It is also useful, for all error types, to examine prior predictive studies.

Finally, it is essential to use the dependability study of the system. In the course of building a failure modes and effects analysis (FMEA), cause trees and consequence trees, we can pick up errors which might prevent the system from fulfilling its function.

Examining procedures, feedback from operating experience, results of simulator tests and previous HRAs; studying system analyses (FMEA, cause and consequence trees): these various approaches have partial overlap, but each one can provide new elements. It is thus desirable to use all of them in parallel to ensure the HRA is thorough.

16.5.2 Selecting significant errors

It is neither possible nor useful to make out detailed analysis of all the potential errors brought out by the preceding phase. Some selection must be made. We will retain only the errors judged to be pertinent to the study. The rigour of selection will obviously depend upon the level of detail chosen for the HRA in its ensemble. The weight to be given to an error depends on several parameters:

- its consequences: does the error lead directly to the 'undesirable event' or must it be combined with failures or other errors? One measure of this parameter is the order of the cut set or the minimal sequence in which the error may intervene.

- the probability of failures or other errors with which the error must be combined in order to come to the 'undesirable event';

- the probability of the error itself.

Depending on the degree of selection desired, we will use one or several of these parameters to sort the errors.

A purely qualitative selection can be made by eliminating from the start all the errors which intervene only in higher-order minimal cut sets. However, the order of an error may be compensated for to a certain degree by the high probability of this error or of the failures with which it may combine. This first method thus will allow us to eliminate without risk only those errors intervening at very high order.

We can also make a rough quantitative selection using two parameters: the order of the cut set and the probability of the failures which might combine with the error. We attribute a probability of 1 to all the human errors, and a realistic probability to hardware failures. We calculate the probabilities of the cut set, and eliminate all the errors which appear in cut sets having a probability inferior to a certain value. This threshold value must be chosen in a manner coherent for the entire overall study.

The most refined approach is, of course, that which uses all three parameters: cut set order, probability of combined failure, probability of the error itself. We proceed as described above, but this time attributing more realistic probabilities than 1 to the human errors, although still over-estimating since the goal of this phase is not to make a refined quantification.

At the end of this stage, we come out with a list of errors which it is judged useful to analyse in detail.

16.5.3 Detailed analysis of significant errors

The next step is now to gather all information which will allow us to estimate the probability of each error. This qualitative analysis is not just a preliminary to making quantifications. It also allows us to bring out which factors may contribute to the occurrence of errors, the weak spots in man–machine interaction. The subsequent introduction of probabilities is simply a way of hierarchizing errors, and thus of facilitating the choice among possible improvements which may be made on the man–machine system.

We must proceed thus to make an ergonometric analysis of each action which could be affected by the errors under consideration. This means familiarizing ourselves with the action; ideally, the analyst should be able to walk through it himself. For each action we will note:

- the characteristics of the action: complexity, time available, time necessary, integration in a sequence, etc.

- the characteristics of the man–machine interface: quality of equipment design; quality of written procedures (in form and in content); legibility of registers (measuring devices, alarms, etc.), arrangement of controls (labelling and distribution of controls on control panel or on the material itself; means of communication, etc.

- characteristics of the environment: heat, noise, lighting, difficulties in access, dangerous work area, protective clothing required, etc.

- organizational characteristics: task distribution, administrative rules (consignment of materials, use of procedures or of tools, checks, etc);

- ways of recovering error: means of detection (alarms, verifications), time limits and means of intervention, etc.

- consequences of the error.

This list recalls the prinicipal 'performance-shaping factors' mentioned in Section 16.3.4.

All this information should be sought out from the persons who know best how to run and to maintain the system: the operators who run and maintain similar systems; those who wrote up the procedures; system specialists.

It is best to examine, if possible, the material in context as well as the man–machine interface on existing installations. In order to pursue the analysis in the best conditions after the end of the visit, it is very useful to have taken detailed photographs.

The level of detail chosen for the analysis is a function of the objectives of the HRA and the quantification models which will be used.

By the end of this phase:

- new error possibilities may have been uncovered; they should be added to the original list;

- the weak spots in the man–machine interaction have been identified;

- the information needed to make quantifications has been gathered.

16.5.4 Integration with system modelling

The system has already been modelled with the help for example of cause or consequence tree methods.

Since the HRA is interwoven with the study of the system, it cannot be independently quantified. Thus the HRA's integration in the cause or consequence trees must be perfected. We use the term 'perfected' because a great number of errors will already be represented in the trees (see Section 16.5.1). Certain errors have been eliminated in phase 2 (selection), others have been uncovered in phase 3 (detailed analysis): the trees must be modified in consequence. The level of detail of representation in the trees is defined as a function of the data available to make quantifications. Thus, it is useless to detail the errors which we can only quantify in a global way.

Last, we must study the dependence between errors themselves and between errors and hardware failures. This is both an important and delicate point: important because dependence is a weighty factor in quantification, and because human errors, highly context-bound, are often dependent upon previous errors or failures; delicate because experience shows that these dependencies are difficult to predict. It appears indispensable to carry out an analysis similar to seeking 'common cause failures'. Thus we must look for common points among errors: actions using the same information, the same means of control, calling on the same reasoning, or the same diagnosis; actions making up part of a sequence, etc. The cause trees must then be modified in accordance with the dependencies which have been found.

16.5.5 Quantification

Once they have been introduced into the trees, errors are no longer distinguished from equipment failures. We must assign them probabilities, then calculate the trees with the help of the usual methods. Models are available which give probabilities of each error as a function of its characteristics, collected in phase 3 (detailed analysis). Some allow dependencies to be taken into account. These models are presented in Section 16.6.

After the overall study has been quantified, it is generally advisable to evaluate the sensitivity of the results to modification in the probability of the human errors carrying the most weight. Indeed, the high uncertainty of the probabilities assigned by current models means that a final result will be of little significance if it is unaccompanied by a sensitivity study or an error factor.

16.5.6 Presentation of the approach used and its results

Human errors depend on so many factors that an HRA will call upon numerous sorts of information, hypotheses and judgements. These must be made explicit in the final report if the HRA is to be verifiable. The report must then include next to each quantified tree the hypotheses, models and qualitative information drawn upon, as well as their sources.

16.5.7 Some remarks

The approach to HRA having been described, some general remarks should be made.

The linear succession of phases in our presentation is schematic. An actual HRA is an iterative process, with several back-and-forth passes over phases.

HRA is situated at the intersection of three domains of knowledge (see Figure 16.2):

- dependability, inasmuch as HRA is intimately linked with the study of the system and must be integrated in its modelling;

- ergonomics, since HRA includes a detailed analysis of tasks and man–machine interface (phase 3);

- system engineering (whether it be electric, mechanical, thermohydraulic, etc): HRA requires understanding the 'why' of each action, task constraints (time limits, etc), the consequences of errors.

This is why HRA must be made by generalists able to discuss with specialists of the three domains. It is desirable to organize the study in a way that will favour such dialogue. In particular, the HRA and the system study

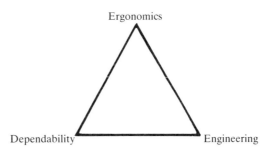

Figure 16.2.

must be led in close conjunction. The persons who are studying the system must at the same time participate in the search for errors; those who are charged with the HRA must follow the system study step by step, to be in a position to integrate the errors.

16.6 QUANTIFICATION MODELS

Quantification models allow us to match a probability to an error as a function of situation characteristics or 'performance-shaping factors' (see Section 16.3.4). The factors most often taken into account are:

- the complexity of the action;
- the time available;
- the design of the interface;
- the operator's stress;
- the operator's experience and training.

We will succinctly present here some characteristic models. For a broader survey, see Hannaman's presentation [56] of 16 models.

With one exception, the models presented here are oriented to the needs of nuclear industry. However, they can certainly be adapted to other industries.

16.6.1 TESEO

TESEO (Tecnica Empirica Stima Errori Operatori) was developed by ENI (an Italian oil company) for the oil industry [57, 58]. The probability of error is represented in the following form:

$$P[E] = K_1 \times K_2 \times K_3 \times K_4 \times K_5$$

The K_i allow us to take into account the following factors:

K_1: complexity of the action
K_2: time available
K_3: experience and training of the operator
K_4: operator's emotion (according to the gravity of the situation)
K_5: man–machine and environment interface.

The value of each K_i is given by a table of this type:

Complexity of the action	K_1
simple, habitual action	0.001
habitual action but requiring attention	0.01
unaccustomed action	0.1

This model is easy to use and takes into account a reasonable number of factors. However, it can be applied only to very quick actions (taking less than one minute). In particular, it seems little adapted to the quantification of complex diagnoses. Finally, it is only very partially validated by experimental data. TESEO should thus be used more as a tool of comparison between different designs of the man–machine system than for obtaining absolute probabilities.

16.6.2 THERP

As indicated above, Swain [25, 54] has authored a complete HRA method. Quantification techniques hold an important place in THERP (Technique for Human Error Rate Prediction).

This approach varies with the type of task considered: the elaboration of a diagnosis or the performance of operating or maintenance actions.

16.6.2.1 Elaboration of a diagnosis

An incident arises. The operators must effect a diagnosis, infer the actions to be taken, and then carry them out. The model gives the probability of failure of the diagnosis as a function of the time available for making it:

$$P[E] = F_i(t)$$

Three curves for $F_i(t)$ are provided (see Figure 16.3): the first is to be used when situation characteristics are favourable, the second when they are average, the third when they are unfavourable. We can thus globally take into account factors other than time.

The probability obtained covers the different ways of failing in diagnosis: absence of diagnosis, erroneous diagnosis. The model applies to the ensemble of an operator team within a nuclear power plant. It presupposes that the situation will be definitely detected as abnormal. It also presupposes that the

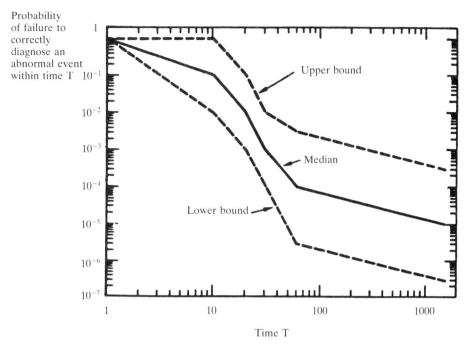

Figure 16.3 Model for diagnosis within time T of one abnormal event by control-room personnel (Swain's diagnostic model).

choice of actions to be carried out is immediate once the diagnosis has been made; the diagnosis dictates the choice of procedure, which in turn dictates action. In cases which deviate from this scenario, we must choose the 'pessimistic' curve or else introduce corrections.

The model is adapted to diverse time spans necessary to make a diagnosis (from one minute to several hours). It is easy to use. Although it is compatible with available experimental data, it is based principally upon the consensus of HRA experts in nuclear industry. It is thus preferable to calibrate the model with some experimental data specific to the context in which it is to be employed

The example given in Chapter 25 uses this method of quantification.

16.6.2.2 Carrying out operation or maintenance

The preceding model does not cover the performance of programmed actions. For tasks of this nature, the approach is quite different. We begin by dividing the task into elementary operations: read an indicator, press a button, open a valve, etc. Then we construct an arborescent structure such as a consequence tree, specific to the task in question. This diagram represents the error combinations possible among the elementary operations. An example of such a structure is given in Chapter 25.

The probability of each elementary error is then estimated:

$$P[E] = P_1 \times K \times P_2$$

where P_1 is the basic probability, function of the characteristics of the operation and of the man–machine interface.
K is a corrective coefficient which allows us to take into account the operator's stress.
P_2 is the probability of non-recovery of the error, in light of the recovery factors available.

These values are given by some twenty tables (probabilities of omission, of misreading, of error in selecting an indicator or a control, of detecting an error or an anomaly, of response to an alarm; stress coefficient). The situational characteristics which do not appear explicitly in the tables are taken into account in a global manner: we choose the upper limit in the table if conditions are unfavourable, the lower limit if they are favourable. Tables 16.2 and 16.3 give two examples.

Once the elementary error probabilities have been thus estimated, we calculate the consequence tree. A quantitative model allows us to introduce the dependencies.

Table 16.2 Human error probabilities (Swain and Guttmann [25], Table 20.7). Estimated probabilities of errors of omission per item of instruction when use of written procedures is specified* (from Table 15.3 [25])

Item†	Omission of item	Probability	Error factor (EF)
	When procedures with check-off provisions are correctly used:‡		
1.	Short list, ≤ 10 items	0.001	3
2.	Long list, > 10 items	0.003	3
	When procedures without check-off provisions are used, or when check-off provision are incorrectly used§:		
3.	Short list, ≤ 10 items	0.003	3
4.	Long list, > 10 items	0.01	3
5.	When written procedures are available and should be used but are not used§	0.05**	5

* The estimates for each items (or perceptual unit) presume zero dependence among the items (or units) and must be modified by using the dependence model when a non-zero level of dependence is assumed.
† The term 'item' for this column is the usual designator for tabled entries and does *not* refer to an item of instruction in a procedure.
‡ Correct use of check-off provisions is assumed for items in which written entries such as numerical values are required of the user.
§ Table 20–6 lists the estimated probabilities of incorrect use of check-off provisions and of non-use of available written procedures.
** If the task is judged to be 'second nature', use the lower uncertainty bound for 0.05, i.e. use 0.01 (EF = 5).

Table 16.3 Human error probabilities (Swain and Guttmann [25], Table 20.13). Estimated HEPs for selection errors for locally operated valves (from Table 14.1 [25])

Item	Potential errors	Probability	Error factor (EF)
	Making an error of selection in changing or restoring a locally operated valve when the valve to be manipulated is		
1.	clearly and unambiguously labelled, set apart from valves that are similar in *all* of the following: size and shape, state, and presence of tags*;	0.001	3
2.	clearly and unambiguously labelled, part of a group of two or more valves that are similar in *one* of the following: size and shape, state, or presence of tags*;	0.003	3
3.	unclearly or ambiguously labelled, set apart from valves that are similar in *all* of the following: size and shape, state, and presence of tags*;	0.005	3
4.	unclearly or ambiguously labelled, part of a group of two or more valves that are similar in *one* of the following: size and shape, state, or presence of tags*;	0.008	3
5.	unclearly or ambiguously labelled, part of a group of two or more valves that are similar in *all* of the following: size and shape, state, and presence of tags*.	0.01	3

*Unless otherwise specified, Level 2 tagging is presumed. If other levels of tagging are assessed, adjust the tabled HEPs according to Table 20.15.

This method is adapted to task execution where there is little thinking or time constraint involved. The probabilities are given for a single operator. Nonetheless, correction factors are suggested for taking into account the role of other team members in recovering errors.

This method is very complete; no equivalent exists. A good number of the models given appear to be experimentally validated to an acceptable degree. Others (dependence, stress) rest essentially upon expert judgement. However, because the method is very analytic, its use is quite laborious and requires numerous choices among tabulated values. We can thus fear that when a complex task is assessed, the final estimations reached by two independent analysts will be noticeably different. It is thus indispensable to consult the user's guide [54] and to attempt to validate, even if only partially, the hypotheses and models chosen.

An illustration of this method is given in the example at the end of Chapter 25.

In the final analysis, THERP models for quantifying diagnosis and manoeuvres constitute a unique package, both complete and well documented, and full of interesting qualitative information. THERP is thus much used for HRA in the nuclear industry.

16.6.3 HCR

HCR (Human Cognitive Reliability) was developed by EPRI (Electric Power Research Institute) [56]. Under this method, the probability of *absence of response* to an incident is expressed:

$$P[E] = F_i \left[t/T_{1/2} (1 + K_1)(1 + K_2)(1 + K_3) \right]$$

where t is the time *available* for the choice and execution of response.
$T_{1/2}$ is the estimation of median time *necessary*.
K_1 is a function of operators' competence.
K_2 is a function of their stress.
K_3 is a function of the characteristics of the man–machine interface.

The K_i factors are given by three tables. Depending on the complexity of the response choice process, we use one of the three F_i curves provided:

- one curve corresponds to the case in which the response choice can be made 'automatically' (as defined under 'skill-based behaviour' in Section 16.3.1);

- a second curve covers those cases requiring 'rule-based behaviour';

- the third curve is used when reflection and analysis are necessary ('knowledge-based behaviour').

Like THERP's diagnosis model, HCR applies to the whole of a nuclear plant operating crew. HCR has the advantage though of taking explicitly into account the degree of complexity of the situation and the principal characteristics of operators and man–machine interface. The normalization of t by $T_{1/2}$ allows it to be used for any given time span. The nature of the task is indifferent. Only its level of complexity counts. This is why the model appears to be as compatible with data obtained from nuclear power plant simulator tests as it is with laboratory data on completely different tasks.

16.6.4 Simulation models

The models presented above were specifically designed for HRA. Also deserving of mention are models developed in a broader context: the simulation of human behaviour by computers. Indeed, though these models were not designed with this single goal in mind, they can be used to quantify human errors.

OPPS (Operator Personnel Performance Simulation) [59], developed by Oak Ridge National Laboratory (ORNL), can serve as an example. OPPS models the activity of a nuclear plant operating crew following an incident. The four stages: detection, information processing, action, and error correction, are broken down into basic operations. Each operation is assigned a distribution of execution time. This distribution can be modified to take into consideration certain situation characteristics (type of signal, type of procedure, etc).

Errors are introduced during the action stage (and only this stage), with a certain probability of occurrence. The probabilities and distributions issue from operating experience and simulator trials.

Operator activity thus having been modelled, we can obtain through Monte Carlo simulation the distribution of execution time for the entire task. We can then infer the probability of success within a given time span. When these simulations are repeated using the distributions corresponding to each case, we can compare several man–machine interface designs.

The results of this model have been compared to operating experience for a type of incident. The median times obtained are very close, but real distribution shows much greater dispersion than is present in the model.

This model has numerous limits: it does not represent complex cognitive processes; it envisages errors only at the level of actions; it takes only a limited number of factors into consideration, etc. Nevertheless, its basic principle of using data from operating experience and simulator trials seems worth notice.

ORNL is carrying out parallel development of a much more ambitious model: simulating at the same time the behaviour of both operators and the nuclear installation itself [60]. Moreover, the operator simulation would be more complete, since even complex operations of situation analysis and decision will be represented. This model uses artificial intelligence techniques.

16.6.5 Conclusions

The models specifically designed for HRA—and THERP in particular—are those most used today by the nuclear industry for quantifying human errors.

All these models, including the simulation models, have an outstanding deficiency: their validation by real or experimental data remains rather weak. They are based upon disparate data, more or less adapted and completed by expert judgements. However, progress is being made, thanks in great part to the developing use of simulators for research aims (see Section 16.7.3).

This limitation signifies that some amount of caution must be exercised in quantifying errors. We must pay attention to the conditions under which the model may be used, the underlying hypotheses, and the data which served to validate them.

16.7 DATA

Our look at quantification models revealed one of the principal problems encountered by HRA: the lack of observation data on human behaviour gathered in real or experimental settings. Prediction thus often remains quite speculative. To resolve this difficulty, various methods are used to gather qualitative and quantitative data.

16.7.1 Data collection

16.7.1.1 Operating feedback

The most natural method of data collection consists of studying what goes on in analogous systems already in existence. We can carry out trials in order to identify intervention times and qualitative information. We can also inventory and analyse operating incidents. This last method is widely used in aeronautics and the nuclear field. A summary report is made of all the incidents which come under pre-established criteria, whether their cause is material or human. The most significant incidents are analysed in detail.

This compendium provides information useful to human factors experts and trainers: what must be improved in man–machine interface, in organization, in training? It is also a source of precious information for HRA: types of errors, modes of reasoning, operators' tendences. Finally some quantitative data can be obtained (mainly reaction times).

Nonetheless, this approach has important shortcomings:

- many errors do not lead to an incident, and thus are never recorded;

- it is often difficult to determine the number of opportunities to which the number of errors should be matched, in order to calculate frequencies;

- we do not have direct access to information, since we do not make real-time observations; thus the data can be altered by the protagonists, whether voluntarily or not;

- serious incidents are fortunately rare. The operating feedback thus gives us few elements on conduct during an accident situation.

To resolve the first of these difficulties, a collection of all events which might provoke an incident has been organized in American aeronautics: a summary report is made spontaneously by the individual involved or by a witness, even if there were no noticeable consequences. This system has been tried out in the nuclear field by EDF (Electricité de France) and INPO (Institute of Nuclear Power Operations) [61–62].

The other shortcomings of incident collections can be overcome by using other means, simulators in particular.

Despite its deficiencies in regard to HRA, the use of feedback from experience is essential. Indeed, human behaviour is highly context-bound. Observations made in a real context are thus indispensable for calibrating the other methods used (simulator or laboratory studies, or expert judgement).

16.7.1.2 Simulators

Simulators are composed of two parts:

- a mock-up (often on a 1:1 scale) of all or part of the control room or cockpit. The operators can act on the controls and follow the evolution of parameters on indicators;

- a computer which drives the mock-up, calculating the evolution of parameters as a function of the conditions defined at the start (a simulated breakdown, for instance) and the manoeuvres carried out by operators.

These simulators allow us to reproduce normal and accident conduct. They are fairly widespread in the nuclear industry and in aeronautics, where they are essential training tools. They can also be used to obtain data on operator behaviour. This use of simulators is growing in the nuclear industry, especially in the United States and in France [42, 44, 63–66].

Since 1980, EDF has organized periodic test campaigns devoted to the observation of operator behaviour. At each campaign, several accidents are simulated, each for up to a dozen operating crews. The crews come from their plants for this precise purpose. Each crew is composed of two operators, a shift supervisor and a shift technical advisor (STA).

How does a simulator test proceed? We explain the objectives to the crew, and ask them to act as they will, in the most realistic manner possible. We do not indicate the type of accident simulated. We guarantee anonymity and feedback on the results of the campaign.

The simulation lasts from one to several hours. During the test, the operators are filmed; the words exchanged are recorded as are several physical values characterizing the state of the reactor. Each of the two operators is followed by an observer who notes mainly the moments at which actions and errors are made. Another observer follows the dialogues and the behaviour of the shift supervisor and STA.

After the simulation, the crew is interviewed to learn what difficulties were encountered and to hear their remarks on the test, the control room, the procedures used. This dialogue also allows the observers to make sure they understood correctly the evolution of the test and the crew's behaviour.

A thorough analysis of these tests allows us then to obtain information on the quality of the control room, procedures, organization, and training. Precious information is also acquired for HRA: diagnosis time, execution time, error types and mechanisms.

As an illustration, Figure 16.4 presents the probability of non-diagnosis before time t (represented by points) as found in simulator tests carried out by EDF in 1984 [66]: the curve shows good agreement with the median curve proposed by Swain [25] (solid line). Note that the 31 simulator trials concerned six different incidents and accidents, and that the correct diagnosis was made in every case (average time: 5 min 30 s; standard deviation: 3 min).

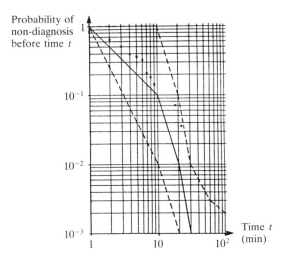

Figure 16.4 Some results of simulator tests made at EDF.

In addition, the errors turned up were closely analysed: about 50% were detected and corrected during the trial, either by their author (30% of the time) or by another member of the team (70%). Figure 16.5 gives the irrecoverability of these errors as a function of time.

Overall, simulators make an excellent source of data, in that they allow us to:

- study accidents which have never happened in reality;

- master certain performance factors;

- carry out fine observation, in real time and in retrospect (video recording);

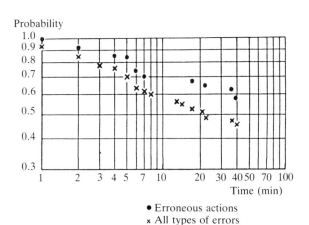

• Erroneous actions
× All types of errors

Figure 16.5 Irrecoverability of errors as a function of time.

- determine the number of error opportunities, for calculating probabilities.

Their drawbacks are:

- economic: each simulator-hour is very expensive, and often priority must be given to training use;

- methodological: the operators' situation during simulation is not perfectly representative of reality, even if the plant operation is very well simulated. In particular, the crew expects to be confronted with an accident. This must be taken into consideration when interpreting the results. To allow this, we advise carrying out systematic comparisons of real and simulated incidents.

16.7.1.3 Expert judgements

Expert judgements have been widely used to elaborate current models and they will continue to be used to fill in the gaps left by other data sources.

Methods have been developed to improve their reliability [67, 68]. These give criteria for the selection of experts, spell out judgement procedures and means of verification. The number of experts to be consulted depends upon the precision desired; in general, it is upwards of eight.

According to Stillwell [68], the methods most adapted to the needs of HRA are 'paired comparison' [69] and 'ranking/rating'. SLIM-MAUD [70], developed for Brookhaven National Laboratory, also deserves mention.

16.7.1.4 Laboratory experiments

Laboratory experiments permit us to study very specific tasks with great precision. But this specificity often renders difficult the direct use of their results for HRA. Nonetheless, Hannaman [56] was able to use experiments of this type to validate the HCR model (see Section 6.3). The results obtained are consistent with results of simulator tests.

16.7.2 Data banks

The information elaborated with the help of the methods presented above must be processed and grouped in data banks in order to be utilizable.

One type of bank provides purely qualitative information, for example standardized error descriptions. Information stemming from different sources is grouped in a homogeneous form, without undergoing statistical analysis.

The value of such a system is that it allows varied users (dependability specialists, trainers) to carry out on demand the specific statistical treatments of interest to them. The OECD Nuclear Energy Agency has put out a standard classification system for reporting errors [37] which could serve to structure such a bank. EDF and INPO are also working in this direction. Table 16.4 gives the principal characteristics used to describe each error by CONFUCIUS, the EDF data base [71, 72].

Table 16.4 Standardized description of human errors. Principal characteristics recorded for each error in CONFUCIUS data base [71].

Identification	Plant, type, data source, date . . .
Description	Free-style description of error
Number of opportunities	
Reactor state	Full power, cold shutdown . . .
Transient initiator	Type of incident during which the error was produced
Activity	Activity of the person making the error (operation, testing, maintenance)
Localization	Locale where the error was produced
Error mode	Mode (non-performed, ill-performed . . .) and operation affected (control, diagnosis, tagging, etc.)
Psychological mechanism	Basic mechanism (unperceived, error of logic, motor clumsiness, . . .), and global mechanism (inadequate image of goal)
Hardware affected	Identification numbers of affected system component
Interface	Error was committed in using which element of man–machine interface (display, control, procedure . . .)?
Interface quality	Characteristics of the interface
Required interface	Was the operator supposed to use an interface element (procedure . . .)? If so, did he?
Dependence	Dependence of the error in regard to other errors
Dynamic factors	Was the operator distracted? Did some previous change favour the error?
Operation characteristics	Complexity, number of parallel operations, time constraints, frequency of the operation
Environment	Temperature, sound level, difficulties in access, . . .
Operator status	Department, function
Operator characteristics	Activation level, motivation, number of hours of work previous to error, experience, training, . . .
Actual consequences	Consequences for equipment, safety and production
Potential consequences	Same categories as above
Detection	Moment of detection, department and function of detector, circumstances and means of detection
Required verification	Was a verification of the operation on which the error took place prescribed by rules of procedures? If so, had verification been made?
Recovery	If error was recovered: when? If not, was it in fact recoverable?
Remedial actions	Actions undertaken to prevent the renewal of the error

The characteristics (about one hundred in all) are entered into the data base via an input program which submits to the user's choice the various options available under each heading (see Table 16.4).

But specialists must also have immediate access to needed quantitative data. Several human reliability data banks have been designed to meet this objective [73]. They often have a very analytical approach (probabilities of error for basic actions such as 'read a gauge' or 'press a button'), like the American Institute for Research (AIR) or OPREDS (American Navy) banks; otherwise they are highly specific to one type of activity (the Aerojet bank for maintenance of the Titan II rocket propulsion system); or else, finally, they lean heavily upon expert judgement (THERP data base).

16.7.3 Range of values

Here we mark out the range of values generally found in estimates of human error probabilities.

The probabilities of error per action generally fall between 1 and 10^{-5}. A common value is 10^{-3}, corresponding to an action of moderate difficulty. Hannaman [55] provides the following distribution:

- 5×10^{-5} to 5×10^{-3} for 'automatic' act;
- 5×10^{-4} to 5×10^{-2} for 'ruled-based' acts;
- 5×10^{-3} to 5×10^{-1} for 'knowledge-based' acts.

The probabilities depend greatly on the nature of the task. Thus, for signal response tasks which call simply upon vigilance, Ablitt [74] gives the following error probabilities (function of available response time):

t(min):	1	5	10	>10
P:	10^{-1}	10^{-2}	10^{-4}	10^{-5} to 10^{-6}

For complex diagnosis tasks, the probabilities become much greater. Swain [26] finds (taken from Figure 16.3, Section 16.6.2.1):

t(min):	1	5	10	20
P:	1	2×10^{-1}	10^{-1}	10^{-2}

Probabilities depend as well on situation characteristics. In particular, stress has considerable impact on performances. Ronan [2, 25] finds a probability of useless or harmful action to be 0.15. In experiments carried out by the American army, Berkun (Lees [74]) finds that 30% of subjects exposed to a simulated attack (in ignorance of the simulation) are incapable of carrying out the action which would allow them to remove the danger.

Some human error probability estimates for simple tasks from one probabilistic risk assessment (US Reactor Safety Study, Chapter 19 [25]) are also given in Table 16.5.

Table 16.5 Human error probabilities used in US nuclear reactor safety study (PRA, 1975)

Human error probability	Activity
10^{-4}	Selection of a key-operated switch rather than a non-key switch (this value does not include the error of decision where the operator misinterprets situation and believes key switch is correct choice)
10^{-3}	Selection of a switch (or pair of switches) dissimilar in shape or location to the desired switch (or pair of switches), assuming no decision error. For example, operator actuates large handled switch rather than small switch
3×10^{-3}	General human error of commission, e.g. misreading label and therefore selecting wrong switch
10^{-2}	General human error of omission where there is no display in the control room of the status of the item omitted, e.g. failure to return manually operated test valve to proper configuration after maintenance
3×10^{-3}	Errors of omission, where the items being omitted are embedded in a procedure rather than at the end as above
3×10^{-2}	Simple arithmetic errors with self-checking but without repeating the calculation by re-doing it on another piece of paper
$1/x$	Given that an operator is reaching for an incorrect switch (or pair of switches), he selects a particular similar appearing switch (or pair of switches), where x = the number of incorrect switches (or pair of switches) adjacent to the desired switch (or pair of switches). The $1/x$ applies up to 5 or 6 items. After that point the error rate would be lower because the operator would take more time to search. With up to 5 or 6 items he does not expect to be wrong and therefore is more likely to do less deliberate searching
10^{-1}	Given that an operator is reaching for a wrong motor operated valve (MOV) switch (or pair of switches), he fails to note from the indicator lamps that the MOV(s) is (are) already in the desired state and merely changes the status of the MOV(s) without recognizing he had selected the wrong switch(es)
~1.0	Same as above, except that the state(s) of the incorrect switch(es) is (are) not the desired state
~1.0	If an operator fails to operate correctly one of two closely coupled valves or switches in a procedural step, he also fails to correctly operate the other valve
10^{-1}	Monitor or inspector fails to recognize initial error by operator. Note: With continuing feedback of the error on the annunciator panel, this high error rate would not apply
10^{-1}	Personnel on different work shift fail to check condition of hardware unless required by checklist or written directive
5×10^{-1}	Monitor fails to detect undesired position of valves, etc., during general walk-around inspections, assuming no checklist is used
0.2–0.3	General error rate given very high stress levels where dangerous activities are occurring rapidly

Table 16.5 (*cont.*)

Human error probability	Activity
$2^{(n-1)}x$	Given severe time stress, as in trying to compensate for an error made in an emergency situation, the initial error rate, x, for an activity doubles for each attempt, n, after a previous incorrect attempt, until the limiting condition of an error rate of 1.0 is reached or until time runs out. This limiting condition corresponds to an individual's becoming completely disorganized or ineffective
~1.0	Operator fails to act correctly in first 60 seconds after the onset of an extremely high stress condition, e.g. a large LOCA
9×10^{-1}	Operator fails to act correctly after the first 5 minutes after the onset of an extremely high stress condition
10^{-1}	Operator fails to act correctly after the first 30 minutes in an extreme stress condition
10^{-2}	Operator fails to act correctly after the first several hours in a high stress condition
x	After 7 days after a large LOCA, there is a complete recovery to the normal error rate, x, for any task

Note LOCA: Loss of coolant accident (break in the primary circuit)

These selected values demonstrate the great variability of error probabilities according to situation. Thus we must always be careful not to employ them without first checking on their precise conditions of use.

REFERENCES

[1] P. M. Fitts and R. E. Jones (1947) *Analysis of Factors Contributing to 460 'Pilot Error' Experiences in Operating Aircraft Controls*, US Air Force, Air Material Command, Memo Rep TSEAA-694-12.

[2] W. W. Ronan (1953) *Training for Emergency Procedures in Multiengine Aircraft*, Am. Inst. Res. Pittsburgh, Pa, Rep, AIR, 153-53-FR-44.

[3] J. Cooper (1961) Human initiated failures and malfunction reporting, *IRE Trans. Hum. Factors Electron*, **HFE-2**, 104.

[4] L. W. Rook (1962) *Reduction of Human Error in Industrial Production*, Report SCTM-93-62 (14) Sandia Corporation.

[5] L. W. Rook (1964) Evaluation of system performance from rank order data, *Human Factors*, **6**, 533.

[6] C. E. Cornell (1968) Minimizing human errors, *Space Aeronautics*, March, 49.

[7] R. L. Scott (1971) *A Review of Safety-related Occurrences in Nuclear Power Reactors from 1967–1970*. ORNL-TM, 3435.

[8] H. L. Williams (1958) *Reliability Evaluation of the Human Component in Man Machine Systems*, Electrical Manufacturing.

[9] D. Meister and G. F. Rabideau (1965) *Human Factors Evaluation in System Development*, New York, Wiley.
[10] D. Meister (1971) *Human Factors: Theory and Practice*, Series in Human Factors, John Wiley & Sons, Chichester.
[11] E. Edwards and F. P. Lees (1974) *The Human Operator in Process Control*, Taylor and Francis Ltd, London.
[12] C. G. Drury and J. G. Fox (1975) *Human Reliability in Quality Control*, Taylor and Francis Ltd, London.
[13] P. Toulouse (1981) Prise en compte de la fiabilité humaine dans la conception des avions civils, Revue Générale Nucléaire, n° 5, septembre–octobre.
[14] D. Meister (1964) Methods of predicting human reliability in man machine systems. *Human Factors*, 621–46
[15] D. Meister (1973) A critical review of human performance reliability predictive methods. *IEEE Transactions on Reliability*, 22, 116–23.
[16] A. D. Swain (1963) *A Method for Performing a Human Factors Reliability Analysis*, Monograph SCR-685, Sandia Corporation.
[17] A. D. Swain (1964) THERP, Sandia lab., Albuquerque, New Mexico, Rep. SC-R-64-1338.
[18] A. D. Swain (1964) Some problems in the measurement of human performance in man–machine systems. *Human Factors*, 687–700.
[19] D. Payne and J. W. Altman (1962) *An Index of Electronic Equipment Operability: Report of Development*, Am. Inst. Res. Rep. AIR, C43-1/62/FR.
[20] J. W. Altman (1962) *A Central Store of Human Performance Data*, Symposium on Quantification of Human Performance, Albuquerque, New Mexico.
[21] L. V. Rigby (1967) *The Sandia Human Error Rate Bank (SHERB)*, Sandia Lab., Albuquerque, New Mexico, Rep. SC-R-67-1150.
[22] A. D. Swain (1970) *Development of a Human Error Rate Data Bank*, Sandia Lab., Albuquerque, New Mexico, Rep. SC-R-70-4286.
[23] Reactor Safety Study (1975) *An Assessment of Accident Risk in US Commercial Nuclear Power Plants*, WASH 1400 (NUREG 74/014) US Nuclear Regulatory Commission.
[24] J. G. Kemeny (1979) *Report of the President's Commission of the Accident at Three Mile Island*.
[25] A. D. Swain and H. E. Guttmann (1983) *Handbook of Human Reliability Analysis with Emphasis on Nuclear Power Plant Application*, US-NRC-NUREG/CR-1278.
[26] *The Man–Machine Interface and Human Reliability: an Assessment and Projection* (1981) NRC/BNL/IEEE Standards Workshop on Human Factors and Nuclear Safety, Myrtle Beach, USA, NUREG/CP-0035.
[27] *Proceedings IEEE Third Conference on Human Factors and Power Plants* (1985) Monterey, USA.
[28] E. M. Dougherty, Jr. and J. R. Fragola (1988) *Human Reliability Analysis—A Systems Engineering Approach with Nuclear Power Plant Applications*, Wiley-Interscience Publications, New York.
[29] J. Rasmussen, K. Duncan and J. Leplat (1987) *New Technology and Human Error*, John Wiley & Sons, Chichester.
[30] J. Rasmussen (1973) *The Role of the Man–machine Interface in Systems Reliability*, Atomic Energy Commun. Res. Est., Risö, Denmark, Rep. Risö-M-1673.
[31] J. Rasmussen (1974) *The Human Data Processor as a System Component, Bits and Pieces of a Model*, Atomic Energy Commun. Res. Est., Risö, Denmark, Rep. Risö-M-1722.
[32] J. Rasmussen (1976) The role of the man–machine interface in systems reliability, in *Generic Techniques in Systems Reliability Assessment* (E. J. Henley and J.

W. Lynn, eds) Noordhoff Leyden, p. 315.
[33] J. Rasmussen and J. R. Taylor (1976) *Notes on Human Factors Problems in Process Plant Reliability and Safety Prediction*, Atomic Energy Commun. Res. Est., Risö, Denmark, Rep. Risö-M-1894.
[34] J. Rasmussen (1977) Man as a system component, in *Man–Computer Research* (H. Smith and T. Green, eds.) Academic Press, New York.
[35] J. Rasmussen (1980) Notes on Human Error Analysis and Prediction. NATO Advanced Study on Synthesis and Analysis Methods for Safety and Reliability Studies. Edited by G. Apostolakis, S. Garribba, G. Volta-Plenum Press.
[36] J. Rasmussen (1982) *Human Reliability in Risk Analysis* in *High Risk Safety Technology*, (A. E. Green, ed.) John Wiley & Sons, Chichester.
[37] J. Rasmussen, O. M. Pedersen, G. Mancini, A. Carnino, M. Griffon and P. Gagnolet (1981) *Classification System for Reporting Events Involving Human Malfunctions*, Commission of the European Communities, Report EUR 7444EN.
[38] C. Lievens (1976) *Sécurité des systèmes*, Cepadues Editions, Toulouse.
[39] J. C. Wanner (1989) Le facteur humain dans la conduite de grands systèmes. *Le Progrès Technique*, No.21, 1981.
[40] J. C. Wanner (1981) L'interface de l'homme et de la machine et la fiabilité humaine, *Revue Générale Nucléaire*, No.5.
[41] J. J. Mira (1983) *Actions Menées par Electricité de France pour minimiser les Erreurs Humaines*, Operational Safety of Nuclear Power Plants, AIEA, Marseille.
[42] M. Legaud, A. Villemeur and A. Oliot (1983) *Operator Actions Following Abnormal Transients: Tests on Simulators Anticipated and Abnormal Transients in Light Water Reactors*, American Nuclear Society, Jackson.
[43] M. Llory and J. F. Chaubaron (1981) *La fiabilité humaine et l'interface homme machine: actions entreprises à EDF*. Conférence à la Belgian Nuclear Society, EDF-DER, HT/13/47/81.
[44] A. Villemeur, J. M. Moroni, F. Mosneron-Dupin and T. Meslin (1986) *A Simulator-Based Evaluation of Operator's Behaviour*, International ANS Topical Meeting on Advances in Human Factors in Nuclear Power Systems, Knoxville, USA.
[45] M. Griffon-Fouco (1980) *Méthode d'analyse des incidents dus à des erreurs humaines dans des installations nucleaires*, Conférence ANS/ENS sur la sûreté des réacteurs thermiques. Knoxville, USA.
[46] J. Leplat (1985) *Erreur humaine, fiabilité humaine dans le travail*, Armand Colin.
[47] J. P. Nicolet, J. C. Wanner and A. Carnino (1990) *Man and Risks. Technological and Human Risk Prevention* Marcel Dekker, New York.
[48] C. Mays and M. Poumadere (1989) Decentralizing risk analysis in large engineered systems: an approach to articulating technical and socioorganizational dimensions of system performance, *Risk Analysis*, **9** (4).
[49] C. Bievenu (1981) L'homme ou l'automate? *Revue Genénérale Nucléaire*, Septembre–October.
[50] M. Reuchlin (1977) *Psychologie*, Presses Universitaires de France, Paris.
[51] B. H. Kantowitz and R. D. Sorkin (1983) *Human Factors*, John Wiley and Sons, Chichester
[52] E. Edwards and F. P. Lees (1973) *Man and Computer in Process Control*, London Inst. Chem. Eng.
[53] M. R. Murphy (1980) Analyis of Eighty Four Commercial Aviation Incidents: implications for a resource management approach to crew training, *Proceeding Annual Reliability and Maintainability Symposium*, USA.
[54] C. R. Bell, A. D. Swain (1983) *A Procedure for Conducting a Human Reliability Analysis for Nuclear Power Plants*, Final report, US-NRC-NUREG/CR, 2254.

[55] G. W. Hannaman and A. J. Spurgin (1984) *Systematic Human Action Reliability Procedure (SHARP)*, Electric Power Research Institute, EPRI NP-3583.

[56] G. W. Hannaman, A. J. Spurgin and Y. O. Lukic (1985) *A Model for Assessing Human Cognitive Reliability in PRA Studies*, Third IEEE Conference on Human Reliability, Monterey, USA.

[57] G. C. Bello and V. Colombari (1980) *The Human Factors in Risk Analysis of Process Plants: the Control Room Model TESEO*, 6th Advances in Reliability Technology Symposium, UKAEA-NCSR-R23.

[58] G. C. Bello (1985) *The Human Operator Failures in Petrochemical Facilities: an Historical Investigation and a Tentative Simulation Model*, International ANS/ENS Topical meeting on probabilistic safety methods and applications, San Francisco.

[59] E. J. Kozinsky et at. (1984) *Safety Related Operator Actions: Methodology for Developing Criteria*, US NRC-NUREG/CR-3515.

[60] Baron et al. (1982) *An Approach to Modelling Supervisor Control of a Nuclear Power Plant*, US NRC-NUREG/CR-2988.

[61] F. Ghertman and P. Dietz (1985) *Human error data collection and analysis program*, International ANS/ENS Topical meeting on probabilistic safety methods and applications, San Francisco.

[62] I. Finlayson (1983) *Nuclear Power Safety Reporting System*, Vols 1 and 2, US NRC-NUREG/CR-3119.

[63] E. J. Kozinsky and K. S. Pack (1982) *Performance Measurement Systems for Training Simulators*, Electric Power Research Institute, NP-2719.

[64] D. E. Crowe et al. (1983) *Criteria for Safety-Related Nuclear Power Plant Operator Actions: 1982 Pressurized Water Reactor (PWR) Simulator Exercises*, US NRC-NUREG/CR-3123.

[65] A. N. Beare, R. E. Dorris et al. (1984) *A Simulator-Based Study of Human Errors in Nuclear Power Plant Control Room Tasks*, US NRC-NUREG/CR-3309.

[66] J. M. Moroni, A. Villemeur, F. Mosneron-Dupin and T. Meslin (1986) *Etude sur simulateurs du comportement des opérateurs de centrales nucléaires en situations perturbées*, 5e colloque international de fiabilité et de maintenabilité, Biarritz, France.

[67] D. A. Seaver and W. G. Stillwell (1983) *Procedures for Using Expert Judgement to Estimate Human Error Probabilites in Nuclear Power Plant Operations*, US NRC-NUREG/CR-2743.

[68] W. G. Stillwell, D. A. Seaver and J. P. Schwartz (1982) *Expert Estimation of Human Error Probability in Nuclear Power Plant Operation: a Review of Probability Assessment and Scaling*, NUREG/CR-2255.

[69] D. M. Hunns (1982) The method of paired comparisons, in *High Risk Safety Technology* (A. E. Green, ed.) John Wiley & Sons, Chichester.

[70] D. E. Embrey, P. Humphreys et al. (1984) *SLIM-MAUD: an Approach to Assessing Human Error Probabilities Using Structural Expert Judgement*, NUREG/CR-3518.

[71] A. Villemeur, F. Mosneron-Dupin, M. Bouissou and T. Meslin (1986) *A Human Factors Data Bank for French Nuclear Power Plants*, International ANS Topical Meeting on Advances in Human Factors in Nuclear Power Systems, Knoxville, USA.

[72] A. Villemeur, F. Mosneron-Dupin and T. Meslin (1986) *Base de connaissances relatives aux facteurs humains pour les centrales nucléaires françaises*, 5e colloque international de fiabilité et de maintenabilité, Biarritz, France.

[73] Topmiller, Eckel and Kozinsky (1983) *Human Reliability Data Bank for Nuclear Power Plant Operations*, Vol. 1: *A Review of Existing Human Reliability Data*

Banks, Vol. 2: *A Data Bank Concept and System Description*, US NRC-NUREG/CR-27744.

[74] Franek P. Lees (1980) *Loss Prevention in the Process Industries*, Vol. 1, Butterworths, London.

17
MECHANICS

17.1 INTRODUCTION

Engineers' concern for the reliability of mechanical components (ball bearings, gears, etc) is far from recent. However, it was only in the 1950s that the studies on mechanical components were first performed. The object of these studies was to predict right from the design stage, the reliability of mechanical components and, in particular, structural components (or parts). NASA (National Aeronautics and Space Administration) played an important role in initiating numerous studies to improve the reliability (then judged inadequate) of mechanical and electromechanical components [1].

Up to this time, engineers had been using the concept of the 'safety margin'. They realized, however, that this provision was inefficient in preventing fractures and, in addition, often led to excessive weights and costs, especially in the space industry. It became clear that concepts like stress and strength could only be defined statistically. Thus was born the probabilistic approach which is used to calculate the reliability of mechanical components by modelling stresses and component strength [2]. This approach is now widespread.

The construction of huge mechanical structures, which are often unique or built in small quantities and whose fracture can entail serious consequences requires the use of methods to predict their reliability; we are mainly referring to structures like the vessels of pressurized-water reactors and structures of offshore oil rigs. Because these structures are very few and must meet high quality standards, it is virtually impossible to use statistics. The only means left for assessing their reliability consists in developing probabilistic models of their mechanical behaviour. A new discipline initially developed by Freudenthal in the late 1950s [3, 4] and called probabilistic fracture mechanics gave rise to numerous studies on nuclear plants and off-shore oil rigs in the 1970s [5].

In France, a large number of studies were carried out in this field; two books were published [6, 7] and can be consulted.

General principles are first presented in this chapter; then the reliability of a mechanical component is calculated using the stress-strength theory. We shall also discuss briefly the statistical approach and probabilistic fracture mechanics in relation to fatigue phenomena. We will end the chapter with a brief description of the general method recommended for dependability assessment as applied to mechanics.

17.2 GENERAL CONSIDERATIONS

Mechanical components have numerous failure modes: wear, seizure, deformation, cracking, fracture, etc. The mechanisms are often complex and involve numerous parameters such as mechanical stresses, corrosion, heat and irradiation. The engineers are preoccupied with the cause of deformations, cracks or fractures of mechanical components.

Mechanical engineers commonly use a 'safety margin' defined in the following way:

$$S = \frac{\text{strength}}{\text{applied stress}} = \frac{S}{s} > 1 \qquad (17.1)$$

Sometimes, a safety-multiplying factor is used to find the strength from the material yield strength. In fact, the strength cannot be found deterministically because materials are heterogeneous, have imprecise dimensions and different methods of manufacture; it can only be represented by a distribution. The same holds for the applied stress which may vary according to the mechanical component environment or use (Figure 17.1).

From Figure 17.1 we can see that, under certain circumstances, s can be greater than S and thus cause the mechanical component fracture.

The reliability of the mechanical component is therefore measured by the probability that S is greater than s.

Let us assume that the safety margin has been calculated from the average values of the distributions. When the spread of the distribution varies, reliability can greatly vary for the same safety margin. Thus a relatively high safety margin can be paired with bad reliability; in the same way, good reliability can go with a low safety margin.

In fact, probabilistic calculation is the only means to relate a safety margin with a reliability level. Note that safety margins can be defined from these

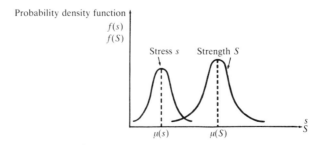

Figure 17.1 Stress and strength: probabilistic approach.

distributions [8, 9]; an example of a safety margin is that based on the 5% bound of the strength and 95% bound of the stress.

17.3 STRESS AND STRENGTH

Reliability assessment consists in calculating the probability that strength will exceed effectively applied stresses during the entire mission considered.

$$R = P[s < S] \tag{17.2}$$

Consider the probability density functions $f_1(s)$ and $f_2(S)$ (Figure 17.2). The probability of having a strength S greater than s_1 is

$$P[S > s_1] = \int_{s_1}^{+\infty} f_2(S) \, dS \tag{17.3}$$

Reliability is therefore written as

$$R = \int_{-\infty}^{+\infty} f_1(s_1) \left[\int_{s_1}^{\infty} f_2(S) \, dS \right] ds_1 \tag{17.4}$$

Given an initial strength S_2, it can be shown that

$$R = P[S > s] = \int_{-\infty}^{+\infty} f_2(S_2) \left[\int_{-\infty}^{S_2} f_1(s) \, ds_1 \right] dS_2 \tag{17.5}$$

Several methods exist for solving this equation:

• **Graphical approach** Mellin transforms, which can be applied to any distribution, are used:

$$F = \int_A^{+\infty} f_2(S) \, dS, \qquad G = \int_A^{+\infty} f_1(s) \, ds$$

We get

$$0 \leq F \leq 1, \qquad 0 \leq G \leq 1, \qquad dG = f_1(s) ds$$

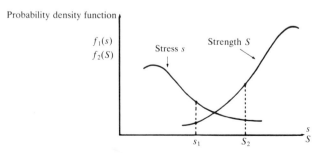

Figure 17.2 Reliability assessment.

Hence

$$R = \int_0^1 F \, dG$$

We can then plot the values of F versus G.

- **Resolution by the Monte Carlo method** A random stress value s is first sampled and then a random strength value S. Reliability is then given as

$$R = \frac{\text{number of times } S > s}{\text{number of samples } (S > s)}$$

- **Analytical solution** This can be applied to numerous known distributions; for example, if the two distributions are considered to be exponential, reliability is

$$R = \frac{1}{1 + \alpha} \quad \text{or} \quad \alpha = \frac{E[s]}{E[S]}$$

Existing publications abound with examples of analytical solving techniques [7, 10].

The previously expounded theory assumes that stresses are static. However, strength and even stress variations can be taken into account, provided fatigue phenomena, which will be discussed now, are not involved.

17.4 FATIGUE

The periodic application of loads causes changes in the material properties; this fatigue phenomenon may result in the mechanical component fracture at stresses which are often lower than the material yield stress. This fracture originates in a crack which progressively extends until the remaining cross-section can no longer stand the applied load.

The probabilistic approach to these phenomena involves a statistical treatment, on the one hand, and what is known as probabilistic fracture mechanics on the other.

17.4.1 Statistical approach

If periodic (sinusoidal) stress cycles of a maximum amplitude s and constant frequency are applied to test specimens, fracture occurs after a number N of cycles. When the maximum stress s is varied, the result is a curve known as the $S-N$ curve or endurance curve or Wöhler curve (Figure 17.3). Three different domains are usually observed on this curve:

- a plastic deformation section corresponding to high stresses: fracture

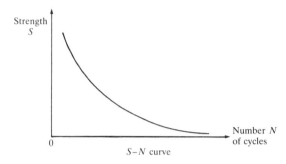

Figure 17.3 S–N curve.

occurs after a small number of cycles and is preceded by a significant plastic deformation;

- a section of limited fatigue or endurance where fracture occurs after a limited number of cycles which increases as the stress decreases;

- an unlimited endurance section or safety section under low stresses and where fracture only occurs after a large number of cycles.

In reality, fatigue is a random phenomenon as evidenced by the spread in the fatigue resistance characteristics of apparently identical components [11–13]. The randomness of the fatigue phenomenon accounts for the variety of S–N curves obtained (Figure 17.4). Thus point (N, S) indicates that the component subjected to stress S has a probability P of breaking at the end of N cycles.

According to the stress, the fracture probability during a test whose duration is limited to a given number of cycles, N, is given by the intersections of a vertical with the different curves. For a given stress, the cumulative distribution function of the number of cycles, N, to fracture is given by the intersections of a horizontal with the different curves.

For example, Figure 17.5 shows what can be observed for common materials; with logarithmic scales, the S–N curve is virtually linear between

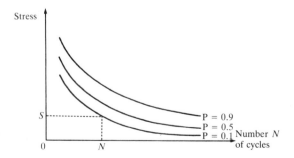

Figure 17.4 S–N curve.

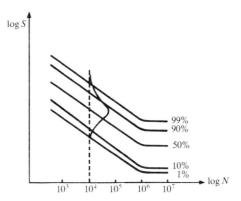

Figure 17.5 Example of S-N curve.

10^3 and 10^6 cycles. The material life can virtually be regarded as infinite below the endurance limit. Moreover, the distribution of s is normal as well as that of $\log N$ in the domain of high stresses. For example, reference [14] gives these characteristics for titanium-base alloys and steel grades used for constructing aircraft.

When the $S-N$ curves are known, we can find the reliability of a mechanical component subject to cyclic stresses or, in other words, the probability that a fracture does not occur before N_i cycles. We assume that the distribution of $\log N$ is a normal distribution with a probability density function $f(x)$. The probability that a fracture will occur before N_i cycles is

$$P(\log N_i) = \int_{-\infty}^{\log N_i} f(\log N)\, d(\log N) \qquad (17.6)$$

Generally, fatigue stresses result from the combination of static and variable stresses. The following classification is usually made:

- *Reversed stresses*. Loads change direction alternately. So far we have had loads of opposite signs and with equal extreme values.

- *Repeated stresses*. The loads are always in the same direction and vary between zero and a positive or negative value.

- *Irregular stresses*. The load varies above or below an average positive and negative non-zero value.

Generally, any periodic stress can be considered to be the resultant of a constant or static load s_m and an alternating stress of amplitude s_a. The $S-N$ diagram which gives the variation of strength versus the number of cycles is applicable to all types of stress; all that needs to be done is to fix a mean stress and study the variations in amplitude of the alternating stress according to the number of cycles to fracture.

Thus, whatever the number of stresses a steel specimen is subjected to, if the maximum difference (algebraic difference between the maximum load and the minimal load) remains below a certain threshold value, fatigue-induced fracture will not occur. Wöhler found a given threshold difference for each mean load value.

We try to represent the variations in the threshold difference when the stress is the combination of a purely alternating stress s_a and a constant stress s_m, both acting in the same direction, whether this stress is a tensile, compression, bending or torsional stress.

One possible representation (Haigh diagram) consists in plotting the values of the endurance limit or the alternating strength S_a for a given number of cycles on the y axis, and the corresponding values of the static or mean strength S_m (Figure 17.6) on the x-axis. A Gerber parabola seems to be quite a valid representation. Just as in Figure 17.4, distributions are adopted instead of deterministic values. The distributions governing strength are considered to be normal for different ratios S_a/S_m.

Now consider the stresses represented in Figure 17.7; how is the structure damaged?

A new approach is required when a mechanical component is subjected to periodic (sinusoïdal) stresses with a no longer zero mean [6, 7, 15, 16]. This

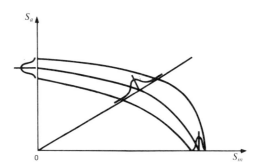

Figure 17.6 Haigh diagram.

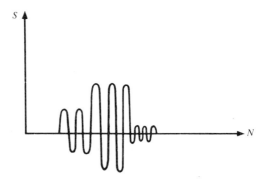

Figure 17.7 Example of stresses.

approach is based on the damage theory. This theory has a physical and a descriptive aspect.

The physical aspect deals with the changes in the physical properties of 'fatigued' materials; the descriptive aspect deals with the quantitative description of the endurance of materials subjected to various stresses.

Miner [17] proposed a simple damage rule based on the following hypothesis: the application of n_i cycles at a given stress level s_i, for which the average number of cycles before fracture is N_i, causes an increase in damage equal to n_i/N_i. Fracture occurs when

$$\sum_i \frac{n_i}{N_i} = 1 \qquad (17.7)$$

Thus, if during a fraction $n_i/N_i = Z$ of its life, a material has been subjected to a certain level of stress s_1, its remaining endurance at another level s_2 is $n_2/N_2 = 1 - Z$. The Miner rule is generally held to be imprecise; it does not take account of the order in which the levels of stress follow one another. Its main advantage, however, lies in its great simplicity.

Other damage rules have been proposed [7]. Various studies have been performed to come up with a probabilistic interpretation of the Miner rule [18–21]; the last reference gives one such interpretation for a given distribution of breaking stresses. Note that the Aerojet Nuclear Systems Company [22] proposed to allow for the fact that fatigue strength is a distribution by modifying the Miner rule as follows:

$$\sum_{i=1}^{m} \frac{n_i}{N_i} = \gamma \qquad (17.8)$$

γ is a normally distributed variable with a mean and a standard deviation of respectively 1 and 0.3, for instance [7].

17.4.2 Probabilistic fracture mechanics

Generally, the ability of materials to resist fracture is lower by several orders of magnitude than their theoretical strength, which is based on the cohesive energy of the crystal lattice. This low value is due to the presence of pre-existing flaws in the materials.

Under certain conditions, the defects propagate under the effect of loads. Thus in a structure subjected to fatigue, a defect whose size is not critical can become critical due to its slow propagation under the effect of stress variation.

According to the theory of fracture mechanics, catastrophic failure occurs when $K_I > K_{IC}$, where

- K_I is the stress intensity factor which is a measure of the stress level in the mechanical component taking the crack into account.

- K_{IC} is the toughness, that is the ability of the material to resist the sudden propagation of a crack.

A general law applies here; fracture occurs when stress is higher than resistance. As an illustration, let us consider an infinite plate to which a stress is applied. This plate exhibits a crack of length a straight through the thickness (Figure 17.8). We get

$$K_I = \sigma \sqrt{\left(\frac{\pi a}{2}\right)}$$

The values of K_{IC} can be determined from tests performed on standardized specimens. The critical size of a defect can also be deduced; catastrophic failure occurs when the size of the defect exceeds the critical size. Obviously, the values of K_I and K_{IC} are not deterministic; in fact, they are governed by distributions.

Fatigue (possibly corrosion or other factors) produces the crack growth. The crack propagation rate is considered to be a function of the variation of the stress intensity factor during cycling. The most commonly used law is that of Paris, where da/dN is used to measure the fatigue crack growth rate per cycle:

$$\frac{da}{dN} = C(\Delta K)^n \qquad (17.9)$$

where

$$\Delta K = K_{max} - K_{min} \qquad (17.10)$$

Constants C and n depend on the material and other factors such as the type of crack growth and the loading mode. Test results show that below a given limit of the stress intensity range the crack stops propagating. In practice, this is very important. Indeed, in certain cases, we can determine an allowable crack size at which a crack will not propagate from initial defects in a structure subjected to fatigue.

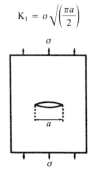

Figure 17.8 Infinite plate with a defect.

The principles which were previously defined can be used to calculate the predicted reliability of a complex structure; however, the following features must generally be known in detail:

- the distribution of initial defects: all the characteristics (size, depth) of initial defects originating from the manufacturing process and present before the structure is first used;

- structure loading conditions: all the loads which the structure will have to bear and which will cause fatigue are inventoried. These loads result from normal operation, incidents and accidents producing high stresses in the structure and from environmental conditions, etc.

- structure inspection conditions: a structure is normally inspected when it comes out of the factory or when it is put into service. Non-destructive tests may be performed to detect and measure defects in the most vulnerable parts of the structure. Since no detection technique is perfect, a probability of detecting defects is also taken into account. These tests once completed and considering the repairs which may have been performed, the defect distribution is no longer the same as immediately before the tests.

Thus, knowledge of these characteristics and the use of the principles defined above make it possible to predict the crack propagation during the life of the structure and the possible structure failure. The probability of structure fracture is deduced.

Three methods can be used to calculate probability:

- **Analytical method [23]** This consists, for each parameter expressed probabilistically, in finding the distribution representing best the collected experimental data and, next, in performing an analytical calculation based on the functions thus defined. Input data concern the amplitude of the stress variation, the number of corresponding cycles, the initial distribution of the defect length and the material toughness.

The data are input as cumulative distribution functions. Then the stress intensity range $\Delta K = \Delta \sigma h(a)$ is computed. The Paris law is integrated over N cycles which correspond to each $\Delta \sigma$:

$$\int_{a_o}^{a_t} \frac{\mathrm{d}a}{[h(a)]^n} = C \sum_{k=1}^{N(t)} \Delta \sigma_k^n \qquad (17.11)$$

Then the fracture criterion for each value of the stresses is applied to the crack size a_t thus determined.

- **Monte Carlo method [24]** The calculation is done in the same way as above; but the Monte Carlo method allows the use of any form of input data.

Its main disadvantage is the long computation time, so that it cannot be used for a calculation involving a high number of random parameters.

- **Histogram method** All the algebraic calculations are performed with variables expressed as histograms. The result is also a histogram. The method consists in performing the requested operation between the histogram blocks which define the parameters and then applying to the result a probability equal to the product of the weighted probabilities of each block [7, 25].

In the nuclear industry, the pressure vessels of pressurized water reactors are closely monitored. The probability of a fracture occurring in one of these vessels is considered to be extremely low considering the safety measures adopted, the acquired experience and the dimensioning: the vessel can operate at pressures above 200 bars, for a nominal pressure of around 150 bars. Despite its serious consequences (core melt), this failure is not taken into account in the reactor design. In 1974, the Advisory Committee on Reactor Safeguards gave a probability of 10^{-7}/year for the fracture of such a vessel: this figure is drawn from a study based on an analysis of the operating experience of pressure vessels used in non-nuclear industries. The study showed the difficulty involved in obtaining such statistics since, for these pressure vessels, the wall thickness, materials and building specifications were very far from those found in nuclear reactor pressure vessels.

Marshall and his team, who adopted an approach based on probabilistic fracture mechanics, found a similar order of magnitude [26, 27]: they also performed sensitivity studies [28–31]. Below are two characteristics of the pressure vessels covered by the study:

- the vessel becomes increasingly brittle in the long run due to the neutron irradiation of its walls;

- the manufacturing defects which cause fracture are most often due to welds (and not to defects in the main shell), and their distributions differ according to whether butt welds (shells and heads) or corner welds (branch connections) are considered.

Other similar studies on vessels [25, 32, 33] or on the large pipes of the primary circuit in PWR reactors [34, 35] can also be mentioned.

In the nuclear industry, certain PRAs have shown that an earthquake can contribute considerably to the risk of a serious accident. To take this risk into account, all the mechanical components and their response to earthquakes must be considered. For the majority of mechanical or electromechanical components of safety systems, a fragility curve is used to represent the probability of component failure versus, say, the earthquake-induced ground acceleration. The theory of probabilistic fracture mechanics is sometimes used for the structural components; thus usual loading conditions and seismic loads [36–39] are considered.

Similar studies have been performed for off-shore oil rigs; one of the major concerns is about the reliability of welded joints exposed to fatigue phenomena mainly due to the wind and the waves. References [40–43] deal with this problem.

The analysis of structural reliability is certainly not an easy task given the large number of factors involved; let us quote among others some of these factors:

- the material quality and characteristics;
- the structure architecture;
- the welding techniques;
- pre-operational tests (the hydrostatic test for a pressure vessel);
- pre-operational inspection;
- the statistical distributions of toughness;
- the variation of toughness versus temperature or other factors;
- initial defect distribution;
- (normal, incident, accident, etc) loading conditions of the structure in service;
- fatigue-induced defect growth;
- in-service inspection;
- structure repairs.

Studies have demonstrated the merits of using probabilistic fracture mechanics to highlight the main hypotheses or the quality factors which affect the fracture probability [44, 45]. The orders of magnitude of these probabilities must be used with care for, as far as we know, very few tests have been performed to prove the validity of this method and its major hypotheses. Studies also show how much the initial distribution and size of defects can affect the results. However, it is not always well known despite the use of sophisticated investigation techniques [44, 45].

Therefore, probabilistic fracture mechanics must be further developed and put to the test before its use can be generalized.

17.5 DEPENDABILITY OF MECHANICAL SYSTEMS

The methods of predictive analysis (see Part 2) are applicable to mechanical systems or to systems with mechanical components in so far as mechanical components can be identified. The failure modes and effects analysis is particularly efficient for identifying relevant failure modes as well as their effects (example: stresses) on other mechanical components.

As a rule, the stresses applied to different components should be determined. When these components undergo the usual stresses and suitable dependability data are available, the 'systems' approach (explained in Part 2) will be used. Note that there are, on the one hand, methods to calculate the predicted reliability of particular mechanical components (bearings, gears, worm screws, springs, etc) [7] and, on the other, data sources (see Chapter 5).

For mechanical structures existing in limited numbers, the use of such an approach is not always possible. The specific mechanical reliability assessment methods described above are especially efficient and valuable in that case. Their use requires a detailed analysis of the applied stresses and of the strength and a knowledge of their statistical distribution. The reliability calculation then takes the specific component characteristics into account as much as possible. The 'stress–strength' approach should be used only when necessary because it is rather cumbersome.

These two approaches seem to complement each other when we consider their advantages and disadvantages [46]. The analyst may therefore use them together but he must restrict the application of the specific methods to those mechanical components they are adapted to, for example to components whose critical role was highlighted by the first approach.

REFERENCES

[1] W. M. Redler (1966) Mechanical reliability research in the NASA 1966, *Annals of Reliability and Maintainability Conference*, IEEE, New York 763–8.
[2] E. B. Haugen (1968) *Probabilistic Approaches to Design*. John Wiley and Sons, Chichester.
[3] A. M. Freudenthal (1956) Safety and the probability of structural failure, *Trans. ASCE* **121**, 1337–75.
[4] A. M. Freudenthal (1961) Reliability and structural design, *J. Struct. Div.* **87**(ST3).
[5] G. O. Johnston (1982) A review of probabilistic fracture mechanics literature, *Reliability Engineering*, **3**(6).
[6] C. Marcovici and J. C. Ligeron (1974) *Utilisation des techniques de fiabilité en mécanique*, Editions Lavoisier.
[7] J. C. Ligeron (1979) *La fiabilité en mécanique*, Desforges.
[8] R. H. Leicester (1977) Load factors for design and testing, *Proc. Conf. Prevention of Fracture*, Melbourne, Australia.
[9] A. H. S. Ang (1974) A comprehensive basis for reliability analysis and design, *Proc. US–Japan Seminar on Reliability Approach in Structural Engineering*, Tokyo.
[10] B. S. Dhillon and C. Singh — *Engineering Reliability: New Techniques and Applications*, John Wiley and Sons, Chichester.
[11] F. Bastenaire (1961) Etude de la rupture par fatigue par les méthodes statistiques, *Bulletin de l'Institut International de Statistique*, 1961.
[12] F. Bastenaire (1971) Aspects probabiliste et statistique sur la rupture par fatigue. *Revue française de Mécanique*, No. 37.
[13] F. Bastenaire (1975) *Estimation et prévision statistique de la durée de vie des matériaux en fatigue*. Conférence faite au stage Fiabilité en Mécanique organisée par ADERA, AFCIQ, AFNOR, Paris.
[14] J. C. Whitaker (1972) *Development of Titanium and Steel Fatigue Variability*

Model for Application of Reliability Analysis Approach to Aircraft Structures, Air Force Materials Laboratory Report AFML, TR 72, 236, Ohio.

[15] D. Kececioglu (1972) *Mechanical Structural Reliability Analysis and the Associated Confidence level*, Tenth Annual Reliability Engineering and Management Institute, University of Tucson, Arizona.

[16] D. G. Kececioglu *et al.* (1974) Sequential cumulative fatigue reliability, *Annals of the Reliability and Maintainability Symposium*, 533-9.

[17] M. A. Miner (1945) Cumulative damage in fatigue, *J. Appl. Mech.*, **12**(3), A159-64.

[18] R. R. Gatts (1962) Cumulative fatigue damage with random loading, *Trans. AMSE J. Basic Engineering*, 403-409.

[19] S. C. Saunders and Z. M. Birnbaum (1967) *A Probabilistic Interpretation of Miner Rule*, Boeing Scientific Research Laboratories, 1967.

[20] J. A. B. Lambert (1973) *ESDU Cumulative Damage Hypothesis*, Agard Lecture No. 62. Fatigue life prediction for aircraft structures and materials.

[21] A. N. Kinkead and P. Martin (1985) An approach to the fatigue reliability of sequentially loaded demountable joints, *Reliability Engineering*, **12**(2).

[22] *Simplified PRE-PDR Techniques for Assessing Component Reliability 'NERVA'* (1970) Aerojet Nuclear System Company reliability calculations for the case of combined stress and fatigue loading.

[23] Nilsson (1977) *A Model of Fracture Mechanical Estimation of the Failure Probability of Reactor Pressure Vessel*. Third conference on Pressure Vessel Technology, Tokyo.

[24] Becher and P. E. Pedersen (1973) *Application of Statistical Linear Elastic Fracture Mechanics to Pressure Vessel Reliability*, Report M 1650, Danish Atomic Energy Commission, Riso.

[25] J. Dufresne (1977) *Approche probabiliste de la mécanique de la rupture: Application à une cuve de réacteurs nucléaires*. CEA-EDF. Cycles de conférences sur la Fiabilité et Disponibilité des systèmes mécaniques et leurs composants, Jouy en Josas, 03/07/10.

[26] W. Marshall (1977) *An Assessment of the Integrity of PWR Pressure Vessels*, A UK Study Group Report, Proc. 4th Int. Conf. SMIRT, California.

[27] W. Marshall (1982) *An Assessment of the Integrity of PWR Pressure Vessels*, Second Report by a Study Group under the chairmanship of Dr. W. Marshall, UK Atomic Energy Authority.

[28] A. B. Lidiard and M. Williams (1977) A simplified analysis of pressure vessel reliability, *Inst. Br. Nucl. Energy. Soc.*, **16**(3), 207-23.

[29] A. B. Lidiard and M. Williams (1978) The sensitivity of pressure vessel reliability to material and other factors, *Reliability Problems of Reactor Pressure Components*, Vol. 1, IAEA, Vienna.

[30] A. B. Lidiard and M. Williams (1978) *A Theoretical Analysis of the Reliability of PWR Pressure Vessels*, C80/78, Institute of Mech.-Eng., London.

[31] A. B. Lidiard (1980) Applications of probabilistic fracture mechanics to light water reactor pressure vessels and piping, *Nucl. Eng. Des.*, **60**, 49-56.

[32] W. E. Vesely, E. K. Lynn and F. I. Goldberg (1978) OCTAVIA: a computer code to calculate the probability of pressure vessel failure from pressure transient occurrences, *Proc. IAEA Int. Symposium*, Vol 1, Vienna.

[33] M. J. Dufresne, J. M. Lanore *et al.* (1980) PWR reactor pressure vessel failure probabilities, *Proc. 4th Int. Conf. Pressure Vessel Technology*, I. Mech E, Paper C13/80, 37-44.

[34] C. K. Chou (1982) A probabilistic assessment of the primary coolant loop pipe fracture due to fatigue crack growth for a PWR plant, *Nuclear Engineering and Design*, **71**(3).

[35] Lawrence Livermore National Laboratory (1981) *Probability of Pipe Fracture in the Primary Coolant Loop of a PWR Plant*, NUREG/CR, 2189.

[36] S. C. Lu and R. D. Streit (1980) Probabilistic assessment of decoupling the effects of large LOCA and earthquake in nuclear power plant design, *Proc. 8th Water Reactor Safety Research Information Meeting*, Gaithersburg, Maryland.

[37] S. C. Lu and D. O. Harris (1981) Probabilistic assessment of decoupling loss of coolant accident and earthquake in nuclear power plant design, *Proc. 6th SMIRT*, Paris, France.

[38] C. A. Cornell (1982) Some thoughts on systems and structural reliability, *Nuclear Engineering and Design*, **71**(3).

[39] P. D. Smith (1982) Seismic Risk Analysis and decisions for nuclear power plants, *Nuclear Engineering and Design*, **71**(3).

[40] P. H. Wirsching (1979) Fatigue reliability in welded joints of off-shore structures, *Proc. Conf. Off-shore Technology*, Paper 3380, 197–206, Houston.

[41] M. K. Ochi (1978) Probabilistic extreme values and their implication for off-shore structure design. *Proc. Conf. Off-shore Technology*, Paper 3161, 987–92, Houston.

[42] M. J. Baker and T. A. Wyatt (1979) *Methods of Reliability Analysis for Jacket Platforms*, 2nd Int. Conf. Behaviour of Off-shore Structures, Paper 84, 499–520, London.

[43] S. B. Jensen, B. Vedeler and E. Wulff (1979) *Risk Assessment. A Study of Risk Levels within Norwegian Off-shore Petroleum Activities*, Royal Norwegian Council for Scientific and Industrial Research Report, 26-27/2.

[44] G. O. Johnston (1983) Probabilistic fracture mechanics (PFM), A meeting report, *Reliability Engineering* **5**, 83–97.

[45] Probabilistic fracture mechanics (PFM) (1985), A meeting report, March 1983, *Reliability Engineering*, **12**, 43–53.

[46] J. F. Barbet, A. Delage and J. C. Ligeron (1982) Méthode d'approche probabiliste de la sécurité des systèmes mécaniques, EDF, *Bulletin de la Direction des Etudes et Recherches*, Série A, Nucléaire, Hydraulique, Thermique, 2, 5–17.

18
SOFTWARE DOMAIN

18.1 INTRODUCTION

Computer systems are present in an ever-increasing number of industrial applications and equipment, ranging from military defence systems, air traffic and banking control systems to commonly used products! Software is becoming more and more a major component, mainly due to its cost, which can by far exceed that of hardware.

Since the beginning of computer system development, reliability has been a major concern, primarily for relay computers designed by Bell Laboratories and Harvard University, then for the first electronic computers such as ENIAC (1946). Already, in the late 1940s, a better understanding of system bugs and the constant search for a means to eradicate them or reduce their effects spurred reliability growth. Further work was carried out, mainly under the impetus of space and military programs and computer and telephone systems manufacturers, resulting in a great number of publications since the early 1970s. Software-related problems have been fully taken into account ever since.

The increasing size of software progams and their presence in almost all industrial sectors have made the question of their reliability a vital one. Over the last 15 years, various studies have been carried out in an attempt to gather statistical data on software cost factors. The majority of this information was collected by a US Army station, the Rome Air Development Center (RADC). In 1976, it set up a centre, the DACS (Data and Analysis Center for Software), specialized in the collection and analysis of software statistical data, which has been running a computer database since 1978. For instance, bugs in about 30 projects were compiled [1, 2]. There was, on average, one software design bug for every 100 instructions for software containing 1000–300 000 instructions. It should be mentioned that there is a great spread of the results; e.g. for programs comprising about 100 000 instructions, there is a ratio of 1 to 100 between the best and the worst of them.

Software designers have therefore developed numerous techniques to produce software that is as reliable as possible from the earliest stages of elaboration. Progressively, design and development procedures have been designed in fields as varied as programming languages, programming methods and specification writing and validation. Since these procedures proved insufficient, exhaustive tests are also performed to improve software quality and reliability.

Despite all the methods devised, it is difficult, if not impossible, to produce bug-free software. Therefore, research has been carried out, together with the above-mentioned approach, in order to assess the 'confidence' or reliability attributable to a program. These problems were first discussed during conferences in 1968 (Garmisch, Federal Republic of Germany) and 1969 (Rome, Italy) [3, 4]. The first software reliability prediction models were developed independently and simultaneously (1971) by Jelinski and Moranda [5] and Shooman [6]; a large number of models are now available. Software reliability modelling and later dependability modelling were extensively studied, particularly in the United States. [7, 8] and in France [9].

In this chapter we shall first define the concepts related to software dependability and then give the main characteristics of the software system. Methods roughly estimating software reliability and those providing predictive assessments are then developed. The question of software dependability is treated afterwards, the software being considered as a system made up of components (modules, for example). Finally, we shall examine the characteristics of fault-tolerant software systems and the principles for evaluating the dependability of computer systems.

18.2 CONCEPTS

Concepts associated with dependability can generally by applied to software. In this field, however, terms and definitions differ somewhat depending on the author. Consequently, our approach is meant to be as consistent as possible with that used for hardware. Below is a commented list of the main concepts.

- *Software system.* Software can be regarded as a system. Its components are, for example, the instructions or the modules it is made up of. A module is then termed 'software component'. One of the main characteristics of a software system is that it is immaterial and abstract.

- *Failure.* This is the termination of the ability of software to perform a required function. The latter is carried out under given conditions, for example the definition of the computer system it was designed for and the conditions included in its specifications.

Software contains bugs (division by zero, for instance) due to human design errors; these are only revealed when an event arises which is linked to the input configuration and implies the use of the erroneous sequence of the program. When the bug is revealed, the program stops functioning correctly although its state is not altered. Some authors consider, therefore, that strictly speaking, there is no such thing as a software failure. Nevertheless, this concept is used here since it is necessary to make a clear distinction between the moment (before the failure) when the software is still capable of functioning correctly and the moment when it no longer is!

In fact, a program is a mapping of an input space (I) into an output space (O) (Figure 18.1) [10]. If the program contains bugs, these will be revealed, causing a software failure, whenever the input data are in a certain space denoted by I_B. Since this space is encountered randomly, software failures occur randomly!

In fact there are two sources of uncertainty:

- input data-related uncertainties: these cannot all be anticipated or tested;

- program-related uncertainties: the life of the software can be regarded as a succession of programs $P(1)$, $P(2)$, ..., $P(n)$, ..., each of them different from the previous one because of the corrections introduced. A succession of spaces $I_B(1)$, $I_B(2)$, ..., $I_B(n)$ ..., of various sizes correspond to them.

Programmers intend to obtain $I_B(j) \subset I_B(j-1)$ ($\forall j > 1$) after debugging but this cannot be guaranteed. Indeed, the correction of bugs often introduces further bugs. Thus the program sequence is itself random.

Because failures occur randomly, probabilistic techniques are justified to assess software reliability. This modelling is interesting since it allows for the distinction between

- behaviour up to the first failure when only the first uncertainty source arises;

- behaviour during the life cycle where the uncertainty sources stem from inputs and programs.

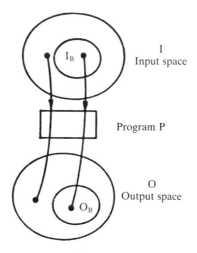

Figure 18.1 Program execution viewed as a mapping.

- **Software bug** The discrepancy between a software characteristic and its required characteristic. A required characteristic can, for instance, concern output data, instructions (content, links, etc).

Obviously a software bug is caused by a human error in software design affecting an instruction or links between modules. Bugs can be extremely varied [2]. They include, for example:

- defects related to input data manipulation (wrong definition of input data, etc.);

- defects affecting the interface between the different modules of a program, i.e. number, format and type of subroutine parameters;

- defects involving operation sequencing;

- semantics defects: typing errors which modify the name of a variable;

- defects affecting the program structure: segments which cannot be reached or from which it is not possible to escape (endless loops);

- defects linked to computational accuracy;

- defects resulting from non-compliance with operating system rules;

- defects related to language use (syntax and semantics);

- defects affecting input–output operations;

- performance defects (real-time, required memory size);

- design defects due to erroneous interpretation of specifications.

Thanks to statistical studies, defects can be categorized. They are generally revealed during testing and are classified as shown in Table 18.1.

Table 18.1 Percentage of defects per category

Categories	Compiler [11]	Real-time software [12]	Real-time software [13]
Computation	6	9	9
Logic	38	26	26
Input/output	2	14	16
Data manipulation	15	18	18
Interface	13	16	17
Data definition	19	3	1
Data base	1	7	4
Others	—	7	9

Logical defects account for 40% of all defects and interface defects for 20%. It should be noted that input–output defects occur most often in real-time applications as these involve numerous input–output functions. Moreover, 2/3 of the defects stem from wrong specifications and 1/3 from incorrect coding [11].

- **Failure mode** The effect by which a software failure is observed.

Generally speaking, failure modes are classified in five different ways, beginning with the easiest to detect and ending with the hardest [2]:

- operating system ceases to function;
- program stops running with clear explanation display;
- halt with no diagnosis;
- program runs but yields inconsistent results;
- program runs, yielding apparently correct results for the chosen input configuration, but which, in fact, are erroneous.

The last two failure modes are serious causes of concern for safety!

- **Fault** The inability of software to perform a given function.

In this definition, the inability is assumed to be taken into account from the moment it is uncovered by input data revealing software bugs.

- **Reliability** The ability of software to perform a given function, under given conditions, for a given time interval.

A measure of this reliability is the probability that the software will operate without failure for a given period of time and under stated conditions. Conventional definitions of reliability and its assessment apply unreservedly in this case: 'stated conditions' must be specified and are characteristic of the software.

- **Maintainability** The ability of software to be restored to a state in which it can perform its required function, when maintenance is performed under given conditions using stated procedures and resources.

Maintainability is often used to mean the ability of software to undergo changes which will extend its possibilities. A failed software product can be repaired using two different techniques:

- program debugging: the bug must first be detected, then identified before it is corrected;

- program re-initiation: the program is restarted in an input data configuration which is different from the one which revealed a bug in the first place. This is not always possible.

Note that when software is corrected, it is brought into another state and becomes, in fact, a new software program. Moreover, software does not need preventive maintenance since it does not normally deteriorate with time.

18.3 MAIN FEATURES OF THE SOFTWARE SYSTEM

In this section are listed the main software system features as regards life cycle stages, testing, quality assurance and reliability, with stress put on differences with hardware systems. The reader might also read Shooman's work [3] on the subject.

18.3.1 Life-cycle stages

The life cycle of the software, like that of any industrial product, can be broken down into several distinct stages:

- **Functional specification definition** Functional specifications must describe the aims of the program and its conditions of use as fully as possible; they must deal with all conditions linked to input–output data. The system is thus cut into large logical interfaced functions. Performance and quality characteristics are also stated and, in most cases, the right balance between cost and expected performance should be sought.

Generally, many software bugs originate from specification errors. This is the reason why it is essential to check whether specifications comply with aims; simulations can prove helpful in this case.

- **Design and development** The software design is based on the above-mentioned functional specifications. It must define the software structure and its division into modules. The internal logic of these modules is then established. In general, one should be able to describe and test modules separately.

Choosing a language is an important task; indeed, it can affect the future reliability of the software. Hence the development, over the past few years, of structured languages in an attempt to reduce language-related error sources.

- **Coding process** This is the next step. Coding is a major bug source, which increases with the size of the program. The code is reviewed before the program is machine tested. It has been shown that debugging is a very long process for programmers; the use of compilers facilitates the detection of many syntax defects before the program is executed.

- **Verification and validation** The conformity of the software with the functional specifications for performance and reliability is verified. Tests are performed since they are a determining factor for the future reliability of the program. On the one hand, software ability to allow for input data variation is verified and, on the other hand choices made in the software design are validated!

Testing all possible input data is often out of the question; therefore tests should be conducted on the most relevant data; i.e. the most likely values or/and most critical values in regard to the functions to be accomplished. Simulation techniques are sometime used to generate input data randomly. It is particularly useful to record any bugs (and their characteristics) discovered during testing; as will be seen later, they provide valuable information for assessing software reliability.

However, we should not forget that, if testing reveals the presence of bugs, it does not reveal their absence! No testing can prove that the program is perfect. The end of this stage, though not well defined, often occurs when the bug number per unit time has become low enough. Further techniques are necessary to evaluate the software reliability; they will be presented in the following section.

In a programmed system, hardware–software interfaces will also be treated during the verification stage, in order to make sure hardware and software are correctly integrated.

- **Operation and maintenance** The term 'operation' refers to activities concerned with the availability and the functioning of the software. 'Maintenance' generally covers different aspects such as (cf. Section 2):

- debugging following bug detection;

- updating consisting in either a modification of functional specifications or an adjustment to a modified environment;

- improvement, i.e. enhanced performance or improved system quality (documentation, modularity, etc).

18.3.2 Software testing

There are several testing techniques available [2]:

- *Static test*. The specifications and code are read before the machine run. The aim is to detect structural, logic and syntax errors and verify compliance with the applicable standards. This reading process can range from a simple exchange of documents (specifications or code) between two programmers to a formal inspection [14, 15].

- *Symbolic test*. Specifications (if they are formal) or the code are executed symbolically by computing variables using their symbolic names. The aim is to analyze the program behaviour along a path.

- *Dynamic test*. The (standard) test, during which the program undergoes a series of tests, and the execution and results are checked. There are three kinds of tests:

  —tests based on the software structure (the so-called structural tests or 'white box');
  —tests based on the definition of the function to be performed (the so-called functional tests or 'black box');
  —tests based on the bugs to be detected.

An important question arises during each testing period. How can one determine exactly when to stop testing? The programmer's intuition (or weariness) is certainly not an adequate criterion. Objective criteria should be set; the main criteria are the following:

- *Criterion based on the test sets*. The number of branches, paths and loops covered by the test sets is evaluated. This brings up the notion of coverage rate. For more sophisticated tests, a similar test coverage notion is determined by counting the number of elementary functions executed, of module calls, etc. It is therefore necessary to provide the program with tools to control inputs into and outputs from modules.

- *Criterion based on the number of bugs detected*. An average bug rate (number of bugs per 1000 lines of code) can be statistically determined according to various parameters (type of application, operators' qualification, etc.). It is also possible to utilize reliability models which yield the number of residual errors in the program (see Sections 18.4 and 18.5). Comparing the number of potential bugs in the program with the quantity of detected bugs gives an indication on when to end testing. This indicator can be further refined by weighing bugs according to their criticality.

- *Criterion based on the marginal detection rate*. The marginal detection rate is the derivative of the number of bugs detected with respect to time. This rate, which is high during the first tests, tends to drop as the testing proceeds. To determine the end of the test, a minimum value for the marginal rate is set. Testing is continued until this value is reached.

- *Criterion based on a target reliability level*. Representative reliability models should be used. Based on the bugs detected throughout the testing procedure, these models provide a reliability measure, the MTTF, for instance. The test is thus stopped once the predicted MTTF reaches an acceptable level.

Miller [16] gives some results of static and dynamic tests performed on software written in PL/1, containing 28 318 instructions and 75 modules; this real-time software was to be used in a control system and involved numerous database-style operations. It should be noted that, because of the criticality of the program, the programmers selected to develop these modules were among the best.

Static tests showed 662 coding violations, i.e. around nine violations per module (example: IF not followed by THEN and ELSE). In general, the bugs detected were not of trivial nature. Dynamic tests implied the use of 937 test sets providing a 87% coverage of the 2612 segments of the software; a segment, here, is defined as a set of logically connected instructions which are always executed together. In this application, each segment included 8 to 15 instructions. It was chiefly through manual comparison of the actual behaviour of the program with, first, the behaviour expected and prescribed in the internal documentation and, then, the specifications available, that bugs were discovered during the dynamic test phase. On the whole, 116 bugs were recorded.

Most bugs observed belong to one of the following categories:

- bugs in the logical diagram (e.g. the symbol = used instead of $\neq$);
- incompatible interfacing between modules;
- obvious non-compliance with the program specifications.

18.3.3 Quality assurance

Software quality is like the quality of any other product; the product must meet the user's needs (general adequacy to its usage) [17]. To assure product quality, one must first define an acceptable quality level, and then determine a mechanism to assure that the chosen level is maintained throughout the manufacturing process and reached at the end of the process; this mechanism is termed 'quality control'. However, it is difficult to define a quality level and its relevant control mechanism, due to two specific features of software products: first, software is most often a unique product; second, it is an essentially abstract object.

In order to clarify the concept of quality for software, to avoid subjective criteria and insert clauses concerning quality in industrial contracts, several trade associations and government agencies have recently worked out or are currently preparing standards on quality assurance for software products.

Some examples are the US Department of Defense [18–20], NATO [21], the Federal Aviation Agency (FAA) [22], the National Bureau of Standards [23], the Institute of Electrical and Electronics Engineers (IEEE) [24–28], the Association Française de Contrôle Industriel et de Qualité (AFCIQ) [17].

At the same time, numerous techniques have been developed over the years to make software products as reliable as possible. For example:

- *Structured programming*. The aim of this is to break the program into modules, each one having its own particular technical specifications, specific programming and tests.

- *Top-down programming*. This technique is now recommended. Originally, major software was programmed module by module and the modules were then integrated to form the program. More reliable programs are achieved through the reverse process: top-down programming. The core of the program is written first, with modules regarded as 'black boxes'; then, these modules are written in the same way as the core.

- *Structured languages*. These are recommended and make it possible to avoid such instructions as GO TO, which are universally considered to be the cause of many bugs! The use of high-level languages also improves program reliability, whereas assembly language programming is less and less advised for highly reliable software.

Despite size and complexity, quality software is produced thanks to the care taken to minimize the amount of bugs by using these different techniques throughout the software design process. It has been shown, however, that software, even if designed according to the most recent techniques, still contains bugs. Thus software quality measurements and reliability assurance are becoming increasingly important. Indeed, analysis techniques to predict software reliability (see Sections 18.4 and 18.5 below) need to be developed since it is impossible to prove the absence of bugs or because of the certainty that some still remain.

18.3.4 Reliability

The problem of reliability is treated differently, for hardware and software systems. For the latter, two scenarios are possible:

- The software is bug-free: this is the case if the software can be tested for each possible input and the test results are satisfactory. No failure will occur, except if inadequate data are entered. Thus, failure probability equals 0 and reliability, 1.

- The software contains bugs: in some data configurations which lead to the detection of bugs, the software failure probability equals 1. Apart from those input data configurations, the failure probability equals 0. Failures occur randomly due to input data-related and software-related uncertainties; the probabilistic methods, which will be discussed later, are therefore justified.

It should be noted that software reliability is not necessarily related to the number of bugs. Software can be highly reliable if the only bugs it contains, however numerous they may be, are located in a seldomly employed module! Inversely, software reliability is low even if it contains a few bugs but they are all located in frequently used paths! The number of software bugs can nevertheless vary during operation:

- it diminishes when a bug is corrected following a failure; reliability then increases;

- it increases when additional bugs are introduced while correcting an observed bug! Reliability then drops.

Note also that software redundancy obviously does not enhance software system reliability—in contrast to what happens in hardware systems—since the redundant software, *a priori*, contains the same bugs; due to this problem, the software programs must be of different design in order to improve reliability.

While methods are available to predict the failure of a hardware system from the failures of its components, there are no such techniques for software systems! Such an approach is presently difficult, if not impossible, probably because of the very high number of possible bugs for each instruction or each module and because of the vast array of failure modes. However, work is being carried out in this direction. Therefore, up to now, only parametric-type models have been developed.

18.4 FIRST INDICATIONS ON SOFTWARE RELIABILITY

Techniques have been developed to inform the designer of the software reliability; they do not give any real assessment of reliability but provide indications which can prove very useful. Techniques can be divided into two categories; techniques aimed at assessing the software complexity and sampling techniques. References [8, 9] should be consulted by the interested reader.

18.4.1 Complexity measurements

Program complexity is measured and an attempt is made to relate these measures to the program reliability measures. These techniques can be used once the program is ready and before the testing period; the most complex programs (or modules) can thus be tested with special care. Three sorts of complexity assessments are usually made:

- a complexity assessment of the program text;
- an assessment of the program structure complexity;
- an assessment of the execution rate of the various parts of a program.

Models to obtain these measures have not been developed to the same extent. Researchers [11] have used these assessments on compilers and thus confirmed that the complexity measures and reliability level were clearly related.

- **Text complexity assessment** Based on computational linguistics and psychometric results, this test indicates the probable number N of bugs present in a program of a given complexity. This technique was developed by Halstead [29]; it assumes that the physical characteristics of programs follow certain rules. Halstead made a list of the latter and of the mathematical relationships between them; hence the name of 'software science' given to these techniques.

First, Halstead evaluated the mental effort W required to make a program. This depends on the 'volume' and the 'length' of the program; formulae were proposed to assess the 'volume' and the 'length' from the quantity of separate operators and operands used. In an experiment, Halstead compared the estimated effort W with the measured effort. The fit between the recorded and the computed results is quite exceptional [8, 9] for programming times ranging from 5 min to 1000 man-years. Moreover data gathered on programs were analysed, and it appears that effort and the amount of observed bugs are closely correlated [9]. One finds, for example, that the number N of bugs is such that

$$N = \frac{W^{2/3}}{3200}$$

Using these techniques, it is possible to obtain an initial assessment of the various parameters of a program and to get a rough idea of the number of bugs in a program.

- **Assessment of the structure complexity** This is based on features of the graph associated with the program. McCabe [30] was the first to propose a complexity measure to detect programs and modules difficult to test or to maintain.

For each program, there is a directed graph G with only one input and one output in which each node corresponds to a strictly sequential set of instructions. A node x and a node y are joined by an arc if and only if the instructions of node x can be executed directly before the instructions of node y. The program complexity is given by the cyclomatic number. Correlations have been found between error rates and the complexity measure. Other studies have been carried out in this field [31, 32].

- **Program complexity assessment as a function of operational behaviour** The procedure consists in analysing not only the program behaviour but also its structure. Program reliability is characterized by an estimation of the exhaustiveness of the tests which were applied [33, 34]. Assessment is based on measurements made on certain classes of paths followed during testing.

18.4.2 Test-related measurements

Methods have been worked out to characterize software reliability using testing programs. The resulting assessments are said to be partial since they cannot predict the reliability of the software. Nevertheless they can give a rough estimate of reliability and help to select the tests to be applied. The main problem met in defining a program is how can the test sequence exhaustiveness be assessed with regard to all potential bugs in the software, on the one hand, and with regard to all input configurations, on the other.

The main techniques are based on samples either from the input data or from the bugs; additional sampling techniques have also been proposed.

- **Sampling the input data** [35, 36]. If following N tests, B bugs were detected, program reliability will be characterized by the following relation:

$$R = 1 - B/N$$

It should be noted that this is not really a reliability measurement. It is used in the following way:

- first, the input domain is partitioned according to the operational use profile of the program; points are drawn randomly in the subsets and the program is run a certain number of times; a reliability measurement is obtained by listing and observing the detected bugs;

- the same procedure can be used for the execution paths of the program.

- **Sampling the bugs** [37]. A set of known bugs (called 'seeded bugs') which, because of their nature and occurrence rate, are representative of the unknown bugs contained in the program (called 'native bugs') are introduced into the program. The number of native bugs is estimated from the number of seeded bugs detected while testing. The conventional method for determining the size of an unknown population is here applied to software. Let us assume that, at any time during the test, the following data are available:

i: number of bugs initially seeded;
j: number of native bugs discovered at a given time;
a: number of seeded bugs discovered at a given time.

It can be shown that, according to the criterion of maximum likelihood, the estimator of the number N_0 of initial bugs is

$$\hat{N}_0 \simeq ij/a$$

If need be, a confidence level can be associated with this estimator [9]. Clearly, one of the difficulties of this technique is to obtain seeded bugs that are representative of unknown bugs.

- **Additional sampling techniques** Various authors have proposed different techniques for sampling instructions, execution paths and, in general, the different modules of a program which are characteristic of its structure and complexity [38–40]. A technique for simultaneously sampling data and bugs also exists [41, 42].

18.5 SOFTWARE RELIABILITY PREDICTION

18.5.1 Introduction to models

Models have been developed to predict the reliability of a software program from the previously recorded data. The assessment is usually carried out once the software has been acknowledged operative. Supposing N failures have been observed (Figure 18.2). t_i denotes the time interval between failures $i - 1$ and i. Prediction, for instance, consists in assessing the next time interval to failure $N + 1$. Schematically, the practical use of these models involves:

- Reviewing all the failures observed while the software was being run or tested; in particular, a record is made of the following:

    —number of bugs detected;
    —intervals of time between detected bugs;
    —cumulative duration of program execution;
    —time to correct each bug;
    —test characteristics; ...

    Data from earlier versions of similar software can be used.

- Predicting a reliability measure (e.g. failure rate) that is related to the random variable T_{N+1}; this characteristic is most often written $f(t, N_0, N, a_i)$ where t, N_0, N and a_i denote respectively the time, the number of initial bugs, the number of failures observed and a specific parameter of the model.

Note that the first proposed models were based on the exponential or the Weibull distributions. In general, the first model is not corroborated by practical experience since it yields a constant failure rate in contradiction with the figures actually recorded. The second model makes it possible to describe

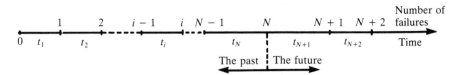

Figure 18.2 Failure chronology.

a decreasing failure rate which is often noticed in reality. However, since the bug detection and correction process is mainly discrete, this kind of model was abandoned for a model producing discrete decreasing patterns. After the first real models were proposed independently and simultaneously by Jelinski and Moranda [5] and Shooman [6] in 1971, numerous models have been constantly developed; more than forty of them are now available [43]. A classification of these models is given below; the most well-known and widely used of them are listed. Model validations problems are treated afterwards.

The models below are based on two fundamental features of software systems:

- the software contains a given number of bugs throughout its existence; the initial number of bugs is considered, for example, before the testing period or once the software's ability to function is acknowledged;

- the bug number can only decrease as time passes, unless other bugs are introduced due to defective debugging.

18.5.2 'Perfect debugging' model

This model was proposed by Jelinski, Moranda and Shooman. After presenting the basic model, we will examine models that have been derived from it.

18.5.2.1 Basic model

The main assumptions are the following:

1. bugs are detected independently from each other and randomly;
2. the failure rate is constant between two successive failures;
3. bugs are corrected immediately following detection;
4. debugging does not introduce additional bugs;
5. failure rate is proportional to the number of residual errors.

For the mathematical model, we use the following notation:

t_i = time between $(i-1)$th and ith failures

N_0 = number of initial bugs

N = total number of failures observed

λ_i = failure rate between $(i-1)$th and ith failures

ϕ = constant of proportionality.

The assumptions used result in the following expression, after the occurrence of the $(i-1)$th failure:

$$\lambda_i = \phi(N_0 - i + 1) \quad \text{(assumptions 4 and 5)} \quad (18.1)$$

The probability density function of the times between two successive failures is

$$f(t_i) = \lambda_i\, e^{-\lambda_i t_i} \quad \text{(assumptions 1 and 2)} \quad (18.2)$$

Failure rate evolution has the shape shown in Figure 18.3. Reliability, and then the MTTF, can be deduced in the following manner:

$$R_i(t) = \exp\left[-\int_0^t \lambda_i\, \mathrm{d}t\right]$$

$$R_i(t) = \exp[-\phi(N_0 - i + 1)t] \quad (18.3)$$

$$\text{MTTF}_i = \int_0^\infty R_i(t)\, \mathrm{d}t$$

$$\text{MTTF}_i = \frac{1}{\phi[N_0 - i + 1]} \quad (18.4)$$

Thus, if it is assumed that N failures were observed, the MTTF is

$$\text{MTTF} = \frac{1}{\phi[N_0 - N]} \quad (18.5)$$

The number N_a of additional errors still to be detected so as to obtain a target MTTF, denoted by M, as well as the corresponding time T_M may also be calculated:

$$N_a = N_0 - \frac{1}{M\phi} - N, \quad T_M = \frac{1}{\phi}\sum_{i=1}^{N_a}(N_0 - N - i + 1)^{-1} \quad (18.6)$$

These formulae are used to determine the moment when the test is considered terminated.

We will now examine the parameter estimation problem. Reliability depends on two unknown parameters N_0 and ϕ; they are estimated using the

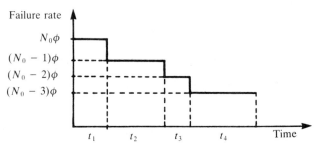

Figure 18.3 Failure-rate evolution.

maximum likelihood and least squares methods (see for example reference [7] in Chapter 4). For example, here are the estimates of $\hat{N}_0$ and $\hat{\phi}$ obtained by the first method:

$$\sum_{i=1}^{N}(\hat{N}_0 - i + 1) - 1 = N\left[\hat{N}_0 - \frac{\sum_{i=1}^{N}(i-1)t_i}{\sum_{i=1}^{N}t_i}\right]^{-1} \quad (18.7)$$

$$\hat{\phi} = N\left[\sum_{i=1}^{N}(\hat{N}_0 - i + 1)t_i\right]^{-1} \quad (18.8)$$

Equation (18.7) yields a value for $\hat{N}_0$ which need only be written into (18.8) to obtain ϕ.

In the previous model, the time intervals between the detection of two successive bugs ($i-1$ and i) were considered. In practice, data are not provided in this form. Often, the number of bugs detected in a time interval is recorded, or the bugs are grouped and, for each group, the time needed for detection is indicated. The model then allows for the interval corresponding to the detection of a bug group [44–46]. Let N_{i-1} denote the total number of bugs detected during the $i-1$ first testing intervals (measured in CPU days, weeks or hours). Hence

$$\lambda_i = \phi(N_0 - N_{i-1})$$

The formulae obtained are similar to the preceding formulae:

$$\text{MTTF}_i = [\phi(N_0 - N_{i-1})]^{-1}$$

Parameters ϕ and N_0 are computed, for example, by maximum likelihood:

$$\sum_{i=1}^{m} m_i(\hat{N}_0 - N_{i-1})^{-1} = m\left(\sum_{i=1}^{m} t_i\right) \cdot \left[\sum_{i=1}^{m}(\hat{N}_0 - N_{i-1})t_i\right]^{-1}$$

$$\hat{\phi} = m\left[\sum_{i=1}^{m}(\hat{N}_0 - N_{i-1})t_i\right]^{-1}$$

where

m = number of test intervals
m_i = number of bugs detected during ith test interval.

Parameters ϕ and N_0 are approximated by another method. Consider two periods where the failure rate is considered to be approximately constant. We note

$$\frac{\Delta T_1}{P_1} = [\phi(N_0 - N_1)]^{-1}$$

$$\frac{\Delta T_2}{P_2} = [\phi(N_0 - N_2)]^{-1}$$

where

ΔT_1, ΔT_2 = length of each period
P_1, P_2 = number of bugs discovered during each period.

Hence

$$N_0 = \frac{\gamma N_1 - N_2}{\gamma - 1} \quad \text{with} \quad \gamma = \frac{\Delta T_1}{\Delta T_2} \frac{P_2}{P_1}$$

$$\hat{\phi} = P_1 \cdot [\Delta T_1(\hat{N}_0 - N_1)]^{-1}$$

The model proposed by Shooman [6, 8, 47–50] is almost identical to the latter. It is indeed based on the same assumptions. The failure rate is given by

$$\lambda = \frac{C}{I}(N_0 - N) \tag{18.9}$$

where

I = number of instructions in the program; it is assumed to be constant during the stated period
C = constant of proportionality.

Let $C/I = \phi$; we find the preceding model again.
Various authors have discussed and used this basic model many times, as the abundant literature on the subject shows [38, 44, 45, 51–56].

18.5.2.2 Derived models

A great many models, quite similar to the previous model, have been proposed. We will just present them briefly and list the main differences.

● **The Schick–Wolverton model** [57, 58] It differs from the basic model because the failure rate is assumed to be proportional to the number of residual errors and to the time elapsed since detection of the preceding bug. Hence,

$$\lambda_i = \phi(N_0 - i + 1)x \tag{18.10}$$

where x = time elapsed since detection of $(i - 1)$th bug.
This model was first developed because its authors noticed that programming teams got better as the project advanced. Thus we have

$$R_i(t) = \exp\left[-\phi\left(\frac{N_0 - i + 1}{2}\right)t^2\right] \tag{18.11}$$

$$\text{MTTF}_i = \left[\frac{\pi}{2\,\phi(N_0 - i + 1)}\right]^{1/2} \tag{18.12}$$

Parameters N_0 and ϕ are estimated by maximum likelihood [8]:

$$\sum_{i=1}^{N}(\hat{N}_0 - i + 1)^{-1} = N\left[\hat{N}_0 - \frac{\sum_{i=1}^{N}(i-1)t_i^2}{\sum_{i=1}^{N}t_i^2}\right]^{-1} \quad (18.13)$$

$$\phi = 2N\left[\hat{N}_0 \sum_{i=1}^{N} t_i^2 - \sum_{i=1}^{N}(i-1)t_i^2\right]^{-1} \quad (18.14)$$

The same authors also proposed a modified version of the model. The failure rate is proportional to the number of residual errors and to a quadratic function of the total time elapsed since the software was first used; this version is proposed to make allowance for a maximum on the curve of the failure rate versus time [58].

- **The Lipow model** [59] The failure rate is proportional, on the one hand, to the number of residual errors and, on the other, to the time necessary to detect previous bugs incremented by a mean detection time in the current time interval.

- **The Moranda model** [53, 60–65] This includes two variants of the basic model:

- in one of them, the failure rate is assumed to decrease geometrically with each bug occurrence [53, 62];

- in the other, the failure rate is assumed to decrease following a geometric progression at dates uniformly distributed over time [53, 62].

- **Bayesian predictions-based model** It is not always easy to estimate N_0 and ϕ; Littlewood and Sofer [66] proposed an evaluation of these parameters using Bayes theorem (see Sections 4.2 and 5.2). It is assumed that the prior distributions (e.g. gamma) of λ and ϕ are known (which implies that N_0 and ϕ are also known). $t_1, t_2, \ldots, t_n$ are found by experimentation. Posterior distributions of λ and ϕ are then derived. Reliability is subsequently calculated:

$$R_{n+1}(t|t_1, \ldots, t_n) = P[T_{n+1} > t | t_1, \ldots, t_n]$$
$$= \iint \text{POSTERIOR}(\lambda, \phi | t_1, \ldots, t_n)$$
$$\times P[T_{n+1} > t | \lambda, \phi] \, d\lambda \, d\phi$$

Clearly the difficulty lies in the choice of the prior distribution of λ and ϕ.

18.5.2.3 Comments

- These models presuppose that debugging does not introduce new bugs. Furthermore, it is implicitly assumed that each bug has the same importance. ϕ is in fact the contribution of each bug to the failure rate. This may not always be true; the first bugs are generally the most liable to occur and the remaining ones (which can be very numerous) have a much lower probability.

- In the initially developed model, calendar time was used; however, it is better to use the CPU execution time when possible.

- If data do not show an increase in reliability, then the model is not applicable; reference [67] specifies the condition the data should satisfy.

18.5.3 'Imperfect debugging' model

These models differ from the preceding basic model since they assume that debugging is imperfect.

18.5.3.1 Parametric debugging model (the Musa model)

This model was proposed by Musa [68–73] in 1975. Compared to the preceding basic model, it assumes that the debugging rate is at all times proportional to the failure occurrence rate. Thus debugging is 'imperfect'. Hence

$$\frac{dN_c}{dt} = B\lambda \quad \text{and} \quad \lambda = \frac{dN}{dt} \qquad (18.15)$$

where

N_c = number of bugs corrected at time t

N = number of failures observed at time t

B = reduction factor

Since

$$\lambda = \phi(N_0 - N_c)$$

It follows that

$$N_c = N_0[1 - \exp[-\phi Bt]]$$

$$\lambda = \phi N_0 \exp[-\phi Bt]$$

$$\text{MTTF} = [\phi N_0]^{-1} \exp[\phi Bt]$$

Let $N_0 = BM$; and $\text{MTTF}_0 = [\phi N_0]^{-1}$.

Hence

$$\phi B = \frac{B}{N_0 \cdot \text{MTTF}_0} = \frac{1}{M \cdot \text{MTTF}_0}$$

$$N = M\left[1 - \exp\left[-\frac{t}{M \cdot \text{MTTF}_0}\right]\right] \quad (18.16)$$

$$\text{MTTF} = \text{MTTF}_0 \exp\left[\frac{t}{M \cdot \text{MTTF}_0}\right] \quad (18.17)$$

This last formula determines the number of additional failures to be observed if a given MTTF_2 is to be obtained from MTTF_1:

$$N = M\left(1 - \frac{\text{MTTF}_0}{\text{MTTF}}\right)$$

$$\Delta N = M \cdot \text{MTTF}_0 \left(\frac{1}{\text{MTTF}_1} - \frac{1}{\text{MTTF}_2}\right)$$

The extra execution time to obtain MTTF_2 from MTTF_1 can also be derived from this formula:

$$\Delta t = N_0 \cdot \text{MTTF}_0 \log \frac{\text{MTTF}_2}{\text{MTTF}_1}$$

In his model, Musa also introduces a factor C ('testing compression factor') to make an allowance for the real conditions of use: it is the ratio of the equivalent operating time under real conditions to the actual operating time under test conditions. In this model, the parameters are difficult to determine; Musa [68] advises making use of the experience gathered from past projects. Factors B and C, in particular, prove very hard to determine. Furthermore, it should be noted that Musa was the first to consider the execution time of a program instead of the calendar time in a model.

18.5.3.2 Probabilistic debugging model

This model was proposed by Goel and Okumoto in 1978 [74–79]. The following assumption is made: an observed bug has only a probability p of being corrected ($0 < p < 1$). Therefore there is a probability q ($p + q = 1$) that the observed bug will not be corrected; q is therefore the probability that the software will be inadequately debugged following a failure of the software.

It can be shown that

$$R_k(t) = P[T_k > t]$$
$$= \sum_{j=0}^{k-1} C_{k-1}^j p^{k-j-1} q^j \bar{F}_{N_0-(k-j-1)}(t) \qquad (18.18)$$

with

$$\bar{F}_i(t) = 1 - F_i(t) = e^{-i\phi t}$$

We can thus calculate the distribution of the time required to obtain a program with a specified number of bugs.

18.5.4 'Random debugging' model

This model accounts for the randomness in debugging which can lead to the removal of a bug and to the introduction of others! It was proposed by Littlewood and Verral [80] and by Keiller *et al.* [81, 82]. The probability density function of inter-failure times is

$$f(t_i|\lambda_i) = \lambda_i e^{-\lambda_i t_i}$$

In this model, the i are assumed independent of one another and gamma distributed. In the Littlewood–Verrall model, the distribution of the λ_i is a gamma distribution with parameters $1/\psi(i)$ and α (see Section 4.3). Hence,

$$f(\lambda_i) = \frac{[\psi(i)]^\alpha \lambda_i^{\alpha-1}}{\Gamma(\alpha)} e^{-\psi(i)\lambda_i} \quad \text{with} \quad \alpha > 0_i;\ \psi(i) > 0$$

so that

$$R_i(t) = \left[\frac{\psi(i)}{t + \psi(i)}\right]^\alpha \qquad \text{MTTF}_i = \frac{\psi(i)}{\alpha} \qquad (18.19)$$

It is usually assumed that $\psi(i)$ is a linear function ($\psi(i) = \beta_1 + \beta_2 i$); parameters are estimated by maximum likelihood. Note that if

- $\psi(i)$ is an increasing function of i: $\text{MTTF}_i > \text{MTTF}_{i-1}$, the software reliability is increasing.

- $\psi(i)$ is a decreasing function of i: $\text{MTTF}_i < \text{MTTF}_{i-1}$, the software reliability is dropping; the observed bugs are inadequately fixed and additional bugs are introduced!

In the Keiller–Littlewood model, the distribution function of the λ_i is a gamma distribution with parameters $1/\beta$ and $\psi(i)$. Hence

$$f(\lambda_i) = \frac{\beta^{\psi(i)} \lambda_i^{\psi(i)-1}}{\Gamma(\psi(i))} e^{-\beta\lambda_i}$$

The proposed distribution for $\psi(i)$ is $\psi(i) = \alpha_1 + \alpha_2 i^{\alpha_3}$. It follows that

$$R_i(t) = \left[\frac{\beta}{t+\beta}\right]^{\psi(i)} \quad \text{MTTF}_i = \frac{\beta}{\psi(i)} \qquad (18.20)$$

Note that reliability increases when $\psi(i)$ is a decreasing function of i. Parameters are also estimated by maximum likelihood.

18.5.5 'Bugs with different occurrence rates' model

This model proposed by Littlewood [83, 84] differs from the first model ('perfect debugging' model) since it assumes that bugs have different probabilities of causing software failure. Let $\phi_1, \phi_2, \ldots, \phi_{N_0}$ be the occurrence rates of the N_0 bugs contained in the software; λ_i is the failure rate of the software after the $(i-1)$th bug has been corrected:

$$\lambda_i = \phi_1 + \phi_2 + \ldots + \phi_{N_0-i+1}$$

$\phi_1, \phi_2, \ldots, \phi_{N_0}$ are assumed to be independent, identically distributed gamma (α, β) random variables: when the program has run for a time t', use of Bayes theorem shows that the remaining rates are distributed gamma distribution $(\alpha, \beta + t')$ random variables.

Therefore the probability density function $f_i(t)$ can be written [83, 84]

$$f_i(t) = \frac{(N_0 - i + 1)\alpha(\beta + t')^{(N_0-i+1)}}{(\beta + t' + t)^{(N_0-i+1)\alpha+1}}$$

Where t' is the time needed to detect and correct $i - 1$ bugs

$$t' = \sum_{j=1}^{i-1} t_j.$$

From this

$$R_i(t) = \left[\frac{\beta + t'}{\beta + t' + t}\right]^{(N_0-i+1)\alpha}$$

$$\text{MTTF}_i = \frac{\beta + t'}{(N_0 - i + 1)\alpha - 1}$$

$$\lambda_i(t) = \frac{(N_0 - i + 1)\alpha}{\beta + t' + t} \qquad (18.21)$$

Figure 18.4 gives the variation of the failure rate versus time. The failure

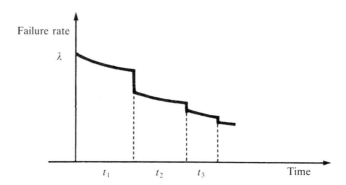

Figure 18.4 Failure-rate evolution.

rate decreases, each time a bug is detected and corrected, by an ever-decreasing quantity. It is implicitly assumed that the most likely bugs are detected and corrected first. The implications of the model are the following:

- a program with numerous bugs can nevertheless be very reliable if the bugs have very low probabilities of being detected;

- the time to achieve a nearly bug-free program can be very long.

The parameters of the model are estimated by maximum likelihood, which is far from simple. It should be noted that gamma distribution was chosen, mainly for reasons of mathematical tractability. Moreover, the model has almost the same limitations as the Jelinski–Moranda model.

18.5.6 Parametric models

Numerous other models have been proposed in which reliability modelling is more parametric than explanatory. Poisson distributions are generally used to describe the number of faults occurring in a time interval. Thus the Duane model [85, 86], which was initially used to describe hardware reliability growth, was also proposed for software. Schneidewind [87, 88] also proposed a parametric model. Studies [89–93] show connections between models based on Poisson distributions and the first model. Note however that, on the whole, these parametric models have not been so widely applied as the models discussed previously.

18.5.7 Model validation

Studies have been conducted to verify and validate the proposed models; it should nevertheless be noted that they are still limited in scope. It is in fact difficult to obtain relevant data in order to validate the models.

Several sets of data were published [53, 68, 69, 71, 82, 93–97]. Reference [9] lists and presents some of them. The main set of data was put together by Musa who gathered failures from 20 different projects developed at Bell Laboratories in the early 1970s [68, 97].

For four of the projects, the MTTF measured in operation was compared to the predicted MTTF after testing; as indicated in Table 18.2, there is a good fit between the values.

However, predictions of the number of bugs to be discovered are not always excellent; one sometimes notes [68] that X bugs remain to be discovered, and, after those X bugs have been discovered, the model predicts that the same number still remain to be discovered!

Since the late 1970s, studies have been carried out to validate the many models proposed using available data sets. Indeed certain models which, with a particular data set, give good predictions, turn out to be inadequate with other sets. Apparently prediction accuracy and quality vary greatly [93] from one data set to another or even inside the same set; no model is able to verify all data sets! Moreover, our understanding of software science is now too imperfect to be able to choose *a priori* one particular model even though we may have an intimate knowledge of the software, its specifications and its conditions of design and use.

Therefore the analyst should find the model which best fits his data set; the quality analysis of model prediction [93] can be of some help to the analyst. It should be remembered that the problem should be considered as follows:

- a set of data is available, which is made up of a sequence of execution times $t_1, t_2, \ldots, t_{i-1}$ between successive failures of the software. These can be regarded as realizations of random variables $T_1, T_2, \ldots, T_{i-1}$;

- the objective is to use $t_1, t_2, \ldots, t_{i-1}$ representing past observations to predict the future unobserved times $t_i, t_{i+1} \ldots$

In practice, the analyst will often have to make do with a measure of this prediction, such as unreliability:

$$[\bar{R}_i(t) = P[T_i < t]]$$

or reliability $[R_i(t) = 1 - \bar{R}_i(t)]$ or the MTTF.

To predict the future $(t_i, t_{i+1}, \ldots)$ from the past $(t_1, t_2, \ldots, t_{i-1})$, one proceeds as follows:

- probabilistic modelling of the distribution of any T_i with unknown parameters $\{\alpha\}$;

- determination of parameters $\{\alpha\}$ from the data;

- prediction for T_i based on the preceding stages.

Table 18.2 The measured and predicted MTTF (in hours)

	Project 1	Project 2	Project 3	Project 4
MTTF measured in operation	14.6	31.4	30.3	9.2
MTTF predicted after testing				
• Maximum likelihood value	20.4	43.5	30.4	14.5
• Confidence interval (50%)	14.5–30.9	> 24.5	> 16.0	7.6–27.9

Probabilistic methods were dealt with in the previous sections; the parameters $\{\alpha\}$ are, in general, determined by maxiumum likelihood or the least squares method or the Bayesian approach among which the latter often proves impracticable. The analyst wonders 'how reliable is my program now?' This question pops up each time a failure is observed and the bug is corrected, with the hope that the successive reliability estimates will indicate a steady improvement!

Let us assume that this estimate is the MTTF. Figure 18.5 shows how these estimates change from the 35th to the 136th failure of a software program [93] modelled using the 'perfect debugging' model (model number 1) and the 'random debugging' model (Littlewood–Verrall model, number 4).

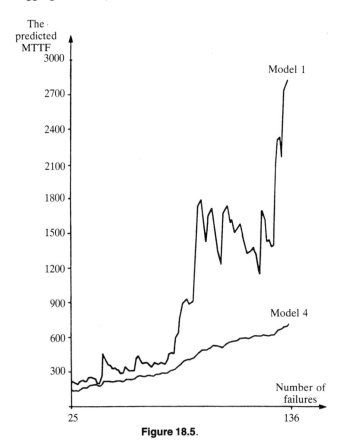

Figure 18.5.

The analyst notices that predictions, which at first are relatively close, later disagree! These predictions agree only on the fact that software reliability increases! The predictions of model number 1 even show momentary drops in reliability. This in fact is due to the effect of parameter estimates.

The prediction quality can be judged by estimating the MTTF for each failure and comparing the estimated MTTF with the real value; the analyst will trust the model if there is a good fit between these values and past failures (past predictions). The authors of references [81, 93] have developed measurement techniques for this purpose. Let's assume that the analyst calculated $R_i(t)$ using the model and data $t_1, t_2, \ldots, t_{i-1}$. When T_i takes the value t_i, $u_i = \bar{R}_i(t_i)$ is calculated.

This calculation is made for a certain number of values of i. If $\hat{\bar{R}}$ is the real distribution of T_i, u_i will be the realization of a random variable in the interval $[0,1]$. If the prediction is accurate, then u_i should be uniformly distributed. Various techniques are available [93] to verify this; the prediction can then be appraised.

The authors of reference [93] compared the performance of the main models for three sets of data; the Bayesian approach does not seem to improve prediction quality much; one notes the following points:

- No one model is always superior to the others.

- Model number 1 gives optimistic results whilst model number 4 yields pessimistic results: this fact probably has a more general impact; indeed, model number 1 overestimates reliability and thus overpredicts the MTTF since it assumes that debugging is always perfect whereas model number 4 underestimates them since it always sees debugging as a random process;

- the number of initial bugs calculated can considerably differ from one model to another. Nevertheless, reliability is the most significant measure, as noted earlier.

18.5.8 Conclusions

Numerous models were proposed to predict software reliability from data on observed failures. Experience has shown the merits of these models but also their limitations. As it is difficult to choose the model best suited to a software program and to its life cycle stages *a priori*, the analyst must endeavour to find the most appropriate model for his data set.

18.6 RELIABILITY OF A SOFTWARE SYSTEM

Software can be looked upon as a set of software components (modules, for example). Studies have been conducted to model a software product from such components [98–102]. Transitions between modules are modelled by

Markov- and semi-Markov-type state graphs. It is generally assumed that the probability of calling a module from a given one depends on these two modules only, or that the time spent in each module is a random variable, characteristic of the source and target modules.

Let us consider a software system comprising n software components whose failure rates λ_i ($i = 1, \ldots, n$) were assessed using the above-mentioned methods. The behaviour of each module and, afterwards, of the software system itself, must be modelled. Figure 18.6 shows the behaviour of a software module which was determined by the state-space method.

In this model, the distribution of the times involved must be specified:

- the inactive and execution times of the component: these periods are random variables with any distribution characteristic of the module and its operating requirements;

- the time to failure: this is a random variable whose characteristics depend on the model chosen (see Section 18.5).

Assuming these times are exponentially distributed, we get a homogeneous Markov-type model (see Figure 18.7).

The notation is:

- η is the demand rate: $1/\eta$ is the mean duration of an inactive period;
- γ is the execution rate: $1/\gamma$ is the mean duration of an execution period;
- λ is the failure rate.

Let $p_i = P[$the component is initially in state $i]$ ($p_1 + p_2 = 1$, $p_3 = 0$). The reliability $R(t)$ of the component is the probability that it is in states 1 and 2. Thus the MTTF can easily be obtained (see Section 14.4).

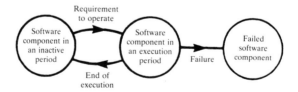

Figure 18.6 States of a software component.

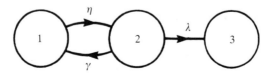

Figure 18.7 Markov-type model of a software component.

$$\text{MTTF} = \frac{1}{\lambda}\left(1 + \frac{\gamma}{\eta} + p_1 \frac{\lambda}{\eta}\right) \tag{18.22}$$

The duration of an execution period is short compared to the mean operating time between two software bugs: thus $\gamma \gg \lambda$, and *a fortiori* $\gamma + \eta \gg \lambda$. Note that no assumption is made on the relation between η and λ. An approximate formula for reliability is

$$R(t) \simeq \exp\left[-\frac{\eta}{\gamma + \eta} \lambda t\right] \tag{18.23}$$

Let a be the ratio of time the component is active to the total failure-free time:

$$a = \frac{1/\gamma}{1/\gamma + 1/\eta}$$

The preceding relations can thus be written

$$R(t) \simeq \exp[-a\lambda t], \quad \text{MTTF} \simeq \frac{1}{a\lambda} \tag{18.24}$$

These relations could have been intuitively obtained if inactive and execution times were considered to be deterministic.

Let us now consider the software system; the transfer of control between the components must be modelled; we will, of course, assume that, at each moment, only one component is being run. Markovian or semi-Markovian models are proposed [101, 102].

As an example, let us take a Markov model. The parameters are:

- γ_i: execution rate of component i; $1/\gamma_i$ is the mean execution time of component i;

- $q_{ij} = P[\text{component } j \text{ is run after component } i \text{ if no failure occurs}]$; $\sum_{j=1}^{n} q_{ij} = 1 \; (\forall i)$.

Figure 18.8 shows the model obtained for a system made up of three components; states 1, 2 and 3 pertain to the failure-free execution of components 1, 2 and 3, state 4 being the failed state of the system.

Let us consider a software system consisting of n components operating in sequence (Figure 18.9).

Execution rates are generally high compared with failure rates ($\gamma_i \gg \lambda_i$). The reliability and the MTTF of the software system can be easily approximated (see Section 14.11.4):

$$\text{MTTF} \simeq \frac{\sum_{i=1}^{n} \frac{1}{\gamma_i}}{\sum_{i=1}^{n} \frac{\lambda_i}{\gamma_i}} \tag{18.25}$$

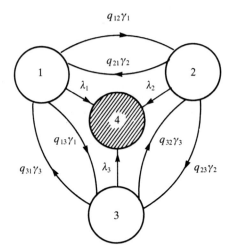

Figure 18.8 Model obtained for a software system made up of three components.

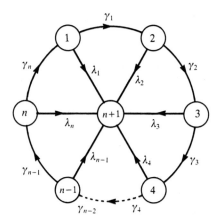

Figure 18.9 Model obtained for a software system consisting of *n* components operating in sequence.

Let a_i denote the proportion of time spent in executing component i (without failure):

$$a_i = \frac{1}{\gamma_i} \sum_{j=1}^{n} \frac{1}{\gamma_j}$$

Hence

$$\text{MTTF} \simeq 1 \bigg/ \sum_{i=1}^{n} a_i \lambda_i, \quad \Lambda(\infty) = \sum_{i=1}^{n} a_i \lambda_i \qquad (18.26)$$

The above assumption is usually verified. A very large number of components can be expected to run before a failure occurs; otherwise, the reliability of the program would be so low it would certainly need improvement!

In the general case of a semi-Markovian model of a software system comprising n components, where the execution rates are assumed to be high as compared to failure rates, Littlewood [101] showed that the system failure rate tends asymptotically to a limiting value.

$$\Lambda(\infty) = \sum_i a_i \lambda_i \quad (18.27)$$

where a_i is the limiting proportion of time spent in component i.

If we, furthermore, assume that some failures are related to interfaces between modules, we have

$$\Lambda(\infty) = \sum_i a_i \lambda_i + \sum_{i,j} b_{ij} \lambda_{ij} \quad (18.28)$$

where

b_{ij} = frequency of transfer from module i to module j

λ_{ij} = failure probability of the interface between modules i and j.

Thanks to its simplicity, this formula is very useful for developing a program from modules; these have been tested and their reliability (λ_i) was predicted using reliability models. The above formula will then be used to calculate the contribution of each module to the program reliability; parameter a_i can be measured by simulating the program. Those in charge of the project can thus deduce the areas that should be improved to reach the required reliability level. It is more difficult to make allowance for interface-induced failures. Parameter b_{ij} is obtained by simulating the program; λ_{ij} is hard to determine as the interface is usually only tested once the modules have been integrated. The analysis of integration tests can be of some help or this rate can be assumed to be independent of the modules and to have been measured on similar programs.

Finally, we should not disregard the importance of the assumption made in Markov or semi-Markovian models: the call rate of module j by module i depends on i and j only and is independent of the previous states of the system (i.e. of the modules executed before i).

18.7 AVAILABILITY AND MAINTAINABILITY OF A SOFTWARE SYSTEM

Studies [8, 101–106] have been conducted in this field, mainly by Laprie. When a software program fails, there are several ways to repair it:

- the software is reinitiated: if the input configuration differs from the configuration which led to the failure, the software will work until the next failure. Otherwise, the software experiences another failure as soon as it is reinitiated and has to be repaired;

- the software is debugged, e.g., off line on a copy which is run on another machine or on the same machine when it is not used.

Debugging is sometimes performed only after a certain number of bugs have been recorded, the corrected software being introduced in the form of a new version. It should be remembered that the software, once corrected, has become another software and thereby a new product. Figure 18.10 below contains a simplified behaviour model.

This model is solely qualitative; to make it more accurate, and quantitative in particular, the processes of failure, debugging, software reinitiation and the introduction of new versions should be examined carefully. Moreover, as seen earlier, the software failure rate varies with each new version. Solving a quantitative model is therefore not easy. We shall only take a relatively simple model as an example: let us consider a software which is rerun following correction only (Figure 18.11).

Such graphs are solved in the framework of the state-space method (see Chapter 14). Clearly the unavailability of the software is the sum of the probabilities of states $E^F(S_i)$. These models may imply complex computations in view of the high number of states and variations in failure and correction rates. To simplify the solving of this problem, the failure (or correction) process is assumed to tend to an asymptotic process, so that when the latter is reached, failures (and corrections) occur at an identical rate (Figure 18.12). The software unavailability is then

$$\bar{A} = \sum_{i=1}^{m} P[E^F(S_i)]$$

$\bar{A}$ can be approximated using a Laplace transform and an asymptotic expansion in λ_i/μ_i ($\lambda_i/\mu_i \ll 1$)

$$\bar{A}(t) \simeq \frac{\lambda_m}{\mu_m} + \sum_{k=1}^{m-1} \alpha_k \exp[-\lambda_k t] - \frac{\lambda_1}{\mu_1} \exp[-\mu_1 t] \qquad (18.29)$$

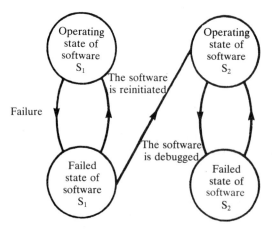

Figure 18.10 Operating and failed states of a software.

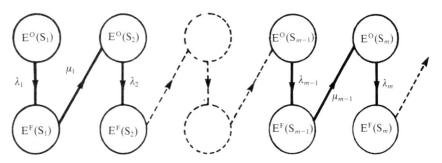

Figure 18.11.

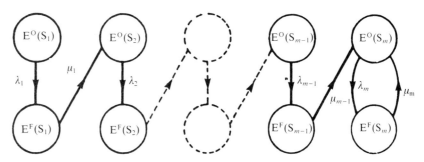

Figure 18.12 Availability and maintainability state graph of a software system.

with

$$\alpha_k = \left(\prod_{i=1}^{k-1} \frac{\lambda_i}{\lambda_i - \lambda_k}\right) \frac{\lambda_k}{\mu_k} + \sum_{j=k+1}^{m-1} \frac{\prod_{i=1}^{j-1} \lambda_i}{\prod_{i=1, i\neq k}^{j} (\lambda_i - \lambda_k)} \frac{\lambda_j}{\mu_j} - \left(\prod_{i=1, i\neq k}^{m-1} \frac{\lambda_i}{\lambda_i - \lambda_k}\right) \frac{\lambda_m}{\mu_m}$$

and

$$\sum_{k=1}^{m-1} \alpha_k = \frac{\lambda_1}{\mu_1} - \frac{\lambda_m}{\mu_m}$$

Let us note that, conventionally, the first term giving the value of α_1 is λ_1/μ_1.

Two cases need to be considered:

- $\lambda_1/\mu_1 < \lambda_m/\mu_m$: the unavailability increases as it approaches its asymptotic value (λ_m/μ_m). This is the case if repair times grow more rapidly than the software MTTFs;

- $\lambda_1/\mu_1 > \lambda_m/\mu_m$: the unavailability exceeds its asymptotic value as indicated by Figure 18.13. The maximum unavailability is

$$\bar{A}_m \simeq \frac{\lambda_m}{\mu_m} + \sum_{k=1}^{m-1} \alpha_k \qquad (18.30)$$

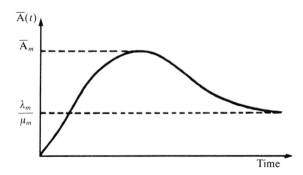

Figure 18.13 Software unavailability.

It is shown that λ_1/μ_1 is an approximate value of $\overline{A}_m$. This value can thus be regarded as an indicator of the maximum unavailability of a software product; in the long run, unavailability tends towards λ_m/μ_m.

18.8 FAULT-TOLERANT SOFTWARE SYSTEMS

In order to improve software reliability, fault-tolerant software systems were developed. We will first describe the architecture of such systems and then discuss the problem of reliability gain obtained with such architectures and give examples of these.

18.8.1 Architecture

The reader can refer to Anderson and Lee's book on fault-tolerant software systems [107]. A technique based on the redundancy principle is used more and more frequently to improve software system reliability. Two techniques are commonly employed:

- *N*-version programming [108, 109]. *N* programs, designed under different conditions and according to different algorithms, are executed simultaneously. The voter compares the outputs, and the common output is validated (Figure 18.14). Whenever outputs differ, the majority output is validated.
  Two-version programming is often used (see Figure 18.15). Two independent versions of the software are produced; outputs are submitted to a comparator and are validated if they agree. If they do not, the operators are alerted by an alarm: indeed the correct result is not necessarily known.

- **Recovery block scheme** Programs, always designed under different conditions and according to different algorithms, are successively executed until a validating condition is met. Thus, if after execution of program P_1, the condition remains unmet, then P_2 is executed, etc. It should be noted that

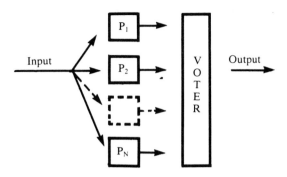

Figure 18.14 *N*-version programming.

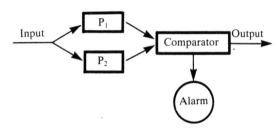

Figure 18.15 Two-version programming.

various acceptance criteria have been proposed [110] as well as hybrid structures [111, 112].

18.8.2 Reliability

What improvement have these techniques brought? There is only little information available on the efficiency of such techniques when applied to software. However, two studies are worth mentioning:

- A first experiment was conducted at UCLA, USA [113, 115]. Students wrote 18 versions of an airport management program, each version containing about 400 instructions in PL/1. All third-order combinations (816) were evaluated. In 27% of the combinations, two correct versions successfully adjust the defective computation of the third faulty version. An erroneous result is obtained in 3% of the cases only; there lies the merit of *N*-version programming.

- Another experiment was conducted at Newcastle University, UK [114, 115]; students developed two versions of a software program for the control of a Royal Navy weapons system. The first version comprised about 8000 lines of CORAL programming; and the second was a fault-tolerant version based on the recovery block scheme technique. A simulator was built to simulate the extremely large variety of input data. The events recorded in the fault-tolerant version were analysed: 'event' stands for any failure observed or any detection of expected bugs during tests.

These 65 'events' fall into the following categories:

—use of recovery blocks and alarm	40
—use of recovery blocks when unnecessary	4
—unsuccessful use of recovery blocks	13
—recovery block-induced failure	4
—failure from various causes	4

It can thus be considered that some 75% of the software bugs (40/53) were eradicated thanks to the use of recovery blocks. It should be stressed that, under certain circumstances, this technique results in the software failure (10% of the events). Moreover the failure rates of the two versions were respectively 1.4/hour and 3.2/hour, that is a gain of about 2.3 for the fault-tolerant version.

Since this study was experimental, the authors think better results would have been obtained, had the use of recovery blocks been better mastered: 90% of the software bugs would have been of no consequence and the increase in the MTTF would have been in the order of 9.

18.8.3 Examples

Multiple software versions have been developed for a number of computer systems, often in the safety field, in order to avoid software-induced failures.

In commercial aviation, navigation, landing and flight stability control are generally computer-aided but the functions of the computer systems are not critical and normally the safety of aircraft does not depend on them. For example, aircraft such as Airbus A310 [116] or Boeing 737-300 [117] use computer-aided flight controls. Two separate versions of the software are used and their results compared; if there be a discrepancy, then the programmed systems disconnect and the crew is warned.

Aircraft projects with reduced stability margins are currently being studied; flight safety for these aircraft would depend totally on computer systems. These programmed systems must therefore be highly reliable. Their failure rate should not exceed 10^{-9} per hour for a 10-hour flight without any maintenance. In this framework, NASA is developing two architecture projects largely based on redundancy, both for the hardware and software systems. These projects are SIFT (Software Implemented Fault Tolerance) [118–20] and FTMP (Fault-Tolerant Multiprocessor) [121, 122]. However, it is difficult both to reach these objectives and to demonstrate that you have reached them.

Another well-known example is the US space shuttle [123, 124]. It is in fact the first aircraft which depends totally on computer systems for its safety during missions. This is the reason why designers gave special care to the reliability of its systems. These include four primary computers executing a normal version of the software in parallel; a voter validates the outputs. A fifth computer runs a different, simplified version. It operates in parallel with the four other computers. It has a back-up role and is able to perform the most vital tasks on request, should the four primary computers fail.

It should be underlined that the first shuttle launch had to be postponed due to a much publicized bug [125], which brought about a lack of synchronization of the computers. This bug, which was not of major consequence, had gone undetected despite an extensive software testing policy. This proves, if proof is still needed, how difficult if not impossible it is to eliminate all software bugs.

18.9 COMPUTER SYSTEM DEPENDABILITY

To model the dependability of such systems, hardware and software component models should be combined. Difficulties result from software modelling and the allowance made for reliability growth. After a software component has been corrected following a failure, it can be considered that a new computer system exists. By way of an example, let us consider a computer system S comprising a hardware and a software component (Figure 18.16).

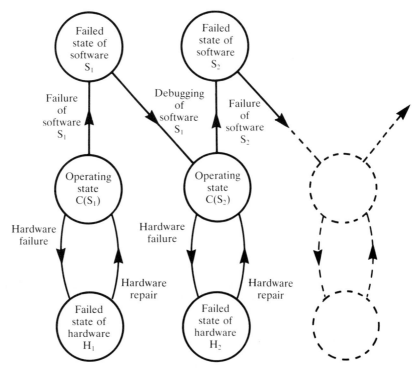

Figure 18.16 Operating and failed states of an elementary computer system.

In this model, it was assumed that the software component (or hardware component) failure could not occur after the hardware component failure, since the system has then stopped operating. Moreover the software is corrected after a failure (no reinitiation).

It is obvious that, in reality, the model must take the operating and maintenance conditions of the hardware and software components into account. The states graphs obtained can then be evaluated using the state-space method.

REFERENCES

[1] Dacs *Data and Analysis Center for Software*, Rome Air Development Center, RADC/ISISI, Griffiths Air Force Base, New York, USA.

[2] J. C. Rault (1984) *Vérification et qualification du logiciel ADI. Sûreté de fonctionnement des systèmes informatiques*, DINOV, Toulouse.

[3] M. L. Shooman (1984) Software Reliability: a Historical Perspective, *IEEE Transactions on Reliability*, **R33**(1).

[4] P. Naur *et al.* (1976) *Software Engineering Concepts and Techniques*, Petrocelli/Charter.

[5] J. Jelinsky and P. B. Moranda (1972) *Software Reliability Research*, McDonnell Douglas Astronautics Co. Huntinghton Beach, CA MADC Paper WD 1808-1971. Presented at conference on statistical methods for the evaluation of computer systems performance, Brown University, 22/23 November 1971. Also printed in statistical computer performance evaluation, W. Freiberger, ed., Academic Press.

[6] M. L. Shooman (1972) *Probabilistic Models for Software Reliability Prediction Presented at Conference on Statistical Methods for the Evaluation of Computer Systems Performance*, Brown University, 22/23 November 1971. Also printed in statistical computer performance evaluation, W. Freiberger, ed., Academic Press.

[7] G. J. Myers (1976) *Software Reliability Principles and Practice*, John Wiley & Sons, New York.

[8] M. L. Shooman (1983) *Software Engineering*, McGraw-Hill Book Company, New York.

[9] Groupe de travail AFCET (1980) Sécurité et disponibilité des systèmes informatique's *Sûreté de fonctionnement des systèmes informatiques*, Edition Hommes et Techniques.

[10] B. Littlewood (1979) How to measure software reliability and how not to, *IEEE Trans. Reliability*, **R-28**, 103–10.

[11] J. L. Albin and R. Ferreol (1982) Collecte et analyse de mesures de logiciels, *Technique et Science Informatiques*, **1**(4).

[12] M. Lipow (1979) Prediction of software failure, *The Journal of System and Software*, **1**, 71–75.

[13] R. W. Motley and W. D. Brooks (1977) *Statistical Prediction of Programming Errors*, RADC-TR-77-175.

[14] M. E. Fagan (1976) Design and code inspections to reduce errors in program development, *IBM Systems Journal*, **15**(3), 182–211.

[15] M. E. Fagan (1977) Inspections software design and code, *Datamation*, **23**(10), 133–44.

[16] E. F. Miller (1979) ACM Sigsoft, *Software Engineering Notes*, **4**(1).

[17] Pour un logiciel de qualité (1980) *Bulletin de l'AFCIQ*, **16**(3), 3–13.

[18] *Software Quality Assurance Program Requirements* (1971), Norme MIL-S-52779 (AD), Department of Defense, USA.
[19] *Military Standard Weapon System Software Development* (1978) Norme MIL-STD-1679. Department of Defense (US Navy), USA, December 1978.
[20] *Military Standard Trainer System Software Development* (1969) Norme MIL-STD-1644 (TD). Department of Defense (Naval Training Equipment Center).
[21] Software Quality Assurance Requirements (1979) Document NATO-AQUAD-13, Draft.
[22] Computer Software quality program requirements (1977) Norme FAA-STD-018, Federal Aviation Agency (FAA).
[23] Guidelines of documentation of computer program and automated data systems (1976) FIPS-PUB38. National Bureau of Standards (Institute for Computer Science and Technology).
[24] Glossary on Software Engineering Terminology (1983) IEEE St 729, IEEE Service Center, 455 Hoes Lane, Pixatway, N.J. 08854.
[25] Software Quality Assurance Plans (1984) IEEE St 730, IEEE Service Center, 455 Hoes Lane, Pixatway, N.J. 08854.
[26] Software Configuration Management Plans (1983) IEEE St 828, IEEE Service Center, 455 Hoes Lane, Pixatway, N.J. 08854.
[27] Software Test Documentation (1983) IEEE St 829, IEEE Service Center, 455 Hoes Lane, Pixatway, N.J. 08854.
[28] Guide to Software Requirements Specifications (1984) Guide 830, 1984. IEEE Service Center, 455 Hoes Lane, Pixatway, N.J. 08854.
[29] M. H. Halstead (1977) *Elements of Software Science*, Elsevier, New York.
[30] T. J. McCabe (1976) A complexity measure, *IEEE Transactions on Software Engineering*, **SE-2**(4), 308–20.
[31] G. J. Myers (1977) An extension to the cyclomatic measure of program complexity, *SIGPLAN Notices*, October, 61–64.
[32] W. J. Hansen (1978) Measurement of program complexity by the pair, *SIGPLAN Notices*, **13**(3), 29–33.
[33] S. Pimont and J. C. Rault (1976) A software reliability assessment based on a structural and behavioral analysis of programs, *Proceedings of the Second International Conference on Software Engineering*, San Francisco, 486–91.
[34] S. Pimont (1977) *Une évaluation de la fiabilité du logiciel s'appuyant sur une analyse de la structure et du comportement dynamique des programmes*, Thèse de 3e cycle. Institut de Programmation de l'Université de Paris.
[35] E. C. Nelson (1973) *A Statistical Basis for Software Reliability Assessment*, TRW Document No. TPR-TR-5.
[36] C. R. Craig, W. L. Hetrick, M. Lipow, T. A. Thayer, J. A. Yoxtheimer and J. A. Whited (1974) *Software Reliability Study*, Rome Air Development Center, US Air Force. RADC-TR-74-250; TRW Report 74-2260-1-9-29; and Report AD-787-784.
[37] H. D. Mills (1970) *On the Statistical Validation of Computer Programs*, IBM Corp. Federal Systems Division, Rept. FSC 72-6015.
[38] S. L. Basin (1974) *Measuring the Error Content of Software*, Science Applications Inc., Palo-Alto, CA.
[39] E. F. Miller (1977) Program testing art meets history, *Computer*, July, 42–51.
[40] D. J. Reifer and S. Trattner (1977) A glossary of software tools and techniques, *Computer*, July, 52–60.
[41] E. Girard and J. C. Rault (1973) A programming technique for software reliability, *IEEE Symposium of Software Reliability*, New York, 44–50.
[42] J. C. Rault (1973) Extension of hardware fault detection models to the verification of software, in *Program Test Methods*, Prentice Hall, Englewood Cliffs, N.J., 255–62.

[43] C. J. Dale and L. N. Harris (1981) *Reliability Aspects of Microprocessor Systems*, British Aerospace Dynamics Group, ST-255358.

[44] A. N. Sukert (1977) A multi-project comparison of software reliability models, *AIAA Conference of Computers in Aerospace*, 413–21.

[45] A. N. Sukert (1977) An investigation of software reliability models, *Proceedings of the 1977 Reliability and Maintenability Symposium*, 478–84.

[46] A. Tal (1976) *Development and Evaluation of Software Reliability Estimators*. University of Utah, Technical Report for Contract F 42600-76-C-0315, Report No. UTEC-SR-77-013.

[47] M. L. Shooman (1972) Probabilistic models for software reliability prediction, *International Symposium on Fault-Tolerant Computing*, 211–15.

[48] M. L. Shooman (1973) Operational testing and software reliability estimation during program development, *Proceedings of 1973 IEEE Symposium on Computer Software Reliability*, 51–7, New York, April/May.

[49] M. L. Shooman (1973) Probabilistic models for software reliability prediction, *Proceedings of the 1973 IEEE Symposium on Computer Software Reliability*, New York, April/May.

[50] M. L. Shooman (1975) Software reliability measurement and models, *Proceedings of the 1975 Annual Reliability and Maintenability Symposium IEEE*, New York.

[51] W. L. Wagoner (1973) *The Final Report on a Software Reliability Measurement Study*, The Aerospace Corp. Report No. TDR-0074 (4112).

[52] T. A. Thayer *et al.* (1976) *Software Reliability Study*. TRW Defense and Space Systems group, Final Technical Report, RADC-TR-76-288.

[53] P. B. Moranda (1977) *A Comparison of Software Error Rate Models*, IEEE Computer Society Repository R77-300.

[54] P. B. Moranda (1978) Comments on reference [58].

[55] M. Lipow and T. A. Thayer (1977) Prediction of software failures, *Annual Reliability and Maintainability Symposium*, 489–94.

[56] L. Lapadula and J. A. Clapp (1973) *Engineering of Quality Software Systems*, MTR.2648, Vol. 1, The MITRE Corporation, Bedford, Massachussetts.

[57] C. J. Schick and R. W. Wolverton (1972) *Assessment of Software Reliability*, 11th Annual Meeting German Operational Research Society, September 1972.

[58] C. J. Schick and R. W. Wolverton (1978) An analysis of computing software reliability models, *IEEE Transactions on Software Engineering*, **SE-4**(2), 104–20.

[59] M. Lipow (1972) *Estimation of Software Package Residual Errors*, TRW Software series and RTW-55-72-09, Redondo Beach, CA.

[60] P. B. Moranda and J. Jelinski (1972) *Final Report on Software Reliability Study*, McDonnel Douglas Astronautic Company, MDC Report No. 63921.

[61] P. B. Moranda and J. Jelinski (1973) Application of probability based model to a code reading experiment, *IEEE Symposium on Computer Software Reliability*, New York, 78–81.

[62] P. B. Moranda (1975) Prediction of software reliability during debugging, *Proceedings of the 1975 Annual Reliability and Maintainability Symposium*, 327–32.

[63] P. B. Moranda (1975) *Probability Based Models for Failure During the Burnin Phase*, Joint National Meeting TIMS-ORSA.

[64] P. B. Moranda (1975) Estimation of *a priori* software reliability, *Proceedings of Computer Science and Statistics 8th Annual Symposium on the Interface* (ed. J. W. Frane), Health Sciences Computing Facility, 364–70.

[65] P. B. Moranda (1977) *Quantitative Methods for Software Reliability Measurement*, IEEE Computer Society Repository R-77-299.

[66] B. Littlewood and A. Sofer (1982) A Bayesian modification to the Jelinski Moranda software reliability growthmodel manuscript. Cité dans mesure et

prévention de la fiabilité du logiciel. Conference on Software Reliability Achievement and Measurement, Paris. DINOV-24-25/04/85.

[67] B. Littlewood and J. L. Verrall (1981) On the likelihood function of a debugging model for computer software reliability, *IEEE Trans. Reliability*, **R-30**, 145–8.

[68] J. D. Musa (1975) A theory of software reliability and its application, *IEEE Transactions on Software Engineering*, **SE-1**(3), 312–27.

[69] J. D. Musa and P. A. Hamilton (1977) *Program for Software Reliability and System Test Schedule Estimation*, Program documentation, IEEE Computer Society Repository R77-243.

[70] J. D. Musa (1977) *Program for Software Reliability and System Test Schedule Estimation*, User's guide, IEEE Computer Society Repository R77-244.

[71] J. D. Musa (1977) Measuring software reliability, *Proc. National Meeting*, ORSA/TIMS.

[72] J. D. Musa (1980) The measurement and management of software reliability. *Proceedings IEEE* **68** (9), 1131–43.

[73] J. D. Musa (1980) Software reliability measurement, *J. Systems and Software* **1**, 223–41.

[74] A. L. Goel and K. Okumoto (1978) *An Imperfect Debugging Model for Reliability and Other Quantitative Measures of Software Systems*, Technical Report 78-1, Department of Industrial Engineering and Operations Research, Syracuse University, Syracuse, New York 13210-1978.

[75] A. L. Goel and K. Okumoto (1978) *Bayesian Software Prediction Models. Bayesian Software Correction Limit Policies*, Rome Air Development Center, Report RADC-TR, 78-155, Vol. 4.

[76] A. L. Goel and K. Okumoto (1978) *Bayesian Software Prediction Models. An Imperfect Debugging Model for Reliability and Other Quantitative Measures of Software Systems*, Rome Air Development Center, Report RADC-TR, 78–155. Vol. 1.

[77] K. Okumoto and A. L. Goel (1978) *Bayesian Software Prediction Models. Classical and Bayesian Inference for the Software Imperfect Debugging Model*, Rome Air Development Center, Report RADC-TR, 78–155, Vol. 2.

[78] K. Okumoto and A. L. Goel (1978) *Bayesian Software Models Availability Analysis of Software Systems Under Imperfect Maintenance*, Rome Air Development Center, Report RADC-TR, 78–155, Vol. 3.

[79] A. L. Goel and K. Okumoto (1979) A Markovian model for reliability and other performance measures of software systems, *National Computer Conference*, 769–74.

[80] B. Littlewood and J. L. Verrall (1973) A bayesian reliability growth model for computer software, *J. Roy Statist. Soc. Ser.* C (Applied Statistics) **22**(3), 332–46.

[81] P. A. Keiller, B. Littlewood, D. R. Miller and A. Sofer (1983) On the quality of software reliability prediction, *Proceedings of NATO Advanced Study Institute on Electronic Systems Effectiveness and Life Cycle Costing*, July 1982, Springer Verlag.

[82] P. A. Keiller, B. Littlewood, D. R. Miller and A. Sofer (1983) Comparison of software reliability prediction, *Proceedings FTCS 13* (13th Annual International Symposium on Fault. Tolerant computing), 128–34.

[83] B. Littlewood (1979) *A Bayesian Differential Debugging Model for Software Reliability*, Report IEEE Workshop on Quantitative Software Models.

[84] B. Littlewood (1981) Stochastic reliability growth. A model for fault removal in computer programs and hardware designs, *IEEE Trans. Reliability*, **R-30**, 313–20.

[85] J. T. Duane (1964) Learning curve approach to reliability monitoring, *IEEE Trans. Aerospace* **2**, 563–6.

[86] L. H. Crow (1977) *Confidence Interval Procedures for Reliability Growth*

Analysis, Tech. Report No. 197, US Army Material Systems Analysis Activity, Aberdeen, Md.

[87] N. F. Schneidewind (1972) *A Methodology for Software Reliability and Quality Control*, Naval Postgraduate School, Monterey, CA.

[88] N. F. Schneidewind (1975) Analysis of error processes in computer software, *Proceedings of the International Conference on Reliable Software*, 337–46, Los Angeles.

[89] N. Langberg and N. D. Singpurwalla (1981) *An Unification of Some Software Reliability Models via the Bayesian Approach*, Tach. Mem. TM-66571, George Washington University, Washington DC.

[90] D. R. Miller (1984) Exponential order statistic models of software reliability growth, Tech. Report T-496/84, George Washington University, Washington, DC.

[91] J. Angus, R. Schafer and A. Sukert (1980) Software reliability model validation, *Proc. Annual Reliability and Maintainability Symposium*, San Francisco.

[92] C. V. Ramamoorthy and F. B. Bastani (1982) Software reliability. Status and perspectives, *IEEE Trans. Software Engineering*, **SE-8**, (4).

[93] B. Littlewood, A. A. Abdel Ghaly and P. Y. Chan (1985) *Evaluation of Competing Software Reliability Predictions*, Centre for Software Reliability, The City University, London.

[94] P. A. Hamilton and J. D. Musa (1978) Measuring reliability of computational center software, *Proceedings of the 3rd International Conference on Software Engineering*, 29–36.

[95] R. C. Tausworthe (1977) *Structured Programming and Software Engineering of Hard Real-time Mini-computer Systems*, 11th Asilomar Conference.

[96] Miyamoto Isao (1975) Software reliability in on-line real time environment. *Proc. 1975. Int. Conf. Reliable Software*, IEEE, New York, No. 75, CHO 940-7CSR, 194.

[97] J. D. Musa (1980) *Software Reliability Data*, Report available from Data and Analysis Center for Software, Rome Air Development Center, New York, USA.

[98] M. Shooman (1976) Structural models for software reliability prediction, *Proc. 2nd Intern. Conf. Software Engineering*, San Francisco, CA, 268–80.

[99] B. Littlewood (1975) A reliability model for Markov structured software, *Proc. 1975 Intern. Conf. Reliable Software*, 204–7, Los Angeles, CA, April 21/23.

[100] B. Littlewood (1970) A reliability model for systems with Markov structure applied statistics, *J. Royal Statist. Soc.* Serie C, **24**, 172–7.

[101] B. Littlewood (1979) Software reliability model for modular program structure, *IEEE Trans. Reliability*, **3**, 28, 241–6.

[102] J. C. Laprie (1983) Evaluation de la sûreté de fonctionnement de logiciels en opérations. *Technique et Science Informatiques*, **2**(4), 233–47. English translation in *Technology and Science of Informatics*, **2**(4), 221–34.

[103] A. K. Trivedi and M. L. Shooman (1975) A many-state Markov model for the estimation and prediction of computer software performance parameters, *Proc. Int. Conf. on Reliable Software*, Los Angeles, 208–220.

[104] A. Costes, C. Landrault and J. C. Laprie (1978) Reliability and availability models for maintained systems featuring hardware failures and design faults, *IEEE Trans. Computers*, **C27**(6), 548–60.

[105] J. C. Laprie (1984) *Models for Software Availability Evaluation*, 2nd Workshop on Fault Tolerant Computer Systems, Melbourne, Australie, Publication LAAS No. 3000.

[106] J. C. Laprie (1984) *Définition, modélisation et évaluation de la sûreté de fonctionnement de logiciels durant leur vie opérationnelle*, Séminaire AFCET 'Approaches Quantitatives en Génie Logiciel', Sophia Antipolis.

[107] T. Anderson and P. A. Lee (1981) *Fault Tolerance Principles and Practice*. Prentice Hall International, Englewood Cliffs, N.J.
[108] J. J. Horning *et al.* (1974) *A Program Structure for Error Detection and Recovery-Lecture*, Notes in Computer Science 16 (ed. E. Gelenbe and C. Kaiser), Springer Verlag, p. 171–87.
[109] L. Chen and A. Avizienis (1978) N-version programming: a fault tolerance approach to reliability of software operation, *Digest of the 8th Int. Conf. on Fault Tolerant Computing*, Toulouse, 3–9.
[110] P. A. Lee (1978) A reconsideration of the recovery black scheme, *Computer Journal*, **21**(4), 306–10.
[111] R. K. Scott *et al.* (1983) Modelling fault tolerant software reliability, *Proc. 3rd Symposium on Reliability in Distributed Software and Database Systems*, Olearwater Beach, 15–27.
[112] M. D. Soneru (1981) *A Methodology for the Design and Analysis of Fault Tolerant Operating System*, Ph. D. Dissertation, Illinois Institute of Technology, Chicago.
[113] J. Kelly and A. Avizienis (1983) A specification oriented multi-version failure software experiment digest, *13th Int. Symposium on Fault Tolerant Computing*, Milan, 120–6.
[114] T. Anderson *et al.* (1985) *An Evaluation of Software Fault Tolerance in a practical System Centre for Software Reliability*, University of Newcastle upon Tyne. Also in *Digest of the 15th Int. Symp. on Fault Tolerant Computing*, Ann Arbor, 1985.
[115] T. Anderson (1985) *Can Design Faults be Tolerated?* University of Newcastle upon Tyne, England. In *Software Achievement and Measurement*, Paris.
[116] D. J. Martin (1982) Dissimilar software in high integrity applications in flight controls, *AGARD Symposium on Software for Avionics*, The Hague, 361.
[117] J. F. Williams *et al.* (1983) *Advanced Autopilot Flight Director System Computer Architecture for Boeing 737–300 Aircraft*, 5th Digital Avionics Systems Conf.
[118] J. H. Wensley *et al.* (1976) The design, analysis and verification of the SIFT fault tolerant system, *Proceedings of 2nd International Conference on Software Engineering*, Long Beach, CA, 458–69.
[119] J. H. Wensley *et al.* (1978) SIFT: design and analysis of a fault tolerant computer for aircraft control, *Proceedings of the IEEE*, **66**(10), 1240–55.
[120] C. B. Weinstock (1980) SIFT: system design and implementation digest of papers, FTCS 10. *10th International Symposium on Fault-Tolerant Computing* Kyoto, 75–7, 1980.
[121] J. H. Lala and A. L. Hopkins (1978) Survival and dispatch probability models for the FTMP computer digest of papers FTCS-8, *8th Annual International Conference on Fault Tolerant Computing*, Toulouse, 37–43.
[122] A. L. Hopkins, T. B. Smith and J. H. Lala (1978) FTMP: a highly reliable fault tolerant multiprocessor for aircraft, *Proceedings of the IEEE*, **66**(10), 1221–40.
[123] J. R. Sklaroff (1976) Redundancy management technique for space shuttle computers, *IBM Journal of Research and Development*, **20** (1), 20–8.
[124] C. T. Sheridaw (1978) Space shuttle software, *Datamation*, **24**(7), 128–40.
[125] J. R. Garman (1981) The bug heard round the world, *Software Engineering Notes*, **6**(5), 3–10.

19
ASSESSING SAFETY

19.1 INTRODUCTION

The notion of safety is often used in a subjective way. It is essential to develop quantitative approaches before it can be used as a functional tool in decision making.

There is no such thing as zero risk for workers, the public or the environment. It is true that we are permanently seeking to reduce incurred risk, but it is equally true that the means for doing so are increasingly expensive. The question therefore is to determine if the gain in safety is worth all the expenses involved.

The problem of safety becomes even more urgent when we consider the fact that, in our technological society, industrial activities have grown in number and have considerably extended, thus giving birth to sometimes completely new hazards. Now, we cannot afford to sit back and wait for an accident to occur before learning from it. Right from the design stage, it is essential to perform risk assessments. This approach raises several questions including that of the acceptable level of risk. Whilst an economic optimization can be used to define the level of reliability to be attained, the acceptable level of risk must be calculated in the light of moral and psychosociological factors which are sometimes difficult to quantify.

If a rational and quantitative approach to safety is to be developed, the following questions must be considered:

- What are the risks generally faced by society? Can one define acceptability thresholds? If yes, which ones?

- What PRA studies have been performed? What impact did they have?

- What are the safety goals set in the different industrial fields?

These different aspects will be discussed in this chapter one after the other.

19.2 THE RISK CONCEPT

The concept of risk needs to be clearly defined. We must first emphasize, however, that it is not an easy concept to define precisely, because of the different meanings attached to it in everyday and technical usage. Besides, people often do not make a clear distinction between this concept and the concept of probability. The reader interested in a definition will find different ones in references [1–3] and a discussion of these definitions in reference [4].

However, often in a vague and uncertain way, the definitions try to associate two aspects of the same event, i.e. its probability of occurrence and its effects or consequences.

The definition proposed below is deliberately as broad as possible.

Risk: risk is defined as the measure of a hazard combining a measure of the occurrence of an undesirable event and a measure of its effects or consequences.

A situation becomes a hazard if it can be harmful to man, the society or the environment. The occurrence of an undesirable event is generally measured by its occurrence probability over a given period or by its frequency (number of events occurring per unit of time), or even by its rate of appearance. Note that these measures are equivalent when the probability of occurrence over a given period is much smaller than 1; they are then used indifferently.

The undesirable event can have various effects or consequences:

- consequences on man: injuries or loss of human lives, diseases, etc

- economic consequences: production loss, cost of repair, etc

- environmental effects: damage to the environment, pollution, etc injured or dead animals ...

These effects are expressed by appropriate measures (C), such as number of fatalities, financial cost, etc. The combination of the two measures can be of a very different nature; often it is the product of the two measures:

$$R = D \times C$$

Risk is then expressed, for example, in monetary unit per time unit, in number of killed per time unit, in probability of death per time unit. Other combinations can be considered; for instance, $R = D \times C^k$ (with $k > 1$) where the combination tends to magnify the importance of events with severe consequences.

Let us consider, for example, the risk in France arising from motor vehicles: roughly 10 000 lives every year for 55 million inhabitants. A measure of this risk is the annual death rate due to car accidents: we thus have

$$R \simeq \frac{10\,000 \text{ fatalities/year}}{55\,000\,000 \text{ inhabitants}} \simeq 2 \times 10^{-4} \text{ death/head-year.}$$

19.3 THE ACTUAL RISKS

19.3.1 Risks from human and natural sources

First of all, we will look at risks linked to disasters. Many disasters are from natural sources or are caused by man [5]. Tables 19.1 and 19.2 give a list of a few particularly deadly catastrophes [6].

Table 19.3 gives the estimated probabilities of disasters which have really occurred or which can be reasonably expected to occur [7]. The compilers of this list used statistics for the most deadly accidents recorded and their expert judgement for potential disasters. Considering the uncertainty involved, these last figures must be handled with care.

We will now examine individual risks. For purposes of illustration, a number of significant data on the American population (approximately 200 million) and on causes of death are presented in Table 19.4. Table 19.5 gives details of the causes of deaths by accident in the American population (1969) [1].

It is sometimes more meaningful to calculate risks per hour of exposure and per exposed individual; this type of calculation is used for risks related to a particular activity (example: mountain climbing) performed by a particular segment of the population. Table 19.6 is an example of what can be obtained [8].

In the UK, Imperial Chemical Industries (ICI) developed the same type of approach. The FAR (fatal accident rate, also sometimes called FAFR: fatal accident frequency rate) or accidental death rate is the average number of deaths due to accidents in 10^8 hours of exposure to a particular activity [9–11]. Note that 10^8 hours correspond to the exposure time of 1000 men

Table 19.1 Catastrophic natural hazard

Hazards	Estimated frequency	Number of fatalities
Landslides	12 over 178 years 6.74 over 100 years	400–4000 (Pérou, 1962)
Riverine flood, tsunami	38 over 102 years 37.3 over 100 years	200–900 000 (China, 1887)
Tornado, hurricane	42 over 112 years 37.5 over 100 years	137–250 000 (Bengale, 1970)
Earthquakes	330 over 100 years	5–700 000 (China)
Volcanic eruptions	2500 over 100 years	1–28 000 (Mont Pelë, 1902)

Table 19.2 Catastrophes caused by man

Hazards	Estimated frequency	Number of fatalities
Drug toxicity (and pollutions)	more than 10 over 20 years	0–6000 (Pakistan, mercuric oxide, 1972)
Dam rupture	more than 14 over 82 years	60–2118 (Italie, Vaiont, 1963)
Fire	more than 40 over 90 years	20–1700 (China, Tchongking, 1949)
Chemical explosion and fire	more than 19 over 156 years	17–1600 (Halifax, 1917)
Mine disaster	more than 27 over 70 years	11–1549 (Honkeiko, 1942)
Ship disaster	more than 25 over 30 years	17–1913 (Titanic, 1912)
Train crash	more than 39 over 63 years	12–800 (Modane, 1917)
Aircraft crash	more than 7 over 22 years	128–570 (Spain, 1976)
Mass panics in stadiums	more than 4 over 14 years	40–400 (Lima, 1974)
Accidents from motor vehicles	250 000 fatalaties and 7 500 000 injured per year (1971)	

Table 19.3 Postulated worst-consequence event

Hazards	Estimated frequency (per year)	Estimated number of fatalities or complete disabilities
Earthquake	10^{-3} to 10^{-4}	100 000–1 000 000
Fire	10^{-6} per high risk site	1 000–3 000
Flood	10^{-2}	200 000–1 000 000
Tidal wave	10^{-2} to 10^{-3}	50 000–500 000
Tornado	10^{-2} to 10^{-3}	1 000–10 000
Dam rupture	10^{-3} to 10^{-4} per high risk site	20 000–750 000
Chemical explosion and fire	10^{-5} to 10^{-7} per site	12 000–25 000
Aircraft crash	10^{-5} to 10^{-7} for a selected site adjacent to a selected airport	6000–20 000
Ship disaster	10^{-6} to 5×10^{-7}	2500–5000
Train crash	10^{-1} to 10^{-2}	200–2000
Mine disaster	10^{-2} to 10^{-3}	500–1000
Nuclear reactor accident	10^{-7} for high risk site	10 000–20 000

Table 19.4

Causes	Individual risks (per year)
All causes	10^{-2}
Cancers	2×10^{-3}
Accidents	6×10^{-4}

Table 19.5 Individual risk of early fatality by various causes

Accident type	Total number for 1969	Individual risk per year
Motor vehicle	55 791	3×10^{-4}
Falls	17 827	9×10^{-5}
Fires and hot substance	7 451	4×10^{-5}
Drowning	6 181	3×10^{-5}
Poison	4 516	2×10^{-5}
Firearms	2 309	1×10^{-5}
Machinery (1968)	2 054	1×10^{-5}
Water transport	1 743	9×10^{-6}
Air travel	1 778	9×10^{-6}
Falling objects	1 271	6×10^{-6}
Electrocution	1 148	6×10^{-6}
Railway	884	4×10^{-6}
Lightning	160	5×10^{-7}
Tornadoes*	118	4×10^{-7}
Hurricanes*	90	4×10^{-7}
All others	8 695	4×10^{-5}
All accidents	115 000	6×10^{-4}

*Tornadoes: average 1953–71.
Hurricanes: average 1901–72.

working 2500 hours per year for a period of 40 years. Figure 19.1 indicates the FAR variation for the different daily activities of an individual who works in the chemical or building industry.

Risks exist for everyone, whatever their way of life. Everyday, our survival depends on our avoiding or overcoming risks like those in Figure 19.1.

In 1971, the chemical industry had a FAR of 4 in the United Kingdom. However, the chemical industry strictly speaking (fires, explosions, release of toxic products, etc) accounts for only half of this figure, the other half being imputable to general causes (motor vehicle accidents, falls, etc). The table below gives a few characteristic examples of FARs for industrial activities in the UK [11–12].

19.3.2 Risk acceptance

The acceptance of risk by people and society is dependent on several factors [13]. Starr [14] has highlighted the most important of these factors. It is

THE ACTUAL RISKS

Table 19.6 Individual risk per hour of exposure

Activity	Individual risk per hour of exposure
Mountain climbing	$4 \times 10^{-5}/h$
Canoeing	$1 \times 10^{-5}/h$
Travelling by motor cycle	$7 \times 10^{-6}/h$
Travelling by air	$2.4 \times 10^{-6}/h$
Travelling by car	$6 \times 10^{-7}/h$
Travelling by train	$5 \times 10^{-8}/h$

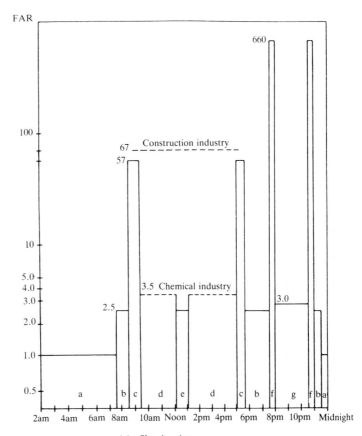

(a) Sleeping time
(b) Eating, washing, dressing, etc. at home
(c) Driving to or from work by car
(d) The day's work
(e) The lunch break
(f) Motor-cycling
(g) Communal entertainment (e.g. pub)

Figure 19.1 Daily risks.

Table 19.7 Fatal Accident Rates (FAR) in different industries and jobs in the U.K.

Industries and jobs in the U.K.	FAR
Clothing and footwear industry	0.15
Vehicle industry	1.3
Chemical industry	4
British industry	4
Steel industry	8
Agricultural work	10
Fishing	36
Railway shunting	45
Construction work	67
Air crew	250
Professional boxers	7000
Jockeys (flat racing)	50 000

related to the willingness or unwillingness to accept the risk; thus people are ready to accept high risks when they are chosen voluntarily. Other factors worth mentioning are [15]: the immediate or delayed effects of a hazard, the presence or absence of alternatives, the degree to which the risk is known, the shared or personal character of the danger, the reversibility or irreversibility of the consequences. Just like many other authors, Starr believes that the statistics of accidents reflect what the public accepts. He defines voluntary and involuntary activities as follows:

- *Voluntary activity*. The person decides of his own free will to engage in a particular activity according to his/her experience, tastes (examples: mountaineering, smoking).

- *Involuntary activity*. The person is subject to a risk he/she has not chosen or has no control over.

In western industrialized countries, the death rate due to diseases is around 10^{-2} per annum; this is a classic example of a high-level risk which is involuntarily accepted by people.

The lowest level of involuntarily accepted risk is that arising from natural events such as lightning, flood, insect bites, etc. It is around 10^{-6}/year. Between these two extremes, the public seems to accept risks to which they are involuntarily exposed according to the benefit derived from the activity [14, 15]. In their survey, Otway and Erdmann [16] propose the following simplified conclusions on the annual levels of individual death risks:

- 10^{-3}/year: this risk level is unacceptable; as soon as a risk gets close to this level, measures are immediately taken to reduce it.

- 10^{-4}/year: the public calls for public expenditure to control and reduce this risk (e.g. traffic, fires, etc).

- 10^{-5}/year: the public is conscious of these risks (drowning, firearms, etc). Advice to reduce this risk is given (for instance, 'don't go swimming alone in the sea').

- 10^{-6}/year: this type of risk does not worry the average individual. The individual is aware of these accidents but feels that they only happen to other people. He feels resigned to such a risk.

The survey authors came to the conclusion that individual death risk of 10^{-7}/year is the highest acceptable risk from accidents in nuclear power plants.

Numerous studies have been carried out by Starr on the public attitude towards accepted or imposed risks and on the relationship between risk acceptance and the expected benefit:

- mortality from disease seems to be the yardstick for the acceptance of a risk;

- accepted risks can exceed imposed risks by three orders of magnitude.

Starr observed the relationship presented in Figure 19.2 for imposed risks [15]. As it is rather difficult to assess the expected benefit, the relationship between risks and benefits has been the object of numerous heated debates [15, 17–19].

19.3.3 Risk perception

Multitudinous factors influence the way individuals perceive risks; readers interested in these factors are advised to consult references [20–24]. In the United States [25] an interesting study showed the relationship between the risk perceived by the public and the actual risk: the public was asked to give its estimate of the annual number of deaths from 41 causes; the number of killed in road accidents in the United States in 1978 (50 000) was the only example given as a reference. Figure 19.3 contrasts this estimate and the actual figure. We can thus observe that:

- The public places deaths from 'diseases' on the same level as those due to road accidents, whereas the first type of death is higher by a factor of ten.

- Risks which contribute significantly to the number of deaths are underestimated.

- Risks which are small contributors are overestimated; thus, exceptional events (tornadoes, floods) are considered to be more dangerous than they actually are.

This survey seems to confirm the fact that, to the public, an activity which claims one human life a day is 'less dangerous' than an activity which claims 300 human lives once a year.

The way in which a risk is perceived by the public depends on numerous moral and psychosociological factors which seem to be difficult to quantify or even explain. Let us examine the following accidents:

- **1. The Flixborough Accident** [5] On Saturday, 1 January 1974, a cloud of chemicals blew the Flixborough (United Kingdom) factory up and killed 28 workers.

The explosion occurred in a sparsely populated area and although 2450 houses in the neighbourhood were damaged, the effect of the explosion on the public was minimal. The accident would clearly have been more serious if it had occurred on a working day, with all the 550 employees on the premises. This accident spurred the Health and Safety Executive (HSE) to decide the creation of the Advisory Committee on Major Hazards. This committee is made up of experts who are supposed to advise the HSE on all aspects of non-nuclear risk management. This committee played a prominent part in setting the priorities for the legislation on major hazards in the United Kingdom.

- **2. The Accident at Seveso** [5] On 10 July 1976, a factory in Meda (Italy) released a cloud of dioxin, an extremely violent and stable poison. Areas were evacuated and people were contaminated. However, the accident was serious neither in the short term nor, apparently, in the long term. It caused

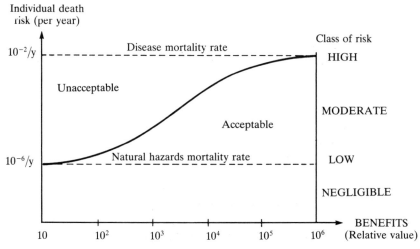

Figure 19.2 Benefit-risk pattern of involuntary exposure.

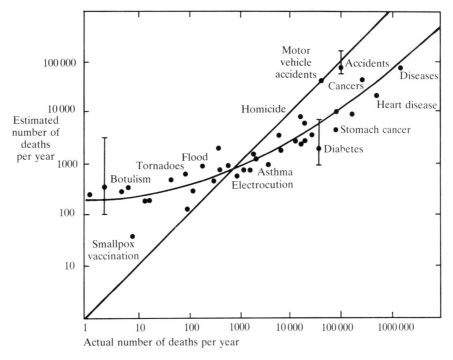

Figure 19.3 Comparison of perceived risk to actual risk.

considerable public turmoil throughout the world. In Europe, it resulted in the so-called post-Seveso directive (enforced in 1985) compelling chemical industrialists to perform detailed safety studies and submit their reports to national authorities.

• **3. The Accident at Los Alfaquès** [5] On 11 July 1978, a tank truck carrying 18 tons of pressurized liquefied propylene exploded near a camp site in San Carlos (Spain). The radiation from the fireball formed after the explosion was so intense that it killed 216 people and burned several hundreds. This accident had little or no impact on international opinion and did not therefore give rise to a noteworthy regulation at the international level.

• **4. Accident at the Three Mile Island Nuclear Power Plant** (1979; see Section 15.3) There were no casualties, but this accident had a considerable impact on public opinion throughout the world, and the American public lost confidence in the nuclear industry. Technically, it had repercussions in almost all the countries where nuclear power plants are operated.

Apparently the public is very sensitive to a new kind of risk and also to the potential or supposed severity such accidents could have in different circumstances.

19.4 PROBABILISTIC RISK ASSESSMENT (PRA)

The first probabilistic risk assessment (PRA) ever performed concerned the Surry 1 and Peach Bottom 2 nuclear power plants in the United States. This US Reactor Safety Study [26], also called the 'Rasmussen Report' after the professor of the Massachusetts Institute of Technology who directed the study from 1972 to 1975, required an effort of some 50 engineer-years and is still now regarded as a reference for the methods it involved as well as for its results.

In the mid-1980s the PRAs are still essentially restricted to the nuclear industry and petrochemistry. The major lessons drawn from reference studies in these two fields are presented below.

19.4.1 Nuclear industry

First, we will describe the main operating characteristics of a nuclear power plant. A nuclear power plant is first made of a nuclear reactor. The most widely used reactors throughout the world, and in particular in the United States and in France, are pressurized-water reactors (PWR) which will be studied here.

The general operating principle of a pressurized water nuclear power plant is the same as in a thermal power plant and is based on the following two elements:

- a heat source—here a nuclear reactor—to vaporize water;

- a steam source (steam generators) to drive a turbine coupled to a generator of electric power.

The plant is essentially made of the following (Figure 15.1):

- a reactor building comprising a concrete structure called the containment because of its protection function;

- a building housing the nuclear auxiliary equipment;

- a building accommodating the electric power generation means.

Inside the containment, we find the nuclear reactor with, among other items, the following:

- The reactor vessel containing the core where the fission of atoms of uranium-235 takes place and releases heat. The core is cooled with ordinary water circulated outside the vessel in loops (2, 3 or 4 according to the reactor power) including primary pipes and each incorporating a heat exchanger and a primary pump. One of the loops is equipped with a pressurizer which regulates the pressure in the primary system; this pressurizer is protected by relief valves which open when the pressure in the primary system—normally some 150 bars—reaches the 162 bars threshold, so that the excess steam is discharged to a so-called relief tank.

- The heat exchangers—also called 'steam generators'—where the primary system water transfers its heat to the so-called 'secondary' system, thus vaporizing the secondary water supplied by the feedwater pumps (secondary system).

The plant safety is ensured by a number of shields called 'barriers' separating the fuel from the public and by elementary systems that must operate in the event of a malfunction. A PWR is provided with a series of three such barriers:

- the cladding enclosing the fuel pellets;
- the primary system;
- the containment.

Let us quote some elementary systems having safety functions (called 'safety-related systems'):

- the reactor trip system made of a group of control rods designed to stop the fission reaction;
- the auxiliary feedwater system (AFS) actuated in the event of a loss of the normal feedwater system;
- the safety injection system injecting cold borated water into the primary system whenever the pressure in this system drops.

The major hazard from a nuclear power plant results from the formation of radioactive products during the fission of atoms. Under normal operating conditions, nuclear power plants release only very little controlled quantities of these radioactive products. However, accidents causing the release of large amounts of radioactive products and hence creating significant risks to the neighbouring populations are not to be disregarded.

Such releases could only occur if the reactor core melted. Therefore, nuclear safety is entirely aimed at preventing such an accident from occurring by providing the plants with numerous safety systems and at mitigating its effects should it nevertheless occur. This core melt may result for instance from a prolonged absence of core cooling.

It should be remembered here that the nuclear fuel continues to give off heat after the nuclear reaction has stopped. If this heat is not removed by a coolant such as water in PWRs, the temperature of the fuel rises until it melts. Radioactive products, including iodine in the gaseous or aerosol forms, are then released.

These products would then be found in the primary system water and in the reactor building air. Complex reactions would occur between them and the reactor constituents (steel, concrete, water, etc). A fraction of these products is normally 'captured' by these constituents, but another fraction may escape from the reactor building, if it is damaged, and be released into the environment (air, water, etc).

The way these products are ultimately dispersed in the environment largely depends on atmospheric conditions. These products may have the following effects:

- immediate death of individuals exposed to very high radiation doses;
- death by cancer of inhabitants exposed to high radiation doses;
- evacuation of the population from contaminated areas.

What were the results of the first PRA (Rasmussen Report)?

- The probability of a core melt in a nuclear reactor was estimated to be around 5×10^{-5} per year and its consequences on the environment are minor (less than one death). This good figure is due to the time available—from several hours to several dozen hours—between the core melt and the release of radioactive products into the environment, which allows the populations at risk to be evacuated. Note that, in this study, the assumed population distribution was based on the average distribution of the 15 million people living in an area of about 40 km around the operating or projected power plants in 1970.

- The hazards to the environment and the populations at risk are very small. Figures 19.4 and 19.5 compare the risks from the operation of 100 nuclear power plants with natural hazards or man-caused events. Thus the risk produced by nuclear power plants does not exceed the risk resulting from meteorites.

The Rasmussen Report conclusions gave rise to a heated argument between the proponents and opponents of nuclear energy. After this study, i.e. from 1975 onwards, the NRC carried out a number of similar studies of limited scope: we can mention the RSSMAP (Reactor Safety Study Methodology Applications Program [27]) for four different reactors (and containments) in order to assess the impact of the different reactor models.

We should also mention the PRAs performed on projects of high-temperature reactor (HTR) or breeder reactor (Clinch River project) nuclear power plants in 1975 and 1978 with methods similar to those used by Rasmussen [28, 29]. A PRA was also conducted for the Biblis nuclear power plant in Germany [30].

In 1977, at the request of the American Safety Authorities—the Nuclear Regulatory Commission (NRC)—a Risk Assessment Review Group [31] was

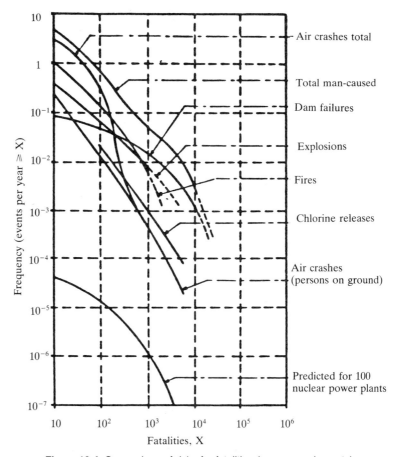

Figure 19.4 Comparison of risks for fatalities (man-caused events).

set up to evaluate the Rasmussen Report. The final report commonly referred to as the 'Lewis Report' (after its chairman) criticized numerous aspects of the Rasmussen report and especially found fault with uncertainty assessments it considered to be underestimated. However, it recommended the extended use of predictive techniques for the safety analyses of nuclear power plants. Thus, in 1979, the NRC rejected the conclusions of the Rasmussen report and nevertheless adopted the above-mentioned recommendation [32]: it concluded that there were sources of both conservatism and non-conservatism in the study but that it was unable to answer the following question: was the risk underestimated, or on the contrary, overestimated?

Among the sources of conservatism and non-conservatism, let us quote the failure to take account of possible human error recovery during accidents and the lack in exhaustiveness of the accident sequence study. Other highly criticized aspects concern the way common-cause failures were dealt with and the inconsistent propagation of uncertainties in the calculations.

Following the Three Mile Island (TMI) accident in March 1979 attitudes towards PRAs changed again. In particular, the NRC realized that the

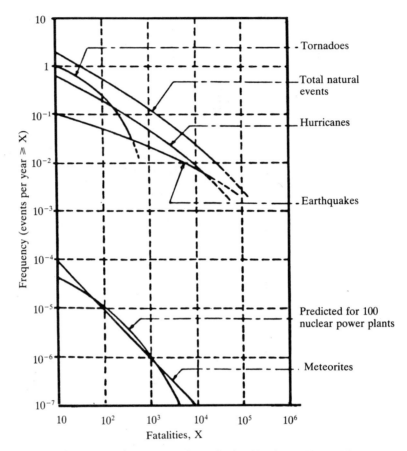

Figure 19.5 Comparison of risks for fatalities (natural hazards).

Rasmussen Report had pointed to accidents beginning with a small break in the primary system and leading to a core melt. There are points of similarity between this accident and the Three Mile Island scenario. The Commission set up by the President of the United States to investigate the accident published its conclusions and again recommended increased use of these methods [33].

In 1980, the NRC undertakes its first research work on probabilistic safety goals to be introduced in the regulation. A provisional draft of these goals was published in 1982 [34–36] and was extensively debated about. A discussion of these safety goals is to be found in Section 19.5.5.

Also in 1980, at the request of the NRC, of the US Department of Energy and of the electric utilities which form the Electric Power Research Institute (EPRI), the drafting of a guide for the application of PRA methods to nuclear power plants was started. It was aimed, first, at describing the procedure for those organizations which wished to perform such analyses and, second, at presenting the available methods which were adapted to the nuclear industry. After a first draft was published in 1981, which was also much discussed, the American Nuclear Society held a conference on the revised version of the guide (Arlington, USA, April 1982). The final version of the Guide was published in 1983 [37–39]. This Guide identifies three main levels in a PRA:

- *First level.* Analysis and identification of accident sequences. They are aimed at assessing the core melt probability. They involve the following main steps:

    —assessment of accident sequences: for instance, use of the consequence tree method;
    —assessment of the probabilities of the undesirable events occurring in accident sequences: for example, use of the cause tree;
    —procedure and human factors analysis;
    —internal and external events: the aim of these analyses is to integrate into accident sequences internal events (for example: flooding, fire resulting from causes inside the nuclear power plant) and external events (for example: aircraft crash, earthquakes, hurricanes, etc).

- *Second level.* Analysis of the physical processes of core melt and of the containment failure modes. This analysis is aimed at inventorying the radioactive products released to the environment and assessing their amounts and at estimating the corresponding occurrence probabilities.

- *Third level.* Analysis of the transport of radioactive products in the environment and its consequences. This analysis is aimed, on the one hand, at characterizing the effects of accidents on the environment and populations (immediate deaths, cancers, surface of contaminated areas . . .) and, on the other, at assessing the probability of such accidents.

An analysis covering only the first two levels is called probabilistic safety assessment (PSA). Between 1979 and 1982, following the Three Mile Island accident, PSAs were performed for five nuclear power plants as part of the IREP (interim reliability evaluation program) at the request of the NRC [40–45]. They were essentially devoted to the assessment of the core melt probability of these plants.

Also in the late 1970s, many nuclear power plant operators engaged in PRAs and PSAs. Indeed, the NRC insisted on the performance of these assessments because, since the Three Mile Island accident, it saw in them a helpful tool for the evaluation of nuclear power plant safety. At the end of 1985 some twenty PRAs and PSAs had been performed [46, 47].

Most of these studies are similar to the Rasmussen Report both in scope (some 20 to 40 engineer-years) and goals. A great many studies compared the results and methods used in these assessments.

Concerning the application of these methods by designers, plant operators and the NRC, the present state of the art has been described in [48]. Let us mention here a few lessons on various aspects of the methods:

- **The results** According to the considered nuclear power plants, the core melt probabilities may range from 10^{-3}/year to 10^{-5}/year. Extreme caution should be exercised when comparing results obtained with sometimes different methods. Nevertheless, the core melt probability may vary by two orders of magnitude although all these nuclear power plants have been designed to comply with the same rules of the deterministic approach. Therefore, this approach clearly does not guarantee consistent safety levels.

  The core melt probabilities obtained generally exceed the values quoted in the Rasmussen report. However, the consequences on the populations and the environment of the most probable core melt accidents are insignificant.

  It appears that initiating events of the external type (flood, fire, earthquake, etc) can significantly contribute to the risk; anyway, the uncertainties in the probabilities of such events exceed by far those associated with internally generated accident sequences.

  The impact on the environment and population depends on the amounts of radioactive products released out of the containment; major uncertainties exist as to the quantities of these products which would remain trapped in the containment. Experts believe that the released quantities would be below those presently anticipated, which is substantiated by experiments currently going on. In this case the immediate death risks would be insignificant.

- **The approach** The performance of a PRA improves our knowledge of nuclear power plants under accident conditions. This deep knowledge makes conventional safety assessment more realistic. Experience shows that conventional assessment, which is essentially qualitative, conservative and deterministic, leaves aside some important aspects that can result in accident scenarios with too high probabilities.

  The probabilistic approach can identify the actually important elements that must be improved. Thus, at the request of the safety authorities, a very large number of alterations were studied at the Big Rock Point power plant, a rather old facility. The introduction of these changes would have resulted in the plant decommissioning because of its no longer being profitable. The PRA performed identified the changes that were really necessary and thus made it possible to continue operating the plant [48].

- **The design and operation** Numerous alterations were introduced as a result of these studies. As regards the design, safety-related systems were modified to eliminate the weak points identified. Sometimes safety systems have been added: this is the case, for instance, for the Oconee nuclear power plant, to reduce the risks of floods brought to light by the PRA [49].

  Concerning operation, these alterations are generally quite numerous and easier to introduce: modification of an accident procedure to reduce the risk of human errors, development of periodic test procedures for safety-related equipment, etc.

  As an illustration, we shall now present the results of the PRA conducted for the Oconee power plant [49]. The reactor in this plant is a PWR and has been operating since 1974, the power plant capacity being 860 MWe. The core melt probability was assessed at 2.5×10^{-4}/year; this is the mean value of a distribution whose 5% and 95% bounds are respectively 7.5×10^{-5}/year and 5.5×10^{-4}/year.

Internal and external initiating events contribute to this probability as follows:

- internal initiating events $\qquad :0.5 \times 10^{-4}$/year

  Among others:

  — loss of auxiliary systems and of cooling
     water system $\qquad :1.3 \times 10^{-5}$/year
  — large break in primary system $\qquad :0.9 \times 10^{-5}$/year
  — small break in primary system $\qquad :0.6 \times 10^{-5}$/year
  — anticipated transients without trip (or rod drop) $\qquad :0.6 \times 10^{-5}$/year
  — rupture of feedwater pipe $\qquad :0.5 \times 10^{-5}$/year
  — loss of instrument air $\qquad :0.3 \times 10^{-5}$/year

- external initiating events $\qquad : 2 \times 10^{-4}$/year

  — flooding of turbine building $\qquad :8.8 \times 10^{-5}$/year
  — earthquake $\qquad :6.3 \times 10^{-5}$/year
  — flood $\qquad :2.5 \times 10^{-5}$/year
  — tornado $\qquad :1.3 \times 10^{-5}$/year
  — fire $\qquad : 1 \times 10^{-5}$/year

The risks to the population are given on Figures 19.6, 19.7 and 19.8.

Figure 19.6 gives an estimate of the annual probability of more than X cancer-induced deaths; the second axis below gives the percentage increase in the resulting number of cancers against the cancers from all causes observed during the same period on the same population. A population of 860 000 living within an area of some 80 km around the power plant site is considered for this estimate. Cancers induced by exposure to radioactive products are assumed to appear over a period of 30 years after a latent period of 10 years. The individual probability of death from cancer amounts to approximately 9×10^{-9}/year. Figure 19.7 gives a calculated annual probability of more than X thyroid cancers: remember that radioactive iodine releases may cause thyroid cancers.

Figure 19.8 gives a calculated annual probability of more than X immediate deaths; by this we mean any death likely to occur during the first year following the radioactive releases and resulting from acute radiation. The individual probability of immediate death approaches 2×10^{-10}/year.

This approach was also used to assess the safety of the transportation, processing and storage of radioactive products; the reader interested in the subject can consult reference [2]. The two international conferences held in 1985 [50, 51], which were solely devoted to the probabilistic safety assessment of nuclear power plants, prove how important and mature these techniques have now become.

522 ASSESSING SAFETY

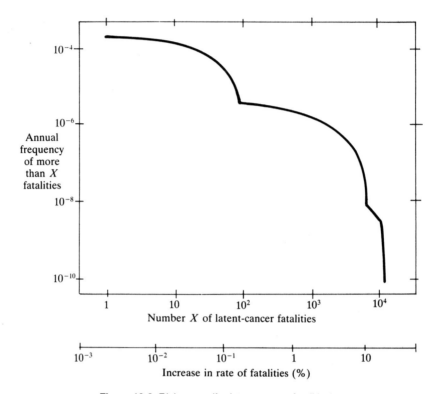

Figure 19.6 Risk curve (for latent-cancer fatalities).

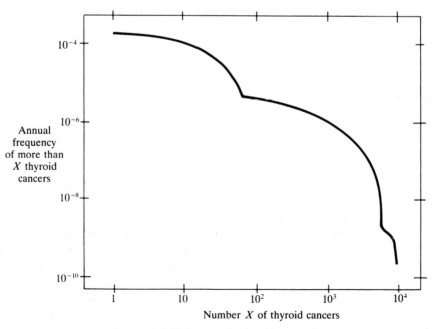

Figure 19.7 Risk curve (for thyroid cancers).

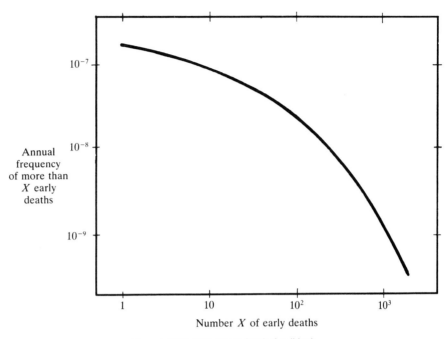

Figure 19.8 Risk curve (early fatalities).

19.4.2 Petrochemistry

The most widely known PRA was performed in the United Kingdom on chemical and oil plants on Canvey island [52]. It was carried out for the HSE by the United Kingdom Atomic Energy Authority.

In 1975, people began to express doubts as to whether it was wise to build an extra oil refinery in this area, especially after the Flixborough disaster in 1974 (see Section 19.3.3). Remember that at Flixborough a cloud of chemicals blew the factory up and killed 28 workers; this disaster occurred in a sparsely populated area, and the impact on the general public was minimal. However, it increased the fear of the inhabitants of Canvey and showed the need for studying potential interactions or chain reactions between different industrial sites. In 1976, it was decided to perform a PRA taking into account all the possible interactions between sites.

The study was published in 1978 and was one of the essential documents presented at the 1980 public enquiry. Over 20 million tons of petroleum products and crude oil are loaded or unloaded each year in the Canvey and Thurrock area; around 35 000 people live in this area. The main industries and activities involved were:

- British Gas Corporation: methane terminal where Liquefied Natural Gas (henceforth referred to as LNG) is imported and stored;

- Texaco Ltd: storage of oil products;

- Mobil Oil Co Ltd: oil refinery, research laboratory;

- Calor Gas Ltd: factory which liquefies petroleum gases (butane, propane, henceforth referred to as LPG);

- Shell Oil UK Ltd: oil refinery;

- Fisons Ltd: manufacturing of ammonium nitrate;

- others: shipping of explosives.

The study of risks was mainly focused on toxic gases (ammonia, hydrofluoric acid) or flammable substances. In 1976 as the chemical industry had little experience of PRAs, the help of UKAEA was sought. The study was supposed to identify the main hazard sources in the plants, assess the associated risks (individual and collective risk) and propose means for reducing them.

The main hazard sources identified were the following:

- oil spill (fire and/or explosion);

- release of an LNG cloud (fire and/or explosion);

- release of an LPG cloud (fire and/or explosion);

- explosion of ammonium nitrate;

- release of a cloud of ammonia (toxic product);

- release of a cloud of hydrofluoric acid (toxic product).

The main methods of predictive analysis used were cause trees and consequence trees. Dependability data were deduced from the analysis of incidents and accidents in the petrochemical sector. The ultimate consequences were assessed in terms of the number of seriously injured persons, no distinction being made between seriously injured persons and the dead as this was not deemed possible or necessary.

In Table 19.8 collective risks are arranged in decreasing order according to the severity of the consequences [53].

Certain characteristics need to be emphasized:

- risks from large quantities of stored LNG are high; they do not however exceed risks from small quantities of stored LPG,

- risks from LNG, LPG and ammonia are equivalent to other risks (petroleum products, hydrofluoric acid, etc),

- the risks change with the severity of the consequences; among major risks (number of seriously injured persons $> 18\,000$), toxic gas-related hazards are high up on the list. A number of interactions between sites were

Table 19.8 Rank order of societal risks for some principal hazards of existing and proposed installations at Canvey

Frequencies in units of 10^{-6}/year

> 10 casualties		> 4500 casualties		> 18 000 casualties	
1. Oil spillage	1366	1. Ammonia vapour cloud	258	1. HF vapour cloud	100
2. LPG vapour cloud (British Gas)	970	2. HF vapour cloud	246	2. Ammonia vapour cloud	73
3. Ammonia vapour cloud	735	3. Oil spillage	156	3. LNG vapour cloud	7
4. LPG vapour cloud (others)	637	4. LPG vapour cloud (other)	96		
5. LNG vapour cloud	497	5. LNG vapour cloud	80		
6. HF vapour cloud	464	6. LPG vapour cloud (British Gas)	80		
7. AN explosion	85	7. AN explosion	17		

Note:
HF: Hydrogen fluoride
AN: ammonium nitrate

identified; for example the threat from ships carrying explosives for LNG storage, those from rotating machines (possibility of missile production) for the storage of ammonia.

One can conclude that:

- the probability of an accident leading to over 10 injured is 31×10^{-4}/year for existing plants;

- the probability of an accident leading to over 10 injured is 16×10^{-4}/year for contemplated extensions;

- the probability of an accident leading to over 18 000 injured is 1.7×10^{-4}/year for present and future plants.

Individual risks have been assessed for various sites; risks vary between 26.3×10^{-4}/year to 2×10^{-4}/year for present and future plants. From this study we learnt that sources of risk could be the methane terminal and the ammonium nitrate factory, both conducting activities which had not been considered by the public to be dangerous.

The report recognized the fact that certain experts found these estimates to be extremely conservative; in fact, it was decided that, when in doubt, a conservative estimate should be chosen and that, therefore, the best estimate should not be systematically used. Changes for reducing these risks were proposed:

- construction of a wall around certain sites to eliminate the risk of oil product spillage (and therefore fire and/or explosion);

- more frequent inspection of the LPG and ammonia storage tanks;

- water sprinkling system for reducing toxic product emissions ...

The effects of these improvements on the risks are outlined in Table 19.9; the probability of an accident leading to more than 10 injured becomes 9×10^{-4}/year for existing plants and 2×10^{-4}/year for future extensions; individual risks from existing and future plants range from 9×10^{-4}/year and 10^{-4}/year. Figure 19.9 shows the collective risks arising from existing and future plants (before and after improvements).

This PRA, which was the first of its kind in a non-nuclear field, was of great interest to the scientific and technical community and, in the industrial sector, it gave rise to numerous debates on the merits and limitations of this approach [54]. In March 1981, a reassessment of the first study performed on the methane terminal was decided by the Health and Safety Executive. This decision was taken because of the criticisms voiced against the methane terminal and its operating conditions and of the high predicted risk level linked to this activity. The second Canvey report was drafted directly as a

Table 19.9 Rank order of societal risks for some principal hazards of existing and proposed installations at Canvey, after improvements suggested

Frequencies in units of 10^{-6}/year

> 10 casualties		> 4500 casualties		> 18 000 casualties	
1. LPG vapour cloud	421	1. Ammonia vapour cloud	71	1. HF vapour cloud	24
2. LNG vapour cloud	316	2. HF vapour cloud	67	2. Ammonia vapour cloud	18
3. Ammonia vapour cloud	240	3. LNG vapour cloud	66	3. LNG vapour cloud	7
4. HF vapour cloud	115	4. LPG vapour cloud	63		
5. AN explosion	8	5. AN explosion	2		

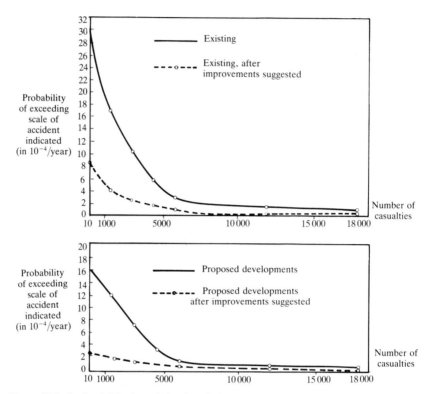

Figure 19.9 Societal risks for existing installations and proposed developments at Canvey.

result of this new study [55]. In fact, since the first study, numerous changes had been carried out and some plants had even been closed down. Moreover, considerable progress had been achieved in modelling the consequences and particularly in calculating the dispersion of dense clouds of heavy gases, this being a great source of uncertainty (and probably of conservatism) in the first study.

The collective risks obtained are much lower, by a factor of 4 for minor accidents and 40 for major accidents (Table 19.10). Individual risk ranges from 4×10^{-6}/year to 67×10^{-6}/year according to the sites.

In May 1982, the public enquiry came to the conclusion, based on the second study, that the methane terminal did not constitute an unacceptable risk and that it could go on operating. The inspector's report contained a number of observations [54]:

- after discussing the respective merits of the so-called 'conservative' approach and of what is known as 'the best estimate', the Inspector chose the first, emphasizing that it was important the risk estimate should always be regarded as an upper limit and not as an underestimate.

- the allowable risk levels in the nuclear industry are lower;

Table 19.10 Societal risks for installations at Canvey—Second PRA

Frequencies in units of 10^{-6}/year N: number of casualties						
$N > 10$	$N > 1500$	$N > 3000$	$N > 4500$	$N > 6000$	$N > 12\,000$	$N > 18\,000$
1005	201	75	34	14	9	5

- the PRA is a decision making aid but is not an end in itself.

Below are other repercussions of these studies:

- plant operators had to pay more attention to safety; numerous detailed safety studies were performed and accident procedures were revised . . .;

- the studies showed the merits and feasibility of this approach, and from this time onwards, numerous recommendations issued by the HSE have been based on these approaches;

- the public was not overly alarmed by major hazards after the publication of these studies.

19.5 QUANTIFIED SAFETY GOALS

19.5.1 Principles of the approach

During the last 10 to 15 years, numerous quantified safety goals have been proposed and used in decision making both for the design and operation of industrial systems.

How are these objectives defined? The three main approaches which can be identified are outlined below.

• **Choice of a goal based on statistical observation** Given the accident statistics, a better record is sought for the future. The goal set is arrived at by dividing the statistical data by a coefficient which represents the gain in safety. This goal must be reasonably feasible from the technical and economical point of view. Examples of this type of approach are the safety goals of the French aeronautical industry (see Section 19.5.2) and those of the British chemical industry (see Section 19.5.3).

• **Choice of a goal based on economic reasoning** The risk is compared with the profit (often assessed in financial terms) for the individual or the community as a whole. Thus the pros and cons of each human activity can be considered. The aim is to find means of optimizing the allocation of financial resources whilst minimizing risk [8, 10, 56, 57]. This type of approach is characterized by recommendations such as: 'as low as reasonably achievable (or possible)'.

Such an approach is known as 'risk–benefit' when the risk is expressed in terms of the benefit the individual or the community gains from the activity in question; it is known as 'cost–benefit' when the cost of an operation aimed at reducing the risk is expressed in terms of the expected benefit.

This approach is mostly used in the nuclear industry, especially in the United States. Assuming that we are seeking to evaluate the usefulness of an additional safety system which would lower the probabilities (or rates) of accident sequences and/or reduce radiological consequences (expressed in rems), the cost–benefit rate is defined as:

$$T = \frac{C}{T_i R_i - T'_i R'_i}$$

where C = discounted cost of the additional safety system (in dollars/year);

$T_i(T'_i)$ = occurrence rate per year of the accident sequences before modification (after modification);

$R_i(R'_i)$ = radiological consequences expressed in man-rem before modification (after modification).

This rate thus represents expenses considered for sparing a man-rem. The American Safety Authorities suggested to limit this rate to $1000 per man-rem spared (see Section 19.5.5). Such rates have been calculated [58] and compared with those in other activites [2, 59]; they show that sometimes there is a difference of up to three orders of magnitude between the expenses incurred to save a human life according to the type of activity. The nuclear industry is one of the sectors where this rate is particularly high [2].

One of the difficulties of this approach is due to the need to place a value on human life. It is often estimated as the marginal cost to save an extra human life. In many fields, the authorities decide the amount of money that has to be spent to save human lives. The judgements on which these decisions are based are implicit; nevertheless the amounts of money spent and the number of lives saved make it possible to estimate the 'value of human life' used for the decision.

This approach highlighted the fact that these values vary widely according to the sector of activity [60–62].

- **Choice of a goal based on acceptable individual and collective risks** Levels of risk acceptability can be determined by studying individual and collective risks incurred by society (see Section 19.3.2). The safety goal is then set accordingly.

At first, often only an acceptable or unacceptable risk threshold was considered.

Now three risk acceptability levels exist [63, 64]:

- *Excessive risk*. This risk level must absolutely be reduced.
- *Average risk*. This is reduced if it is reasonable (or possible) to do so.
- *Negligible risk*. Reducing such a risk is tantamount to wasting resources.

This type of approach is used for instance in the nuclear industry (see Section 19.5.2).

In practice, these approaches are sometimes used simultaneously either implicitly or explicitly. Just as important as the safety goal itself is the way in which it is used by those who manage risk (designer, plant operator, administration, public, etc). These safety goals are generally used in system design. Design is only considered to be satisfactory when the safety goal is achieved. As shown in Figure 19.10, this method is necessarily iterative so that the designer generally proceeds by successive approximations. Note that such goals have sometimes been used for defining systems operation (see Appendix 5).

We will now look briefly at the most common domains where quantified safety goals are used either officially or semi-officially.

19.5.2 Aircraft

The method used by the French aeronautical industry will be explained below.

Let us look first at international accident statistics [65, 66]. The rate of fatal accidents for regular flights was 1.5×10^{-6}/hour of flight in 1977. The distribution of fatal accidents between 1970 and 1976 shows that the causes are closely related, for 23% of fatal accidents, to airworthiness (failure or fire of engines or fires, fire in the cabin or lavatories, failure of other systems, etc) and for 77% of fatal accidents, to the operational aspect (meteorological problems, crash on the runway, etc). It follows that some 10^{-7} of fatal accidents per hour of flight can be directly ascribed to the design.

Initially, aeronautical safety regulations were mainly deterministic. Quantified safety goals were used for the first time on the Concorde project [67]. In fact, a more general and rational approach to safety based on a detailed predictive analysis and probabilistic safety goals needed to be developed since the Concorde project represented a considerable technological leap forward both in terms of use and systems. The French regulation which was developed includes these safety goals; it has been among others applied to the Airbus planes.

These goals are the result of an agreement between the designer, SNIAS (Société Nationale des Industries Aéronautiques et Spatiales) and the technical authorities responsible for checking compliance with the regulation (Service Technique Aéronautique du Ministère de l'Air). Recall that the Gathered Fault Combination Method was developed during the Concorde project and that it is recognized as an acceptable licensing method (see Chapter 11). The safety goals are defined in Table 19.11.

532 ASSESSING SAFETY

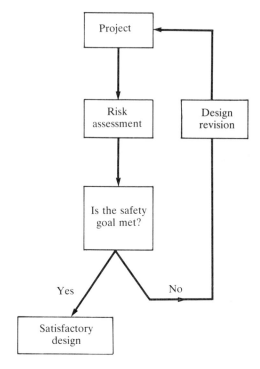

Figure 19.10 Design and safety goal.

The fault effects are divided into four classes:

- *Minor effects.* At most, these faults lead to a slight deterioration of the characteristics or a slight increase in the crew's workload.

- *Significant effects.* These faults can lead to a perceptible reduction of safety margins, to a marked deterioration of the airworthiness or to an appreciable increase in the crew's workload.

- *Critical effects.* These faults may result in a dangerous reduction of the safety margins, in an excessive workload or physical stress, which might prevent the crew from performing all its tasks or cause serious injuries.

- *Catastrophic effects.* These faults lead to the loss of the aircraft and/or numerous fatalities.

In addition, the probabilities have been divided into four classes:

- An event is probable if it can happen from time to time in the service life of each plane (for example, a failed engine). The probability then exceeds 10^{-5}/hour of flight.

Table 19.11 Safety goals in French aeronautical industry

Effects \ Probabilities	10^{-5}/hour	10^{-7}/hour	10^{-9}/hour	
	probable	rare	extremely rare	extremely improbable
Minor effects				
Significant effects	▨			
Critical effects	▨	▨		
Catastrophic effects	▨	▨	▨	

- An event is considered rare if it does not happen to each aircraft, but can occur several times in the life of a fleet of planes. The probability then ranges from 10^{-5} and 10^{-7}/hour of flight.

- An event is considered extremely rare if it does not generally occur in the life of a fleet but must however be regarded as possible. The probability is lower than 10^{-7}/hour of flight.

- An event is considered extremely improbable if we can reasonably be sure that such an event will not occur. The probability must then be lower than 10^{-9}/hour of flight.

The probability–effects couple for each fault configuration is satisfactory if we are outside the hatched area of the preceding table. Note that, for the Concorde project [67], in addition to the preceding goals, two extra criteria were imposed concerning the sum of catastrophic fault probabilities as well as the sum of critical fault probabilities. The first was supposed to be lower than 10^{-7}/hour and the second below 10^{-5}/hour. For this project, about a hundred catastrophic global faults whose occurrence probabilities were at most equal to 10^{-9}/hour have been identified: the total probability of catastrophic faults was thus lower than 10^{-7}/hour. Recall that only those faults which can be ascribed to the airplanes systems are taken into account here, and that human errors during operation are not quantified. Moreover, these goals do not apply to the structure.

Thus with the goal of 10^{-7}/hour, a new plane must be as safe as, if not safer than, the planes of the preceding generation.

On the other hand, it is interesting to note that, in the United States, the Federal Aviation Administration (FAA) suggested a goal of 10^{-9} for the probability of catastrophic events occurring during a flight (including take-off and landing); a big controversy then arose about the desirability of such a goal and the adequacy of its value. Moreover, the main groups involved in risk management do not see eye to eye on the subject. The debate is still under way, and it seems that the FAA has reiterated its request for projected planes [68].

19.5.3 Chemical plants

We will present the method used by the British company, Imperial Chemical Industries (ICI). It is based on the concept of FAR (fatal accident rate) already explained in Section 19.3.1.

In the early 1970s, the FAR of the chemical industry was 4 in the United Kingdom. ICI then decided never to exceed 2 for the most exposed operators. Any risk higher than this goal must be eliminated or reduced in existing or projected plants. A FAR of 1 is considered an average value to attain. Moreover, a particular risk, calculated separately, independently of other risks, must not exceed 10% of the FAR (of 1971) that is 0.4 for an operator; it is thus assumed that each chemical plant involves approximately five important chemical hazards. This last criterion can be adapted if necessary to particular plants [69].

ICI justifies its approach by the need to spend money first and foremost to reduce the most serious risks, namely those above average—and not to waste resources by reducing the risks which are already low. Actually, the FAR measures individual risk. Let us consider plant A and plant B, controlled respectively by an operator and 100 operators. Imagine an explosion in each plant which kills all the staff. The FAR derived from this accident is the same for the two plants: this example demonstrates that collective risk criteria should be defined.

After the accident at Flixborough (see Section 19.3.3), the Advisory Committee on Major Hazards [70] published three reports [71–73] recommending the use of predictive risk analysis methods. It placed particular emphasis on quantitative assessments, the feasibility and merits of which [74] have been demonstrated by the Canvey PRAs (see Section 19.4.2).

In the third report [73], the committee does not set quantified risk goals but recommends PRAs only as an aid for assessing risk. Note that, in its first report [71], the committee underlined that a catastrophic accident should not occur more than once every 10 000 years. It is generally accepted that an accident leading to the death of more than 30 people is catastrophic. Even though this is not an official definition, it seems to be the guiding principle for designers in the chemical industry.

19.5.4 Off-shore oil rigs

We will now describe the approach developed by the Norwegian Petroleum Directorate (NPD) for licensing off-shore oil rigs operating in the Norwegian sector of the North Sea.

Major accidents such as the blow-out (uncontrolled gushing of hydrocarbon) on the Bravo oil production platform, helicopters falling on off-shore platforms and more recently the destruction of the Alexander L. Kielland oil-rig, have made operators, authorities and the public aware of the safety problems of such structures. For example, the blow-out occurrence probability in the North Sea is estimated at some 1.7×10^{-3} per well drilled, and at 4×10^{-4}/year per operated well [75].

In 1981, the NPD issued recommendations for assessing the safety of off-shore platforms right from the design stage [76]; they included recommendations to the effect that quantified safety goals be used.

First of all, the designer must compile a list of accident situations which may cause significant losses of production or human lives. We can quote in this respect: blow-outs, fires, objects falling, collision between ships, helicopter crashes, earthquakes, extreme weather conditions and combinations of the above.

Three main safety functions are considered. Briefly, they are the following:

- *Areas of refuge*. These must remain intact several hours after the occurrence of an accident until a totally safe evacuation can be carried out.

- *Emergency exits*. At least one emergency exit must remain undamaged during at least an hour, providing a way out from the centre of the oil-rig.

- *The main structure*. In case of an accident, the main structure must continue supporting its load for a given period of time.

The quantified safety goal is defined as follows: the improbable accidental events are not considered in the design if the total probability of each type of accidental event does not exceed 10^{-4}/year for any of the safety functions. The designer [77–79] follows the procedure described below:

- identification of all probable accidents and their initiating events;

- assessment of their occurrence probabilities based on statistical studies;

- development of accident sequences for each type of accident; assessment of their probabilities;

- evaluation of their consequences, especially on their safety functions;

- classification of the accident sequences—which jeopardize the safety functions—in a list of so-called residual accidental events. If the total probability of each class of residual events exceeds 10^{-4}/year and per safety function, the design is revised; events which were previously considered to be residual then become design-basis events. Otherwise the design is considered to be satisfactory.

Design accidental events are thoroughly examined in order to ensure that their effects are taken into account in the design.

Note that nine different classes of design accidental events are considered in the design, and that the total probability of residual accidental events is calculated for each class and each safety function. The result is that the total probability of failure of a safety function is at most 9×10^{-4}/year. The probability of events jeopardizing at least a safety function is therefore at most 2.7×10^{-3}/year for the whole of the off-shore platform.

The goal of 10^{-4}/year [77] is an arbitrary value, that is the product of the compromise between the extremely improbable risk of an aircraft crashing on an oil-rig (10^{-8}/year) and the quite probable risk of a blow-out ($> 10^{-4}$/year). The NPD intends to revise its recommendations in order to emphasize the need for a regular safety assessment of oil-rigs and to make allowance for the production availability in the acceptance criterion.

In France [80], there are no quantified safety goals. Note, however, that a study of the blow-out risk was performed for the projected drilling at a depth of 1800 m in the Mediterranean Sea, some hundred kilometres off the city of Marseille [81].

19.5.5 Nuclear power plants

To begin with, we should stress that the first comments were made by Farmer [82] of the United Kingdom Atomic Energy Agency in the mid 1960s. At the time, it was suggested to adopt a maximum allowable probability for serious accidents which would be a decreasing function of their severity. Thus in the probability–severity plane, a line separates an acceptable area from an unacceptable one (Figure 19.11); the values are given as logarithmic co-ordinates and the accident severity can be expressed in terms of financial cost or of the amount of radioactive products released into the environment.

19.5.5.1 Nuclear power plants in the United States

After some preliminary work [83, 84], the Nuclear Regulatory Commission (NRC) proposed both quantitative and qualitative safety objectives in May 1983 [85]. These objectives are aimed at laying the foundations of safety regulations which help the public to understand and accept risks and serve as an aid to decision making in nuclear power plant design and operation.

In May 1983, the NRC considered that it would take some two years to assess the content of these objectives and their repercussions. During this period, organizations and experts were largely consulted. At the end of this time interval, the NRC believed that these objectives, after a few minor alterations, could be used in the regulatory procedure [86]. We shall now examine these proposals.

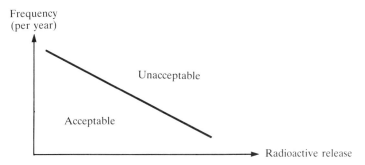

Figure 19.11 Accident release frequency limit line (the Farmer curve).

The qualitative safety objectives are the following:

- 'The public in the vicinity of the facilities should be adequately protected against potential consequences so that no individual might incur a significant additional risk to his life or health due to nuclear power plant operation.'

- 'The collective risk to life and health arising from nuclear power plant operation should be comparable to or below the risks from power production by alternate feasible techniques and should not contribute to increase other collective risks significantly.'

The following quantified safety objective have been set:

- 'The individual immediate death probability in the neighbourhood of a plant site due to nuclear power plant accidents should not exceed one thousandth of the cumulative immediate death probabilities from other accidents the American population is generally exposed to.'

- 'The individual cancer-induced death probability (in the population living near a nuclear power plant site) which may result from the plant operation should not exceed one thousandth of the cumulative immediate death probability imputable to cancers induced by all the other causes.'

- 'The benefits for society from the additional reduction in mortality should be compared with the radiological and financial cost saving on the basis of 1000 dollars saved per man-rem'.

- 'The probability of a nuclear power plant accident resulting in an extensive core melt should normally remain below 10^{-4} per reactor-year'.

Some explanation is required to understand these quantified goals better.

- The individual immediate death probability is defined for the population living in the neighbourhood of a nuclear power plant site, that is 1 mile around the plant. The American population at large is generally exposed to an individual risk of 5×10^{-4}/year; the risk goal is therefore 5×10^{-7}/year.

Note that immediate death means any death likely to occur during the first year after the radioactive releases, that is death resulting from acute radiation.

- The individual cancer-induced death probability is defined for the population living near a nuclear power plant site, that is 10 miles around the plant. This risk is calculated by dividing the total number of cancer-induced deaths in this population by the number of individuals in this population. The American population at large is generally exposed to an individual risk of some 2×10^{-3}/year; the risk objective will, therefore, be 2×10^{-6}/year.

- The cost–benefit objective is aimed at contributing to an efficient distribution of the resources devoted to safety improvement so that the anticipated risk reduction be in keeping with the safety improvement cost. The costs associated with the accident should incorporate all the costs resulting from the accident, including the decontamination, the power plant dismantling costs as well as the cost of the energy supplied to replace the power plant energy, for the most severe accidents.

- The NRC considers that the risks from the normal operation of nuclear power plants (for instance, controlled releases of small amounts of radioactive products) are so insignificant that there is no purpose in demonstrating each time or for each site that these risks meet the safety requirements.

The NRC believes these objectives will contribute to improve the assessment of nuclear power plant safety. Conscious of the complexity of the methods used for the PSA and of the numerous associated uncertainty sources, the NRC proposes that these objectives be regarded by the designers, the operators and the Safety Authorities as guidelines.

How will these guidelines be used? Take, as an example, the guideline for core melt. Assume the median core melt probability (due to causes internal to the nuclear power plant) is P:

- $P < 10^{-5}$/year. Normally the NRC should not propose additional improvements. Nevertheless, it can require that improvements be studied to reduce the probability if an accident sequence has a probability close to 10^{-5}/year. For this study, a cost–benefit approach will be used.

- $P < 10^{-3}$/year. Normally the NRC should ask for improvements to reduce the probability.

- 10^{-5}/year $\leq P \leq 10^{-3}$/year. Normally the NRC should carefully examine the accident scenarios which significantly contribute to this probability; improvements should be required to reduce the sequence probabilities.

Given the significant uncertainties on the core melt probability due to external events, such as earthquakes or floods, the safety goals will not apply to these events. However, the NRC will examine why some of these events may result in a core melt probability above 10^{-5}/year and can ask for improvements. Moreover when individual risk objectives are not met, the NRC should normally demand improvements based on a cost–benefit analysis.

In fact these safety goals will not be used as a strict acceptance criterion but as guidelines to assist the specialists in the global assessment as any other element they normally use.

The sensitivity of these objectives to some assumptions has been studied [87]: thus, as an illustration, the objectives concerning the core melt probability and the individual immediate death risk are normally the most exacting; if these objectives are met, the individual cancer risk criterion is also satisfied.

A decisive change in this approach took place in 1986 [88]: only the first two quantified safety objectives were kept. However, all these objectives, though not official, still influence the experts.

19.5.5.2 Nuclear power plants in France

In France, quantified goals and probabilistic assessments are used in quite a different way from what we have seen above.

Nuclear power plant design is deterministic and is based on such concepts as barriers, defence in depth, the single failure criterion (see Section 19.4.1). In the design and dimensioning phase, operating conditions are subdivided in four classes according to their criticality (Table 19.12); design and construction provisions are thus defined to control these conditions, to limit their occurrence probability as well as their consequences on the facility and environment.

Malfunctions with probabilities lower than 10^{-6}/year are not taken into

Table 19.12 Design–basis conditions

Annual frequency	Design–basis conditions
$\geqslant 1$	Normal operating
10^{-2} to 1	Operating transients (class 2)
	Example: loss of the external electrical supplies
10^{-4} to 10^{-2}	Accidental conditions (low frequency: class 3)
	Example: small break in the primary circuit
10^{-6} to 10^{-4}	Hypothetical accidents (class 4)
	Example: large break in the primary circuit

account in the design. All possible external events are studied. Some of them (like an aircraft crash) are taken into account in the design when the probability of unacceptable consequences exceeds 10^{-7}/year.

We shall now examine the introduction of probabilistic assessments and their main impact on the nuclear power plant design and operation; we will merely quote the research programmes reflecting a change in the dependability assessment status.

As soon the nuclear programme was launched in 1974, EDF intensified its efforts to develop dependability assessment of safety systems (around 15 at the time) as well as the associated methods; this programme, applied to the Fessenheim 900 MWe power plant from 1976 to 1978, had merely research purposes and not regulatory ones.

Numerous lessons were derived from these studies on the significance of safety-related systems and on their reliability level; they helped the designers to improve the dependability of safety systems in the nuclear power plants built afterwards. Moreover they induced the designers to refine their approach and to find counteractions to certain accidents (called additional conditions or beyond design basis events) not listed among the design basis events in the regulations. Consider for instance the initiating event 'loss of the emergency power supplies': in the late 1970s the probability of a core melt accident initiated by this event was found to be 10^{-5}/year. Additional counteractive provisions were defined and introduced in all the nuclear power plants. They have considerably lowered the probability of this family of accidents.

For the Paluel nuclear power plant, the first of the 1300 MWe series, the safety authorities asked EDF to perform dependability assessments for all the safety-related systems (1981–83); the methods described in this book were used to this end. These studies belong to the procedure required by regulations to obtain the operating licence of nuclear power plants.

For the 1400 MWe nuclear power plant series due to be commissioned in the early 1990s, EDF conducted a series of dependability assessments of accident scenarios from 1983 to 1986 to demonstrate that scenarios initiated by complementary events meet the set goals. The assessments thus help to validate the design choices. More precisely, the probability of a core melt accident starting by a complementary design basis situation (e.g. loss of emergency power supplies, loss of the normal feedwater system and of the auxiliary feedwater system) should not exceed 10^{-7}/year [89]. These studies are a new step in the introduction of probabilistic assessments in the design of power plants which from now on will be both deterministic and probabilistic (see Section 23.2).

Moreover since the early 1980s, probabilistic assessment studies have been systematically performed to define rational operating rules for the instances of partial unavailability of safety systems. An agreement was reached with the Safety Authorities as to the safety approach and goals. These are described in Appendix 5. The operating rules must be presented to the safety authorities and documented before a nuclear power plant can be commissioned.

In recent years, quantified objectives have been used for design and operation. Although their scope is still limited, they reflect a fundamental change in the approach to nuclear power plant design and operation.

We must now mention an approach based on a promising principle. The dependability assessment of the cooling system reliability in a gas–graphite reactor (UNGG-type reactor) is updated every year based on the data collected and analysed for the incidents recorded on the equipment required for the cooling function. The reliability, which is assessed with a model based on the state-space method, gives a measure of the nuclear unit safety each year. Such safety indicators help the plant operator to appraise the safety in this unit using objective safety measures.

Note finally that two PSAs were completed in 1990 [90–92]: one carried out by technical services of the safety authorities and the other by EDF. The first concerns a 900 MWe power plant and the second the Paluel power plant. The PSA performed by EDF extensively relied on all the methods discussed in this book and the most recent developments in computers. Beside the conventional objectives, this PSA was aimed at devising a true risk assessment tool (the LESSEPS software) that can be easily updated to take account of the equipment dependability data and the information brought to light by the incidents (see Section 20.8). The core melt frequencies are respectively 5×10^{-5}/year and 10^{-5}/year for 900 and 1300 MWe nuclear reactors. Accident scenarios were identified for all the reactor states (power operation, shutdown); however such external events as fires and floods were not taken into accout. Numerous and essential lessons were derived from these two PSAs both for the design and operation of French nuclear reactors [93].

Generally speaking, the probabilistic assessments tend to be periodically revised during the design and operating phases in order to check and guarantee that models correspond to reality.

REFERENCES

[1] E. J. Henley and H. Kumamoto (1981) *Reliability Engineering and Risk Assessment*, Prentice-Hall, Englewood Clifts, N.J.

[2] N. J. McCormick (1981) *Reliability and Risk Analysis*, Academic Press, New York.

[3] J. Bessis (1984) *La probabilité et l'évaluation des risques*, Masson, Paris.

[4] G. W. E. Nieuwhof (1985) Risk: a probabilistic concept, *Reliability Engineering*, **10**(3).

[5] A. Lagadec (1981) *Le risque technologique majeur*, Collection futuribles, Pergamon.

[6] P. Tanguy and J. F. Guyonnet (1978) *La prévision rationnelle des grands risques*, Le progrès technique, 11–12.

[7] K. A. Solomon, C. Whipple and D. Okrent (1978) Insurance and catastrophic events: Can we expect de facto limits on reliability recoveries? *Probabilistic Analysis of Nuclear Reactor Safety*, Vol. 2, p. IV.7.1, American Nuclear Society, La Grange Park, Illinois, 1978.

[8] D. Okrent (1971) *A General Evaluation Approach to Risk-benefit for Large Technological Systems and its Application to Nuclear Power*, UCLA-ENG-7777.

[9] T. A. Kletz (1971) Hazard analysis. A quantitative approach to safety, *I. Chem. E. Symposium*, Ser. 34, 75.

[10] S. B. Gibson (1976) Risk criteria in hazard analysis, *Chemical Engineering Progress*, **72**(2), 59.

[11] F. P. Lees (1980) *Loss Prevention in the Process Industries*, Butterworths, London.

[12] R. F. Griffiths and T. A. Kletz (1981) *Dealing with Risk*, Manchester University Press.

[13] W. W. Lowrance (1976) *Of Acceptable Risk*, Kaufman, Los Altos, California.

[14] C. Starr (1969) Social benefit versus technological risk, *Science*, **165**, 1232.

[15] C. Starr, R. Rudman and C. Whipple (1976) Philosophical basis for risk analysis, *Ann. Rev. Energy*, **1**, 629.

[16] H. J. Otway and R. C. Erdmann (1970) Reactor safety and design from a risk view point, *Nucl. Eng. Design*, **13**, 365.

[17] H. J. Otway and J. J. Cohen (1975) *Revealed Preferences: Comments on the Starr Benefit Risk Relationships*, International Institute of Applied Systems Analysis Rep. IIASA RM-75.

[18] T. H. Conolly and A. Mazur (1970) The Risks of benefit-risk analysis. *Proc. 6th Annual Health Physics Society, Midyear Symposium*, Richmond. Washington.

[19] W. D. Rowe (1977) *An Anatomy of Risk*, John Wiley & Sons, New York.

[20] A. Tversky and D. Kahneman (1974) Judgment under uncertainty, *Science*, **185**, 1124.

[21] E. L. Zebroski (1975) *Attainment of Balance in Risk–Benefit Perceptions, in Risk Benefit Methodology and Application*, Some papers Presented at the Engineering Foundation Workshop, September 22–26, 1975, Asilomar, California, UCLA-ENG, 7598, p. 666.

[22] E. W. Lawless (1977) *Technology and Social Shock*, Rutgers Univ. Press, New Jersey.

[23] B. Fischhoff, C. Hohenemser, R. Rasperson and R. Kates (1978) *Handling hazard, Environment*, **20**, 16.

[24] C. Starr and C. Whipple (1980) Risks of risk decisions, *Science*, **208**, 1114.

[25] P. Slovic, B. Fischhoff and S. Lichtenstein (1980) Facts and fears: understanding perceived risk, in *Societal Risk Assessment: How Safe Is Safe?* (R. Schuring and W. A. Albers Jr., eds), Plenum, New York.

[26] *Reactor Safety Study (1975) An Assessment of Accident Risks in US Commercial Nuclear Power Plant*, WASH 1400 (NUREG 74/014) US Nuclear Regulatory Commission.

[27] *Reactor Safety Study* (1981, 1982) *Methodology Applications Program*, Sequoyah 1 PWR Power Plant. Oconee 3 PWR Power Plant. Calvert Cliffs 2 PWR Power Plant. Grand Gulf 1 BWR Power Plant. US-NRC-NUREG/CR, 1659, 4 volumes.

[28] *HTGR Accident Initiation and Progression Analysis Status Report* (1976) General Atomic, GA A13617.

[29] *CRBRP Safety Study (1977) An assessment of accident risk for CRBRP* (Clinch River Breeder Reactor Plant).

[30] German Risk Study (1981) Main Report. *A Study of the Risk due to Accidents in Nuclear Power Plants* EPRI NP, 1804, SR.

[31] Risk Assessment Review Group Report to the US Nuclear Regulatory Commission (1978) NUREG/CR-0400.

[32] NRC Statement on Risk Assessment and the Reactor Safety Report (1979) (WASH-1400) in Light of the risk assessment review group report, US-NRC.

[33] J. G. Kemeny (1979) Report of the President's commission on the accident at Three Mile Island.

[34] J. F. Mallay (1981) NRC. Industry cooperative effort sets state for probabilistic

risk assessment studies, *Nuclear News*, September.
[35] *Safety Goals for Nuclear Power Plants: a Proposed Policy Statement* (1982) US-NRC.
[36] *Safety Goals for Nuclear Power Plants: a Discussion Paper* (1982) NUREG-0880 for comment. Office of Policy Evaluation, US-NRC.
[37] *PRA Procedures guide* (1981) *A Guide to the Performance of Probabilistic Risk Assessments for Nuclear Power Plants*, NUREG/CR-2300, Review Draft, US-NRC.
[38] *PRA Procedures guide* (1982) *A Guide to the Performance of Probabilistic Risk Assessments for Nuclear Power Plants*, NUREG/CR-2300, Revision 1, vols. 1 and 2. Review Draft, US-NRC.
[39] *PRA Procedures guide* (1983) *A Guide to the Performance of Probabilistic Risk Assessments for Nuclear Power Plants*. Final Report, vols. 1 and 2. NUREG/CR-2300.
[40] *Crystal River 3 Safety Study* (1981) US-NRC-NUREG/CR-2515.
[41] *Interim Reliability Evaluation Program* (1982) *Analysis of the Arkansas Nuclear one, Unit 1 Nuclear Power Plant*, US-NRC-NUREG/CR-2787.
[42] *Analysis of the Browns Ferry, Unit 1, Nuclear Plant* (1982) US-NRC-NUREG/CR-2802.
[43] *Analysis of the Milestone point Unit 1 Nuclear Power Plant* (1983) US-NRC-NUREG/CR-3085.
[44] *Analysis of the Calvert Cliffs Unit 1 Nuclear Power Plant* (1984) US-NRC-NUREG/CR-3511.
[45] *Interim Reliability Evaluation Program Procedures Guide* (1983) US-NRC-NUREG/CR-2728.
[46] V. Joksimovich (1984) A review of plant specific PRAs. *Risk Analysis*, **4**(4), 255–66.
[47] L. V. Konstantinov (1985) *Probabilistic Safety Assessment (PSA) in Nuclear Safety International Developments*, Seminar on Implications of probabilistic risk assessment, IAEA, Blackpool, United Kingdom.
[48] Probabilistic Risk Assessment (PRA) Reference document (1984) US Nuclear Regulatory Commission, NUREG-1050, Final Report.
[49] *Oconee PRA* (1984) *A Probabilistic Risk Assessment of Oconee Unit 3*, NSAC, Electric Power Research Institute, Palo Alto, California, NSAC-60.
[50] ANS/ENS International Topical Meeting on Probabilistic Safety Methods and Applications (1985) San Francisco, California.
[51] *Seminar on Implications of Probabilistic Risk Assessment* (1985) IAEA-Blackpool, United Kingdom.
[52] CANVEY (1978) *An Investigation of Potential Hazards from Operation in the Canvey Island/Thurrock Area*, Health and Safety Executive, HMSO.
[53] F. P. Lees (1980) *Loss Prevention in the Process Industries*, Butterworths, London.
[54] A. C. Barrell, J. N. Edmondson and P. L. Holden (1985) *Canvey Island. A Case Study of the Application of Probabilistic Risk Assessment*, Seminar on Implications of Probabilistic Risk Assessment, Blackpool United Kingdom.
[55] CANVEY (1981) *A Second Report*, Health and Safety Executive, HMSO.
[56] J. Bowen (1975) *The choice of criteria for individual risk, for statistical risk and for public risk*, in *Risk–Benefit Methodology and Application* (D. Okrent, ed.) Some papers presented at the Engineering Foundation Workshop, September 22–6, 1975, Asilomar, California, School of Engineering and Applied Science, University of California, UCLA-NEG-7598, 581–90.
[57] D. Okrent (1979) Risk–Benefit evaluation for large technological systems, *Nuclear Safety*, **20**, 148.

[58] E. P. O'Donnel and J. J. Mauro (1979) A cost benefit comparison of nuclear and non nuclear health and safety protective measures and regulations, *Nuclear Safety*, **20**, 525.
[59] B. L. Cohen (1980) Society's evaluation of life saving in radiation protection and other contexts. *Health Phys.* **38**, 33.
[60] A. Sauvy (1977) *Coût et valeur de la vie humaine*, Hermann, Paris.
[61] M. Le Net (1979) *Le prix de la vie humaine*, La documentation française.
[62] *Living with uncertainty* Risks in the energy scene (1981) Scientific and Technical Studies.
[63] H. J. Dunster and W. Vinck (1979) The assessment of risk. Its value and limitations. *Nuclear Engineering International*, 23.
[64] D. R. T. Lowe (1980) *Major Incident Criteria. What Risk Will Society Accept?* Eurochem'80, Birmingham.
[65] P. Toulouse (1979) La sécurité du véhicule aérien. Congrès Sûreté des systèmes électriques et électroniques, Toulouse.
[66] P. Toulouse (1981) *Prise en compte de la fiabilité humaine dans la conception des avions civils. Dans la Fiabilité Humaine*, Revue Générale Nucléaire, 5, September–October 1981.
[67] C. Lievens (1976) La sécurité des systèmes, Cepadues Editions, Toulouse.
[68] Y. D. Matteson, J. F. McDonald and A. M. Smith (1984) *Commercial Aviation Experience of Value to the Nuclear Industry*, EPRI-NP-3364.
[69] T. A. Kletz (1984) Hazard analysis. A review of criteria. *Reliability Engineering*, **3**(4).
[70] F. P. Lees (1980) Accident Fatality Number. A Supplementary Risk Criterion. *Proceedings of the 3rd International Symposium on Loss Prevention and Safety*. Promotion in the Process Industries, Swiss Society of Chemical Industries, 6–426.
[71] Health and Safety Commission (1976) *Advisory Committee on Major Hazards*, First Report.
[72] Health and Safety Commission (1979) *Advisory Committee on Major Hazards*, Second Report.
[73] Health and Safety Commission (1984) *Advisory Committee on Major Hazards*, Third Report.
[74] A. F. Ellis (1984) *H.S.E. Policy for the Control of Major Hazards*, A Seminar on the use of reliability data for Major Hazards Evaluation, Eurodata, October 17th, Paris.
[75] A. Leroy (1982) *Utilisation des méthodes probabilistes pour assurer la sécurité des installations et des opérations pétrolières en mer*, Revue de l'Energie, No. 346.
[76] *Guidelines for Safety Evaluation of Platform Conceptual Design* (1981) Norweigian Petroleum Directorate.
[77] O. Berg and O. Thuestad (1984) N.P.D. *Requirements Relating to Safety Evaluation of Platform Conceptual Design*, A Seminar on the use of reliability data for major hazards evaluation, Eurodata, Paris.
[78] T. Gjerstad and A. P. Pyman (1984) *Application of Concept Safety Evaluations*, A seminar on the use of reliability data for major hazards evaluation, Eurodata, Paris.
[79] O. Michelsen (1984) *Reliability and Reality: Major Accident Frequencies Based on Component Data*, A Seminar on the use of reliability data for major hazards evaluation, Eurodata, Paris.
[80] A. Leroy (1981) Analyse du risque de perte du riser lors d'un forage par 1 800 m d'eau, 9ᵉ congrès national du pétrole, Vichy.
[81] J. P. Signoret and A. Leroy (1985) The 1 800 m Water Depth Drilling Project: Risk Analysis, *Reliability Engineering*, **11**(2).

[82] F. R. Farmer (1967) Reactor safety and siting: a proposed risk criterion. *Nuclear Safety*, **8**(6).

[83] USNRC Advisory Committee on Reactor Safeguards (1980) *An approach to quantitative safety goals for nuclear power plants*, NUREG 0739.

[84] *Safety Goals for Nuclear Power Plants: A Discussion Paper* (1982) US Nuclear Regulatory Commission, NUREG-0880.

[85] *Safety Goals for Nuclear Power Plant Operation* (1983) US Nuclear Regulatory Commission, NUREG-0880.

[86] T. E. Murley (1985) *Implementation of Safety Goals in NRC's Regulatory Process*, International Topical Meeting on Probabilistic Safety Methods and Applications, San Francisco, USA, EPRI NP-3912-SR.

[87] R. M. Blond, R. P. Burke and T. S. Margulies (1985) *Safety Goal Evaluation: Sensitivity Studies*, International Topical Meeting on Probabilistic Safety Methods and Applications, San Francisco, USA, EPRI NP-3912-SR.

[88] *Safety Goals for the Operations of Nuclear Power Plants* (1986) Policy Statement, Nuclear Regulatory Commission, 10 CFR Part 50.

[89] N. Schektman, A. Villemeur, J. P. Berger, J. Boisseau, A. Dubreuil-Chambardel and F. Mosneron-Dupin (1985) *Nuclear Power Plant Design: Contribution of Probabilistic Methods and their Application to a Total Loss of Electric Power Supplies*, International ANS/ENS topical meeting on Probabilistic Safety Methods and Applications, San Francisco, USA, EPRI NP-3912-SR.

[90] A. Villemeur, J. P. Berger, A. Carnino, J. M. Moroni, J. M. De Guio, C. Ancelin and A. Ellia-Hervy (1990) *Probabilistic Safety Assessment of a 1300 MWe Nuclear Pressurized-Water Reactor Unit: Methods, Results and Lessons*, International Symposium on Reliability and Maintainability, Tokyo.

[91] *Probabilistic Safety Assessment of the Paluel Nuclear Power Plant* (1990) Synthesis Report, EDF.

[92] *Probabilistic Safety Assessment of the Paluel Nuclear Power Plant* (1991) Abstract, EDF.

[93] A. Villemeur, J. P. Berger, J. M. De Guio, J. Brisbois and J. M. Lanore (1991) *Probabilistic Safety Assessment of French 900 and 1300 MWe Nuclear Power Plants*—Revue Générale nucléaire.

Part 4

COMPUTERIZED METHODS

20
COMPUTER CODES FOR DEPENDABILITY ASSESSMENT

20.1 INTRODUCTION

Computer programs for dependability analysis were introduced in the mid-1960s. The early programs used success diagram models and computed success paths. The Automatic Reliability Mathematical Model, ARMN [1], which was developed by the North American Aviation Company for the US Air Force as early as 1966, is an example of such a program. Its interest is now mainly historical. Cause tree evaluation programs were also rapidly developed, the calculation method was usually based on a Monte Carlo simulation [2]. However, considering the state of computer technology at the time, large trees were virtually impossible to compute: a cause tree with 300 gates took 275 hours to compute on the UNIVAC 1107! However, only 17 minutes were required for simulations with variance reduction [3]!

The first efficient cause tree evaluation program (PREP and KITT) was already created at the close of the 1960s. With the 1970s came other such programs based on different algorithms as well as programs for uncertainty analysis. Some of these programs were also used for consequence tree analysis since programs specifically adapted to this method were only developed in the early 1980s. There is still a very limited number of such programs.

It was also in the mid-1970s that the first state graph evaluation programs and common-cause failure analysis programs were developed.

In this chapter, a brief description will be given of the major programs (listed alphabetically) used in the following fields:

- cause tree evaluation;
- consequence tree evaluation;
- state graph evaluation;
- common-cause failure analysis;

- uncertainty analysis

We have also included programs which do not belong to any of these categories but which are worth mentioning. In addition, the next chapter deals with the automatic dependability assessment programs. Particular attention has been paid to the Monte Carlo-type simulation programs, and a description is given of the Monte Carlo Method.

The author does not, however, claim to have drawn a comprehensive list of all the programs used throughout the world!. The programs listed here are the most commonly known which have already been studied in publications and often implemented and/or are representative of a given state of the art. These programs can normally be purchased on the market.

20.2 CAUSE TREE ANALYSIS CODES

These programs now abound. They can have several functions; four categories are considered below:

- *Qualitative analysis codes*. These solely perform the qualitative analysis of cause trees by determining their minimal cut set. Required input for the code consists of a description of the cause tree structure (logic gates) and of basic events. Most of the time, the algorithm is based on Boolean equation reductions.

- *Quantitative analysis codes*. These solely perform the quantitative analysis of cause trees; the minimal cut sets are entered into the code, and their probabilities are then computed. Importance measures are generally computed (see Appendix 3).

- *Qualitative and quantitative analysis codes*. These are used for qualitative and quantitative analyses. However, the results of the qualitative analysis (minimal cut sets) can also be obtained.

- *Direct evaluation codes*. These codes are used for both quantitative and qualitative analyses. Due to the methods used and the interdependence of the analyses, the results of qualitative analyses (minimal cut sets) are not available.

Normally, programs based on the determination of minimal cut sets involve long computer times and have quite extensive storage requirements. This makes the processing of very large cause trees prohibitive! In fact, the number of cut sets can increase exponentially with the number of gates and may easily reach millions or billions! Moreover the number of gates and basic events is not always representative of the complexity of the cause tree or the number of minimal cut sets. Thus, it is often difficult to predict the storage requirements and running time. Numerous techniques have been introduced in recent years to facilitate the processing of large trees. Some examples are:

- elimination of cut sets whose order exceeds a number fixed by the user;

- elimination of cut sets with probabilities lower than a specified value;

- tree modularization before finding the cut sets; a module is a collection of components (or events) which behaves like a 'macro-component'; an intermediate event of the cause tree corresponds to such a module if none of the basic events contained under this gate appear in other branches of the tree;

- optimization of memory space through bit handling.

The last two methods are particularly efficient. Moreover, a method for determining minimal cut sets based on a Monte Carlo simulation is sometimes used. The method consists in randomly selecting basic events and their combinations and then verifying whether the top event has occurred or not; a cut set is obtained if the result is positive. This method, though sometimes more rapid, is less precise. It is certainly not the best method for finding all the minimal cut sets.

Let us examine now the different codes in each category.

20.2.1 Qualitative analysis computer codes

DEFAIL 84

This code [4] was developed by Electricité de France (EDF, France). DEFAIL is used to find the minimal cut sets of a cause tree with AND- and OR-type logic gates. DEFAIL uses a top-down algorithm. The latest update of this code (1984) processes cause trees containing several hundreds of basic events and logic gates. For quantitative analyses, DEFAIL has to be used together with FIABC 84. It is written in Pascal for IBM 3090 computers.

ELRAFT

ELRAFT (Efficient Logic Reduction of Fault Trees [5]) is used to find the minimal cut sets of a cause tree with AND and OR logic gates. It uses the factorization property of integers into prime factors. Cause trees are computed from the bottom up, and cut sets at the different levels of cause trees are represented by the product of the numbers associated with inputs. This method is not well adapted to the processing of large cause trees. ELRAFT is written in Fortran IV for CDC 6600 computers.

FATRAM

FATRAM [6] is used to find the minimal cut sets from trees with AND and OR logic gates. The FATRAM algorithm is similar to that of MOCUS but it uses less core and less computation time due to an optimization of logic gate processing. The output consists of minimal cut sets up to an order specified by the user. FATRAM is written in Fortran IV for CDC Cyber 76 computers.

FAULTRAN

FAULTRAN [7] finds the minimal cut sets of cause trees with AND and OR logic gates. The reduction technique is based on the binary coding of events, and bit handling.

FTAP

FTAP (Fault Tree Analysis Program) [8] was developed by the University of California, Berkeley (United States). It is used to find the minimal cut sets of cause trees with AND, OR, K-of-N and NOT gates. FTAP offers the user the choice of three reduction methods: top-down, bottom-up and Nelson. The top-down and bottom-up approaches are conventional methods. The third method, Nelson, is a prime-implicant algorithm that is applied to trees containing complement events. FTAP uses particular techniques for processing cause trees (modularizing independent portions of the tree), thereby increasing the code efficiency. FTAP is thus able to generate minimal cut sets of high-order minimal cut sets. The code is written in Fortran and assembly language for CDC 6600/7600 and IBM 360–370 computers.

MOCUS

MOCUS [9, 10] was written in 1972 to replace PREP and serve as input for SUPERPOCUS or KITT1 and KITT2. MOCUS uses a top-down method to determine minimal cut sets from cause trees with AND- and OR-type logic gates. The code is written in Fortran IV for IBM 360/370 and CDC 7600 computers.

PREP

PREP [11] has essentially a historical value because it is one of the first computer codes used for cause tree computation. From 1970 onwards, it was used in the first Probabilistic Risk Assessment of a nuclear power plant (Rasmussen Report). PREP uses two different methods to determine minimal cut sets. The first method works in the following way: it generates basic events, their combinations in pairs, then in groups of three basic events, etc., to find which combinations cause the undesirable event, the user specifying the highest order of the cut sets to be computed. The second method uses the probability of events to find the most likely minimal cut sets from a Monte

Carlo simulation. The main disadvantage of the first method is that it requires a prohibitive amount of computer time. On the other hand, the second method is not guaranteed to find all the minimal cut sets. PREP can be used to compute cause trees containing up to 2000 basic events and 2000 logic gates. PREP is very efficient for obtaining minimal cut sets of up to, say, order 3. It is used with KITT which performs the quantitative analysis. It is written in Fortran IV for IBM 360/370 and CDC 7600 computers.

TREEL and MICSUP

TREEL and MICSUP [12] were developed by the University of California (United States). They are used for processing cause trees with AND and OR logic gates. TREEL and MICSUP are similar to MOCUS except that MICSUP (Minimal Cut Set Upward) works upwards. TREEL determines the maximum number and order of cut sets. Thus, cut sets of a higher order than the limit specified by the user can be eliminated, thereby reducing computer time and storage space. TREEL and MICSUP are written in Fortran IV for CDC 6400 computers.

20.2.2 Quantitative analysis computer codes

FIABC 84

FIABC 84 [13] was developed by Electricité de France (EDF, France). It performs the quantitative analysis based on the minimal cut sets determined by DEFAIL 84. Probabilities are computed as a function of time. A Monte Carlo-type simulation is used to compute uncertainty from the lognormal distribution. FIABC is written in Pascal for IBM 3090 computers.

FRANTIC

FRANTIC (Formal Reliability Analysis including Normal Testing, Inspection and Checking [14–16]) was developed by the Nuclear Regulatory Commission (United States). This program computes time-dependent unavailability of any system modelled by a cause tree. It can assess the effects of test policies and component maintenance on system unavailability. The input to FRANTIC consists of a system model equation or the minimal cut sets from a cause tree as well as the failure rates of the components and their repair and test characteristics. FRANTIC assumes exponential component failure distributions.

A second version of the program, FRANTIC II, has been developed to enhance the capability of the code to model the time-dependent unavailability for basic events; the effects of early failures as well as wearout failures can be modelled by the bathtub model and the Weibull distribution. The FRANTIC and FRANTIC II programs are written in Fortran IV for IBM 360/370 computers.

KITT

KITT1 and KITT2 [17–19] were created together with PREP. They are based on the 'Kinetic Tree Theory' developed by Vesely (see Section 9.8). This theory approximates availability or reliability in cause trees where basic events are independent of one another. Components may be repairable or non-repairable. KITT1 and KITT2 compute the following measures: unavailability, reliability, failure rate and expected number of failures as well as importance. KITT1 assumes that components have constant failure rates and constant repair times; KITT2 also assumes that failure and repair rates are constant, but these parameters can be modified to take phased-mission systems into account. The minimal cut sets generated by PREP and MOCUS can be directly input in KITT. Both KITT1 and KITT2 are written in Fortran IV for CDC 7600 and IBM 360/370 computers.

SUPERPOCUS

SUPERPOCUS [20] was developed by the University of Tennessee (United States) to perform quantitative analyses based on minimal cut sets. It can read the cut sets directly from the MOCUS or PREP output. SUPERPOCUS calculates time-dependent availability or reliability characteristics, the expected number of failures as well as importance measures. SUPERPOCUS uses approximation methods; failure and repair rates are assumed to be constant. SUPERPOCUS is more efficient than KITT. It accepts the initial unavailability of components. The program is written in Fortran IV for CDC 7600 and IBM 360/370 computers.

20.2.3 Computer codes for qualitative and quantitative analyses

ALLCUTS

ALLCUTS [21] was developed by the Atlantic Richfield Company (United States). It is very similar to MOCUS and, like this code, ALLCUTS handles cause trees with AND- and OR-type logic gates. It can compute cause trees with up to 175 basic events and 425 gates. Up to 1000 minimal cut sets can be generated and classified in descending order of probability. ALLCUTS is written in Fortran IV for IBM 360 370 computers and in COMPASS (assembly language) for CDC 6600 computers.

AWE 1

AWE 1 was developed by the European Economic Community Research Centre in Ispra (Italy); it is considered to be an extension of the previous AWE program [22, 23]. AWE 1 is used for the qualitative and quantitative analyses of cause trees with AND- and OR-type logic gates. It generates all the combinations of basic events (up to order 5) and inventories those which cause the undesirable event. The cut sets are then reduced in order to obtain the minimal cut sets. AWE 1 computes the probability of each minimal cut set as well as the expected number of failures. Importance measures are computed as well. AWE 1 is written in Fortran IV for IBM 370/165 computers.

CADI

CADI [24] was developed by the European Economic Community Research Centre in Ispra (Italy). It processes cause trees with up to 500 basic events and 2000 AND- and OR-type logic gates. CADI operates in much the same way as PRET and KITT; all the failure combinations of up to order 5 are generated, then cuts sets are determined and reduced into minimal cut sets. The probabilities of minimal cut sets and of the top event are calculated versus time. An importance measurement is also performed. CADI is written in Fortran IV for IBM 370/165 computers.

DICOMICS

DICOMICS [25,26] was developed by the European Economic Community Research Centre in Ispra (Italy) in collaboration with other organizations. DICOMICS finds the minimal cut sets of cause trees with AND and OR logic gates by using a top-down algorithm. Then the probability of minimal cut sets as well as the expected number of failures are computed. DICOMICS allows for a specified probability threshold. It is written in Fortran IV for IBM 370/165 computers.

FAUNET

FAUNET [27] was created by the Danish Atomic Energy Commission (Denmark). It performs the qualitative and quantitative analyses of cause trees; it uses a Boolean algorithm (and modularization) for determining minimal cut sets. FAUNET can also be applied to network reliability evaluation. It is written in Fortran IV and has been optimized for installation on minicomputers PDP8/PDP11/Borroughs 7800.

KARLSRUHE CODE FOR BINARY COMPONENTS

This program [28] was created by the Kernforschungszentrum in Karlsruhe (Federal Republic of Germany). It processes cause trees with up to 256 events and with AND- and OR-type logic gates. Minimal cut sets are determined using a top-down algorithm. The program computes minimal cut set probabilities as a function of time. It should be noted that it can handle up to 2000 events when probabilities need not be computed versus time. Moreover it can process two-phased mission systems. The program is written in Fortran IV for IBM 3033 computers.

MUSTAFA

MUSTAFA [29, 30] was developed by the Kernforschungszentrum in Karlsruhe (Federal Republic of Germany). It processes cause trees containing up to 1000 basic events and AND, OR, NOT and K-of-N logic gates. Components may have several failed states. The algorithm used works from top to bottom (Nelson algorithm). Then the probabilities of minimal cut sets are calculated. A probability threshold may be specified in the program. The program takes statistical dependencies between components into account. It is written in Fortran IV for IBM 3033 computers.

MUSTAMO

This program [31] was created by the Kernforschungszentrum in Karlsruhe (Federal Republic of Germany). It processes cause trees containing up to 800 basic events and 1000 AND, OR, K-of-N and NOT logic gates. Components may have several failed states. The algorithm starts from the top and works down to the bottom to find the minimal cut sets. Then the probability of minimal cut sets is computed. MUSTAMO is written in Fortran IV for IBM 3033 computers.

PATRICK

PATRICK was developed by the Commissariat à l'Energie Atomique (CEA) in France. It is an extension of the PATREC program (see Section 20.2.4); it contains an option for obtaining minimal cut sets with a top-down algorithm. The program is written in PL/1 for IBM 370/168 and IBM 3081 computers.

PHAMISS

PHAMISS [32, 33] was developed by the Netherlands Energy Research Foundation ECN. It is made up of 4 sub-programs FAULTREE, PROBCAL, IMPCAL and COMMODE. FAULTREE is used to determine the minimal cut sets of cause trees with AND, OR and K-of-N logic gates. The algorithm for determining minimal cut sets works from top to bottom, but also from bottom to top for particular intermediate gates. This program has been specially optimized to process large number of gates and events (up to a total of 4095) within a relatively short computing time. This high efficiency is due to the direct handling of 'bits' in the central memory; thus a single bit is required for the representation of a basic event or a logic gate. PROBCAL computes the probabilities of minimal cut sets; it can also compute the success probabilities of phased-mission systems from the cause tree of each phase and the time limits of each phase. The components are assumed to be non-repairable. 10 is the maximum number of phases, and the number of systems likely to fail is limited to 3. Moreover, the program accepts different types of components: non-repairable components, tested components, etc.

The importance measures are calculated by IMPCAL, and COMMODE identifies common-cause failures (see Section 20.5). PHAMISS offers an option for the quantitative analysis of a consequence tree (see Section 20.3). PHAMISS is written in Fortran IV for CDC Cyber 175 and IBM 3090 computers.

PL MOD

PL MOD [34–36] was developed by the Massachusetts Institute of Technology (United States). It performs the qualitative and quantitative analyses of cause trees with AND, OR, K-of-N and NOT logic gates. The PL MOD algorithm for finding minimal cut sets is based on fault-tree decomposition and modularization. PL MOD separates all replicated events from the rest of the tree, modularizes the independent portions of the tree and then finds Boolean relations between the replicated events and the modules. The Boolean relation is reduced to obtain the minimal cut sets based on these modules. The code MODCUT [36] is used to generate the minimal cut sets of the initial cause tree. Thus, PL MOD determines cut sets of the order specified by the user. Then the probabilities are calculated. An option (PL-MODT) calculates system unavailability at different times. Fussell-Veseley-type importance calculations can be performed for all sorts of events (basic events, module). A Monte Carlo simulation option is available for uncertainty analysis. PL MOD is written in PL/1 language for IBM 360/370 computers.

RALLY

RALLY [37] was developed by the Gesellschaft für Reaktorsicherheit (Federal Republic of Germany) and has been used for the probabilistic risk assessment of the German Biblis nuclear power plant. RALLY is capable of evaluating cause trees with up to 1500 basic events and 2000 AND, OR, K-of-N and NOT-type logic gates. It is made up of numerous subroutines TREBIL, CRESSC, CRESSEX, CRESSCN, FESIVARM, KARI, RISA). TREBIL is similar to PREP. Qualitative and quantitative analyses as well as uncertainty calculations are performed. The code can also be used for two-phased mission systems. RALLY is written in Fortran IV for IBM 360/370 computers.

RAS

RAS (Reliability Analysis System [38]) was developed by the EGG company (United States). It is an integrated package of computer codes such as MOCUS, POCUS, KITT1, SRTPAN and COMCAN. RAS can be used for the qualitative and quantitative analyses of a system modelled by a cause tree with up to 5 different operating phases. RAS is written in Fortran IV for the CDC 7600 computer.

SALP-3

This computer code [39, 40] was developed by the European Economic Community Research Centre in Ispra (Italy) in collaboration with other organizations. SALP-3 can be used for the qualitative and quantitative analyses of cause trees with AND and OR-type logic gates. The algorithm for finding minimal cut sets works from top to bottom. It computes the top event probability as a function of time, the probability of each minimal cut set as well as the expected number of failures for each minimal cut set; importance calculations are also performed.

SALP-3 accepts limits on probabilities or on cut set orders; it can handle trees with NOT gates. The code is written in PL/1 for IBM 370/165 computers.

SALP-MP

This computer code [41] was developed by the European Economic Community Research Center in Ispra (Italy) in collaboration with other organizations. It performs the qualitative and quantitative analyses of cause trees with AND, OR and K-of-N logic gates. The algorithm for finding minimal cut set works from top to bottom. SALP-MP calculates the probability of the top event as a function of time, the probability of each minimal cut set as well as the expected number of failures for each minimal cut set. The code can handle phased-mission systems. It is written in Fortran IV for IBM 370/165 computers.

SETS

SETS (Set Equation Transformation System [42, 43]) was developed by Sandia National Laboratories (United States) and has been used in several probabilistic risk assessment studies conducted on nuclear power plants in the United States. SETS is a computer code designed to manipulate Boolean equations used for analysing cause trees or consequence trees. It generates minimal cut sets of any order for cause trees with AND, OR, NOT or special logic gates represented by any user-specified Boolean equation. The algorithm works from top to bottom; it also identifies independent branches of the cause tree and creates modules. SETS can be used to obtain prime implicants. This is very useful for non-coherent systems or for the processing of consequence trees. SETS can handle 8000 basic events and logic gates; it is thus capable of dealing with very large cause trees. Moreover, the Boolean equations can be reduced by placing a limit on cut sets order or probability. SETS can be used with SEP for probability, uncertainty and importance calculations. The capabilities of this computer code have been extended to handle consequence trees (see Section 20.3).

WAMCUT and WAMCUT II

These computer codes [44–46] were developed by the Electric Power Research Institute (EPRI) in the United States and have been applied to several probabilistic risk assessment studies of nuclear power plants. WAMCUT determines and quantifies minimal cut sets. It handles cause trees with AND, OR, K-of-N and NOT logic gates as well as others. WAMCUT consists of two subroutines WAM and CUT. WAM reads the cause tree and checks for syntax errors. CUT finds the cut sets of each gate, working from the bottom to the top of the tree; then it calculates their probabilities. WAMCUT can process large cause trees with up to 1500 logic gates and 1500 basic events. It is possible to use a probability threshold for the minimal cut set calculation. WAMCUT II [45] is an advanced version of WAMCUT; it analyses cause trees and restructures their logic to obtain and replace independent portions by modules; cut sets are then obtained with a top-down algorithm, thus considerably reducing computing time. SPASM (System Probabilistic Analysis by Sampling Methods), a computer code for uncertainty analysis, can be used with WAMCUT and WAMCUT II. The two codes are written in Fortran IV for CDC 7600 and IBM 370 computers.

20.2.4 Computer codes for direct analysis

GO

For a description of this code, see Chapter 21.

PATREC

PATREC [47–49] was developed at the Commissariat à l'Energie Atomique (CEA) in France. The code uses an original method for processing cause trees. The code recognizes structures (subtrees) in the cause tree, calculates them and replaces them by 'equivalent leaves'. The tree is thus progressively reduced to a single 'leaf' which contains the final result. Thanks to this method, the tree only takes up a minimum of memory space. Moreover, the result precision does not depend on the value of basic probabilities and thus PATREC can perform correct analyses even with large probabilities. The code can evaluate cause trees containing events and their complements as well as K-of-N logic gates. The code uses Bayes' theorem to solve the problem of replicated events; if there are n replicated events, then 2^n elementary calculations must be performed. Despite the rapidity of the method and possible approximations, PATREC still requires a prohibitive amount of computer time to process a cause tree with more than 20 replicated events. This constitutes the main shortcoming of the code. The PATREC RCM version includes an algorithm option for finding minimal cut sets. The PATREC MC version uses a Monte Carlo simulation to take uncertainties in input data into account and thus to deduce the uncertainty in the final result.

The input data distribution is lognormal. The code is written in PL/1 language.

SIFTA

SIFTA (Simple Fault Tree Analysis [50]) was developed by the Canadian Atomic Energy Control Board. It is used for the qualitative and quantitative analyses of cause trees with AND, OR and K-of-N-type logic gates.

The algorithm is based on a cause tree pattern recognition technique which restructures the tree and reduces it progessively. Quantitative analysis starts with the calculation of independent branches followed by a calculation of branches made independent through the identification of common events. If the cause tree cannot be fully reduced, the residual tree is processed by simulation. SIFTA is written for HP 1000 computers.

WAM-BAM

WAM-BAM [51] was developed by the Electric Power Research Institute (EPRI) in the United States. It performs the qualitative and quantitative analyses of cause trees. It consists of three subroutines: WAM, WAMTAP and BAM. WAM and WAMTAP are input preprocessors for BAM (Boolean Arithmetic Model). WAM generates numerical input data for BAM from the cause tree and the basic event probabilities. WAMTAP allows for easy modification of input data for sensitivity or common-cause failure studies. The BAM evaluation subroutine combines GO and cause tree analysis concepts; it uses the GO computational algorithm but cause tree logic gates are used. The probability of the top event is calculated by creating a truth table. WAM-BAM is very easy to use; however the new version of WAM-CUT may be faster and more efficient. WAM-BAM is written in Fortran IV for CDC 6600 computers.

20.3 CODES FOR THE CONSEQUENCE TREE METHOD

Few codes are specifically adapted to this method; the first of these (SETS) is in fact a code for cause tree processing which has been improved upon due to its use for the consequence tree method. The complex nature of the processing stems mainly from the different types of dependences (functional dependences, common component-related dependences) which can exist between generic events. Thus codes which are being currently developed cover particular aspects.

ISA

ISA was developed by Electricité de France (EDF, France) [52]. It computes the probability of accident sequences modelled from a consequence tree; it can take account of sequential-type dependences between systems (or generic events) which are observed in accident sequences. The reliability of each system can be approximated by the sum of exponential functions, and the probability of the initiating event disappearing can be allowed for. ISA can also be used to compute the probabilities of accident sequences whose duration must not exceed a specified time (see Chapter 12). Moreover, the initial event, which is the result of the loss of a repairable system, can be modelled by a semi-Markovian states graph (constant-time transitions). ISA is written in Pascal for IBM 3090 computers and in Basic for TEKTRONIX microcomputers.

SETS

A description of this code for cause tree evaluation has already been given in Section 20.2.3. It was used for the Probabilistic Risk Assessment of nuclear power plants in the United States as part of IREP (Interim Reliability Evaluation Program) and, in particular, for processing consequence trees. This latter application has led to its improvement [53, 54]. It was created by Sandia National Laboratories for the US Nuclear Regulatory Commission.

For each accident sequence, a number of cause trees are merged. By 'merge' we mean putting all the cause tree top events under the same AND logic gate. SETS creates a maximum number of modules in the different cause trees, then restructures the new tree wherefrom it derives the minimal sequences. An accident sequence may involve the operation of a system so that the top event of the corresponding cause tree will not occur. SETS selects among the preceding cut sets only those which suppose the operation of the systems in question. To make calculations easier, a limit is placed on the minimal cut set probabilities. Minimal sequences are calculated assuming that basic events are independent. SETS is written in Fortran IV for CDC 7600 computers.

PHAMISS

The use of this code for cause tree processing has already been discussed in Section 20.2.3. PHAMISS can also process certain types of consequence trees which bring several systems into play during phases which last for a limited amount of time. After modelling each system failure by a cause tree, the probability of each accident sequence is then computed by using the subroutine for phased-mission systems. The computation is based on the assumption that components are not repairable during the operating phase. PHAMISS is capable of handling the failure of three systems at most.

RELOSS

RELOSS [55] was developed by the Electrowatt Engineering Service for the Nuclear Installation Inspectorate (UK). It is capable of computing cause trees with statistically independent branches by identifying minimal cut sets for each sequence of events. RELOSS is written in Fortran IV for PRIME 250/11 computers.

20.4 CODES FOR THE STATE-SPACE METHOD

These codes started to be developed during the mid-1970s, mainly in the nuclear field. They calculate the time-dependent probabilities of different states and deduce the availability or reliability of repairable systems. State graphs are generally assumed to be Markovian. The user selects one of the following methods:

- step-by-step numerical integration of a linear system of first-order differential equations;
- calculation of the eigenvalues of the transition matrix;
- exponentiation of the transition matrix.

Required input for these codes includes the matrix of transition rates. Generally these codes are used only to process state graphs with some hundred states at most. The reader is reminded that a system made up of 10 components has 1024 states, assuming that each component has 2 states. It is easy to imagine how cumbersome this processing can be with a transition rate matrix of 1024×1024!

GRAFFITI

GRAFFITI [56] was created by Electricité de France (EDF, France). It computes the reliability of a repairable system modelled by a Markov or semi-Markov state graph (constant-time transition). The method is based on a compilation of the state sequences in the graph, and probability calculation uses approximations explained in Chapter 14. GRAFFITI is written in Pascal for IBM 3090 computers and in Basic for TEKTRONIK microcomputers.

MARKAN

MARKAN [57] was created by two Italian companies, ENEA and NIRA. It computes the availability or reliability of a repairable system from success paths. MARKAN is based on numerical integration and can handle up to 17 components. It can also handle phased-mission systems. The code is written in Fortran 77 for CDC CYBER 170/835 computers.

MARK-GE

MARK-GE [58], created by the Commissariat à l'Energie Atomique (CEA) in France is used for computing the availability and the reliability of repairable systems modelled by Markovian state graphs. The algorithm used computes the eigenvalues of the matrix representing the transitions between the various states, then diagonalizes and exponentiates this matrix. This code has been used in conjunction with a Monte Carlo simulation similar to that of PATREC in order to derive result uncertainty from data uncertainties. It is written in Fortran for IBM 3081 computers.

MARK-EXD

MARK-EXD [59] was developed by ELF (France). It computes Markov graphs using an algorithm called 'direct exponentiation' which gives it particularly interesting capabilities. The time-dependent probabilities of different states as well as the average state residence time can be obtained. MARK-SMP is an extended version of MARK-EXD. It handles repairable phased-mission systems by computing the graphs of the various phases in sequence. The code is written in Fortran for IBM 3090 computers.

MARKOM

Developed by the European Economic Community Research Centre in Ispra (Italy), MARKOM [60] computes the time-dependent state probabilities; the code inputs are the differential equations or the transition matrix. The Adams method is used for equation solving and the code can handle up to 30 states. It is written in Fortran IV for IBM 370/165 computers.

MARKOV

MARKOV [61] computes the time-dependent probabilities of the different states of a Markov graph through numerical solution of the state equations. It is written in Fortran for FACOM M.190 computers (IBM 3033 compatible).

SURF

Developed by the Centre National de la Recherche Scientifique (CNRS, France), SURF [62, 63] computes the availability or the reliability of repairable systems modelled by Markovian state graphs. More precisely, the probabilities of the various states of the graph are obtained as a function of time considering the initial probability of each of these states. The algorithm used is based on the calculation of the matrix eigenvalues which represent transitions between different states. SURF is written in PL/1 language for IBM computers.

20.5 CODES FOR COMMON-CAUSE FAILURE ANALYSIS

COMCAN, the first of this type of codes, appeared in the mid-1970s in the nuclear field. Since then, with the awareness that common-cause failures may significantly contribute to system failures, an increasing number of similar codes have been created. The codes are used for processing cause tree. They identify minimal cut sets corresponding to the occurrence of a common-cause generic event. Such a cause is known as a common-cause critical event, and the minimal cut set as the common-cause candidate. A brief description of these codes is provided below.

BACKFIRE

The BACKFIRE code [64–66] was developed by the University of Tennessee (United States); it uses a similar method as COMCAN. However it permits more than one location to be specified for the failure of components such as pipes and electrical cables. BACKFIRE has similar input and output characteristics to COMCAN. It is written in Fortran IV for CDC 7600 and IBM 360/370 computers. The MOCUS-BACKFIRE code [65], obtained by merging MOCUS and BACKFIRE, is written in Fortran IV for IBM 360/370 computers. A new version, BACKFIRE II [66], has been developed.

COMCAN

COMCAN [67–69] was developed by the Aerojet Nuclear Company (United States). It is used to identify potential common-cause failures in a system modelled by a cause tree using minimal cut set analysis. A minimal cut set is regarded as a 'common-cause candidate' under the following conditions:

- all the basic failures in a cut set share a special condition which can trigger a simultaneous failure; an example of such a common-cause susceptibility is a single team of repairmen for servicing all the components affected;

- all the failures in the cut set share susceptibility to the same failure cause; the affected components share a common physical location.

Required input consists of the minimal cut sets to be analysed with the potential causes of secondary failure and applicable special conditions for primary failures. Common-cause dependences cannot be obtained in several systems studied and inputting the numerous cut sets can be very cumbersome.

COMCAN is written in Fortran IV for IBM 360/370 computers. COMCAN II [68, 69], an improved version of COMCAN, uses the FATRAM code for identifying minimal cut sets before the COMCAN analysis. COMCAN II is written in Fortran IV for CDC CYBER 7600 computers.

PHAMISS

PHAMISS has already been described in Section 20.2.3. It can be used (COMMODE subroutine) for analysing common-cause failures. It identifies minimal cut sets corresponding to the occurrence of a common-cause event. In PHAMISS, each primary failure is labelled for instance 'T', indicating the temperature effect, or 'R1', indicating the location.

SETS

SETS, already described in Section 20.2.3, can be used for analysing common cause failures. The method of analysis is similar to that of COMCAN; the generic cause susceptibilities are indicated for each primary failure. SETS inventories the minimal cut sets which are common cause candidates; it can handle very large trees.

WAMCOM

The WAMCOM code [70] was worked out by the Electric Power Research Institute (EPRI) in the United States for analysing common cause failures in large and complex cause trees. It can handle cause trees with up to 2000 logic gates and basic events. SETS is one of the numerous subroutines which constitute WAMCOM. Required inputs include cause trees and a table of component susceptibilities to common cause events. Several types of results can be presented:

- *Critical common-cause events.* These events trigger the occurrence of a minimal cut set which then becomes a common cause candidate.

- *Significant common-cause events.* These events can cause enough failures to result, when combined to another failure, in the undesirable event.

WAMCOM is written in Fortran IV for CDC 7600 computers.

20.6 CODES FOR UNCERTAINTY EVALUATION

These codes are designed to calculate the probability distribution of an event (e.g. system failure) by using the distribution of the basic events which produce this event (see Appendix 4). Codes for uncertainty analysis normally use Monte Carlo simulation, and rarely analytic formulae for combining distributions. The user normally inputs the failure function which links the variables corresponding to the basic events or the equation of minimal cut sets obtained through the qualitative analysis codes. Numerous codes for the quantitative analysis of cause trees use uncertainty analysis in varying degrees. This section deals only with codes specially designed for uncertainty analysis and which handle a large variety of distributions.

COVAL

Created by the European Economic Community Research Centre in Ispra (Italy), COVAL [71] computes uncertainties about the probability of a cause tree top event or of minimal cut sets. The distributions of basic events may be exponential, normal, lognormal or Weibull. The mean, median, standard deviation as well as the bounds of a confidence interval are obtained. The code is written in Fortran IV for IBM 370/165 computers or in Basic for microcomputers.

SAMPLE-MOCARS

SAMPLE [72] was used in the first PRA study of a nuclear plant in the United States; it allows for normal and lognormal distributions for the failure rates. The output distribution is assigned estimates of the mean, the median and of the upper and lower bounds of the confidence interval. SAMPLE is written in Fortran IV. MOCARS [73] is similar to SAMPLE; it can however handle a larger variety of known distributions (binomial, Poisson, exponential, normal, lognormal, Weibull and Gamma) as well as empirical distributions.

SCORE

SCORE [74, 75] is a creation of the European Economic Community Research Centre in Ispra (Italy). It computes the uncertainties in the probability of a cause tree top event or of minimal cut sets. The distributions of basic events can be exponential, normal, lognormal, gamma, uniform or empirical, etc. The mean, median, standard deviation as well as the bounds of a confidence interval can be obtained. The code is written in PL/1 language for IBM 370/165 computers.

STADIC II

STADIC [76, 77] was used in 1978 for the PRA studies of a nuclear power plant with a high-temperature helium-cooled reactor. STADIC II [77] is an improved version. It allows for normal, lognormal and loguniform distributions; the user can also select any of the above distributions for any variable. STADIC II is much faster than SAMPLE because of the computational method adopted. The code is written in Fortran IV for the UNIVAS-1180, CDC Cyber 7600 and PRIME computers.

20.7 MONTE CARLO SIMULATION CODES

Monte Carlo simulation can be interesting for the evaluation of the dependability of systems, especially when they are too large or too complex to be analytically solved. For example, this is an efficient means of assessing the reliability of such systems as the large electric power systems of various countries [78–80].

In the 1960s, before the introduction of efficient analytical codes, the Monte Carlo simulation was widely used for calculating success diagrams and cause trees.

20.7.1 Principle

Monte Carlo simulation consists in simulating the behaviour (e.g. operating or failed state) of the components of a system in order to find out the dependability of the system.

Let us consider a first example of a system made up of a repairable component whose mean unavailability is being sought. Clearly

- operating time follows an exponential distribution with a mean MTTF of 200 hours,
- repair time follows a normal distribution with a mean of 10 hours.

About a hundred 5000 hours' periods in the system life are simulated. For each period, a first operating time is sampled (indication of how the sampling is done is given later), then a first repair time, then a second operating time, etc, until the total of 5000 hours is reached. Then the entity unavailability is calculated for the given period. The mean unavailability is then calculated from the hundred or so simulations.

As a rule, let X be a random variable (associated with a system) which depends on a number of random variables $X_1, X_2, \ldots, X_n$ (associated with components c_i) whose distributions are known; the aim is to find out the distribution of this random variable or to evaluate some of its parameters.

The MTTF of the system (assumed to be non-repairable for simplicity's sake) is analysed as follows:

- The system is considered at time $t = 0$:
  - for each component c_i, its time to failure is randomly sampled. Let t_1 be the shortest operating time;
  - the system is considered at time t_1. The system is either in an operating or in a failed state; system reliability models are used to determine the state of the system (success diagrams, cause trees). If the system is in a failed state, the operating time is t_1. If the system is still operating, the sequence of operating times is studied, and the operating time t_k after which the system is failed, according to the model logic, is determined.

- the previous step is repeated, thus giving realizations t_i of the operating times. The mean of these values gives an MTTF estimate

$$\widehat{\text{MTTF}} = \frac{1}{n} \sum_{i=1}^{n} t_i$$

We shall now look at the techniques for sampling random variables.

20.7.2 Distribution generation

No Monte Carlo simulation can be carried out without the random sampling of operation and repair times. Random sampling is based on the generation of uniformly distributed random numbers. Numerous methods are available to generate such random numbers; an example is the 'congruential' method: the random variable R_{n+1} is obtained from a random number R_n by using the following recurrence relation:

$$R_{n+1} = aR_n \text{ (modulo } m\text{)} \tag{20.1}$$

where a and m are positive integers and $a < m$.

The relation (20.1) means that R_{n+1} is the remainder when aR_n is divided by m. There are numerous satisfactory combinations for these parameters [81–84].

Example [83]:

$$a = 455\,470\,314 \qquad m = 2^{31} - 1 = 2\,147\,483\,647$$

$$R_0 = \text{a whole number between 1 and } (m - 1)$$

If random numbers between 0 and 99 are needed, the computer can be instructed to take the two digits of the generated random number. In fact, these numbers, which are generated by the recurrence relation, can be reproduced and are therefore not random strictly speaking; for this reason they are called pseudo-random numbers.

Let us now consider the distribution function of a random variable X; with each realization x, a value u such that $u = F(x)$ can be associated. The values of u are the realizations of a random variable U. Therefore U is a random variable which is uniformly distributed between 0 and 1.

Let us perform a random sampling of values of u between 0 and 1 and then calculate their value:

$$x = F^{-1}(u)$$

The values x are therefore realizations of the random variable X whose distribution function is $F(x)$.

Thus distributions are generated with the help of pseudo-random numbers by inversion of the distribution function.

20.7.3 Simulation characteristics

Obviously, the result accuracy depends on the number of simulations performed. Let us consider again the example of the MTTF calculation; the MTTF is the mean of realizations t_i. According to the central limit theorem (see Section 4.3), the asymptotic estimate of the MTTF follows a normal distribution $N\,(\widehat{\text{MTTF}}, \sigma/\sqrt{n})$, σ being the standard deviation associated with realizations t_i.

The confidence interval associated at a confidence level of $1 - \alpha$ is deduced:

$$\text{MTTF}_{\text{upper}} = \widehat{\text{MTTF}} + U_{\alpha/2} \frac{\sigma}{\sqrt{n}} \qquad (20.2)$$

$$\text{MTTF}_{\text{lower}} = \widehat{\text{MTTF}} - U_{\alpha/2} \frac{\sigma}{\sqrt{n}}$$

If $F(u)$ is the distribution function of the standardized normal distribution, then u_α is such that

$$1 - \alpha = F(u_\alpha)$$

The formulae (20.2) give the confidence interval for the MTTF estimate.

The Monte Carlo simulation is sometimes difficult to use because it requires the performance of a high number of simulations, which demands tremendous computer time. Computer time required for estimating measures related to rare events can be prohibitive.

Variance reduction techniques can be used to solve this problem. They are aimed at reducing the simulation time while increasing accuracy without having to increase the number of simulations. References [19] and [84] deal with these techniques.

20.7.4 Application

Monte Carlo methods can be readily applied to the simulation of even relatively complex systems; they are adapted to different models (cause trees, state graphs, etc). A specific model such as the Petri net seems particularly well suited to simulation-based calculations [85]. Monte Carlo simulation is also very useful for calculating uncertainties (see Section 20.6). Note, however, that this technique only provides estimates and that new simulations have to be performed for the sensitivity studies.

Finally, it must be remembered that the calculation can take a very long time when the model to be analysed involves extremely rare events.

20.8 MISCELLANEOUS PROGRAMS

This section deals with various programs which do not fall under the preceding categories but are, nevertheless, noteworthy.

AVACOM-ETARP

This code [86–88] was developed by the European Economic Community Research Centre in Ispra (Italy). It computes the availability of repairable components whose failure and repair rates follow different distributions. It can handle exponential, normal, lognormal, Weibull and gamma distributions. The computation method is based on a numerical integration method.

CAFTS

CAFTS (Computer Aided Cause Tree Synthesis [89–91]) was developed by the European Economic Community Research Centre in Ispra (Italy). CAFTS offers interactive assistance for the construction of cause trees. The technique is based on the use of two different types of component models which are stored in files:

- macro-component models; available for known components (example: pump) and inventory the macro-component failure modes;

- logic models; used for modelling the way the variations of the main parameters are related to the component failure modes.

CAFTS is used in the following way:

- first, by using a description of the system and logic models, the system is modelled as a cause tree;

- second, the accuracy of the previous cause tree is increased by introducing macro-component models.

CAFTS is written in PL/1 for IBM 3033 computers.

EBAI

EBAI (Expressions Booléennes Automatiquement Interprétées) [92] was developed by Electricité de France (EDF, France). The code allows the results of the Gathered Fault Combination Method to be utilized. From the expressions for global gathered faults and their constituents, EBAI deduces their expression in terms of the internal and external gathered faults by identifying them with Boolean expressions. Thus we obtain, for each global gathered fault, a cause tree whose basic events are internal and/or external gathered faults (or failure modes); next, DEFAIL code reduces the cause tree (see Section 20.2.1). The AIDE code was also developed to facilitate the writing of fault expressions, especially when an expression is modified or a new one is added. These codes are written in Pascal for IBM 3090 computers.

FIABC

FIABC [93] was created by Electricité de France (EDF, France). It computes the reliability of repairable systems. The systems are modelled by success diagrams or cause trees. The method of minimal operating states is used (see Chapter 14). FIABC finds the minimal cut sets or success paths. For quantitative analyses, FIABC generally assumed that the basic events are independent; however, it has been extended to certain dependences (an operating component and two standby components).

FIABC is particularly efficient for processing repairable systems, small or even large, provided they are simple to model. An uncertainty analysis is also performed. The code is written in Fortran for IBM 3090 computers.

GSI

GSI (Génération de Séquences par Inférences) was developed by Electricité de France (EDF, France) [94]. It is used for automatically generating and quantifying event sequences corresponding to the system passage from an operating state into a failed state. GSI uses an expert-system-type approach where the system operating and failed states are modelled as production rules.

Thanks to GSI, the reliability of repairable systems involving consideration of a large number of states can be studied. Reliability computation, based on an inventory of state sequences, uses approximations discussed in Chapter 14. The program is written in APL and in Fortran for IBM 3090 computers.

ICARUS

ICARUS [95] was developed by the Argonne National Laboratory in the United States. It is capable of calculating average unavailabilities, an optimal test interval as well as the contributions of testing and maintenance to unavailability. ICARUS handles only basic events. The user inputs the various failure rates and characteristics of the component test and maintenance and chooses between different test schemes. ICARUS is written in Fortran for IBM 360/370 computers.

IMPORTANCE

IMPORTANCE [96] was developed by the Livermore National Laboratory (University of California, United States) and has been used in PRA studies of nuclear plants in the United States. It ranks basic events and minimal cut sets according to various importance measures. The main importance measures are calculated (see Appendix 3). Cut sets generated by FTAP and SETS can be directly processed by IMPORTANCE. IMPORTANCE is written in Fortran IV for CDC 7600 computers.

EXCON

EXCON [97] was developed by JBF Associates (United States); it is used in conjunction with a cause–consequence analysis for the PRA studies of systems or industrial plants. Relevant inputs are used for this type of analysis and the code searches for all accident sequences which have the required characteristics. It then calculates the probabilities of the resulting accidents. The analyst can then determine which accident sequences are the major contributors to overall risk. EXCON is written in Fortran IV for CDC 7600 and IBM 360/370 computers.

LESSEPS

This program [98] was developed at Electricité de France (EDF, France) to compute complex dependability models and to be able to perform sensitivity studies to data easily and rapidly. It was worked out in the framework of the Probabilistic Safety Assessment of a French nuclear unit (Paluel power plant—see Section 19.5.5.2). The models used are based on the cause tree method, consequence tree method and state-space method. The LESSEPS program comprises a data base specific to the models used. Sensitivity studies to the data are automatically performed when the data are changed, and the program then optimizes the computation sequence.

To compute the models, LESSEPS uses the PHAMISS (cause tree), ISA (consequence tree) and EXPRESS (cause tree built by an expert system) programs described in this chapter.

This program is run on IBM 3090 and IBM PS 286 computers and on the SUN 4 workstations.

RIKKE

RIKKE [99] was developed by the RISO Research Center (Denmark). It is based on the method of the cause–consequence diagram. It can be applied to system risk assessments. Component failure models and circuit diagrams (with oriented information flow) can be input via a terminal. The analyst may choose to use the cause tree by inputting the undesirable event: the code then searches for the causes. The cause tree, once constructed, is processed by FAUNET. The code can also generate the consequences of an initiating event. It is written in Fortran IV for PDP11-VAX computers.

REFERENCES

[1] C. W. McKnight et al. (1966) *Automatic Reliability Mathematical Model*, North American Aviation, Dawney, California, NA 66-838.
[2] D. F. Haasl (1965) *Advanced Concepts in Fault Tree Analysis*, System Safety Symposium, University of Washington and Boeing Company, Seattle, Washington.

[3] C. Lievens (1976) *Sécurité des systèmes*, Cepadues Editions, Toulouse.
[4] D. Mulet-Marquis and A. Dubreuil-Chambardel (1984) *DEFAIL 84, FIABC 84: un ensemble de logiciels pour l'évaluation qualitative et quantitative de la fiabilité d'un système*, EDF-DER-HI/4737-02.
[5] S. N. Semanderes (1971) ELRAFT A computer program for the efficient logic reduction analysis of fault trees, *IEEE Trans. Nucl. Sci.*, **NS-18**(1), 481–7.
[6] D. M. Rasmuson and N. H. Marshall (1978) FATRAM. A core efficient cut set algorithm, *IEEE Trans. Reliability*, **R-37** (4), 250–3, October 1978 (program available from EGG-Idaho Inc.).
[7] D. B. Wheeler et al. (1977) Fault tree analysis using bit manipulation, *IEEE Trans. Reliability*, **R-26**(2), 95–9.
[8] R. R. Willia (1978) *Computer Aided Fault Tree Analysis*, Report ORC-78-14. Operations Research Center, University of California, Berkeley, August 1978 (program available from Operations Research Center, University of California, Berkeley, USA).
[9] J. B. Fussel and W. E. Vesely (1972) A new methodology for obtaining cut sets for fault trees, *Trans. ANS*, **15**, 262.
[10] J. B. Fussel, E. B. Henry and N. H. Marshall (1974) *MOCUS: A Computer Program to Obtain Minimal Sets from Fault Trees*, USAEC Report ANCR, 1156, Aerojet Nuclear Company, NTIS (program available from Argonne Software Center, Illinois, 60439, USA).
[11] W. E. Vesely and R. E. Narum (1970) *PREP and KITT Computer Codes for the Automatic Evaluation of a Fault Tree*, Idaho Nuclear Corporation, Idaho Falls, Idaho, IN-1349, 1970 (program available from Argonne Software Center, Argonne, Illinois, 60439, USA).
[12] P. K. Pande, M. E. Spector and P. Chatterjee (1975) *Computerized Fault Tree Analysis: TREEL and MICSUP* Operational Research Center, University of California, Berkeley, ORC 75-3, April 1975 (program available from Operational Research Center, University of Calfornia, Berkeley, USA).
[13] D. Mulet-Marquis and A. Dubreuil-Chambardel (1984) *DEFAIL 84, FIABC 84: un ensemble de logiciels pour l'évaluation qualitative et quantitative de la fiabilité d'un système EDF-DER*. HI/4737-02 (program available from the same organization).
[14] W. E. Vesely and F. F. Goldberg (1977) *FRANTIC: A Computer Code for Time Dependent Unavailability Analysis*, US Nuclear Regulatory Commission, NUREG 0193, October 1977 (program available from Argonne Software Center, Argonne, Illinois, 60439, USA).
[15] F. F. Goldberg and W. E. Vesely (1977) *Time Dependent Unavailability Analysis of Nuclear Safety Systems*. National Conference on Reliability, Nottingham, England, September 1977.
[16] W. E. Vesely et al. *FRANTIC II: A Computer Code for Time Dependent-Unavailability Analysis*, USNRC Report NUREG/CR, 1924, 1981 (program available from Argonne Software Center, Argonne Illinois 60439, USA).
[17] W. E. Vesely (1969) *Analysis of Fault Trees by Kinetic Tree Theory*, Idaho Nuclear Corporation Idaho Falls, Idaho, INC-1330.
[18] W. E. Vesely and R. E. Narum (1970) *PREP and KITT Computer Codes for the Automatic Evaluation of a Fault Tree*, Idaho Nuclear Corporation, Idaho Falls, Idaho IN-1349 (program available from Argonne, Software Center, Argonne Illinois 60439, USA).
[19] E. J. Henley and H. Kumamoto (1981) *Reliability Engineering and Risk Assessment*, Prentice-Hall, Englewood Cliffs, NJ 07632.
[20] J. B. Fussel, D. M. Rasmuson and D. Wagner (1977) *SUPERPOCUS: A Computer Program for Calculating System Probabilistic Reliability and Safety*

Characteristics, NERS-77-01, Nuclear Engineering Department, University of Tennessee, Knoxville, 1977 (program available from the same organization).

[21] W. J. Van Slyke and D. E. Griffing. (1975) *ALLCUTS: A Fast, Comprehensive Fault Tree Analysis Code*. Atlantic Richfield Hanford Company, Richland, Washington, ARH-ST-112, July 1975.

[22] A. G. Colombo and G. Volta (1974) *Euratom* JRC, Ispra Italie, EUR 5260-1974.

[23] M. Astolfi (1976) *Euratom*, JRC, Ispra Italie, EUR 5638-1976.

[24] A. G. Colombo (1973) *CADI, a Computer Code for System Availability and Reliability Evaluation*, Report Euratom, JRC, Ispra Italie, EUR 4940.

[25] S. Garriba, P. Mussio, F. Naldi, G. Reina and G. Volta (1977) Euratom, JRC, Ispra Italie, EUR 5481.

[26] M. Astolfi and F. Naldi (1977) Euratom, JRC, Ispra Italie, EUR 5811e, 1977.

[27] O. Plats and J. V. Olsen (1976) *FAUNET: A Program Package for Evaluation of Fault Trees and Networks*, Report RISO, 348, Danish Atomic Energy Commission.

[28] L. Caldarola and A. Wickenhauser (1977) The Karlsruhe computer program for the evaluation of the availability and reliability of complex repairable systems, *Nuclear Engineering and Design*, **43**, 463–70.

[29] L. Caldarola (1978) *Fault Tree Analysis with Multistate Components*, ANS topical meeting, Los Angeles.

[30] L. Caldarola (1980) Coherent systems with multistate components *Nuclear Engineering and Design* **58**, 127–39.

[31] L. Caldarola (1981) *Kernforschungszentrum*, Karlsruhe, RFA, KFK 3190 (EUR 7056e).

[32] K. Terpstra (1984) *Phased Mission Analysis of Maintained Systems: a Study in Reliability and Risk Analysis. Netherlands Energy Research Foundation*, ECN.

[33] K. Terpstra and N. H. Dekker (1988) *PHAMISS: a Reliability Computer Program for Phased Mission Analysis and Risk Analysis*, Users Manual, Netherlands Energy Research Foundations ECN (program available from the same organization).

[34] J. Olmos and L. Wolf (1979) *A Modular Approach to Fault Tree and Reliability Analysis Department of Nuclear Engineering*, Massachusetts Institute of Technology, MITNE-209.

[35] M. Modarres, N. C. Rasmussen and L. Wolf (1988) *Reliability Analysis of Complexe Technical System Using the Modularization Technique*, MITNE.228, Department of Nuclear Engineering, Massachusetts Institute of Technology, Cambridge, Massachusetts.

[36] M. Modares, N. C. Rasmussen and L. Wolf (1980) *A User's guide for MODCUT and PL-MOD MC Computer Codes Fault Tree Analysis*, USNRC Report NUREG/CR 1461-1980 (program available from Argonne Software Center, Argonne, Illinois, 60439, USA).

[37] *Gesellschaft für Reaktorsicherheit* (1978) Deutsche Risikostudie Kernkraftwerke: Eine Untersuchung zu dem durch Störfalle in Kernkraftwerken verursachten Risiko (German Risk Study), Verlag TÜV, Rheinland, Federal Republic of Germany.

[38] D. M. Rasmuson, N. H. Marshall and G. R. Burdick (1977) *User's Guide for the Reliability Analysis System (RAS) ERDA Report TREE*, 1168, E.G.G. Idaho, Inc., Idaho Falls, Idaho 1977 (program available from Argonne Software Center, Argonne, Illinois, 60439, USA).

[39] M. Astolfi, G. Mancini, G. Volta, S. Contini and S. Garribba (1978) *SALP (Sensitivity Analysis by List Processing): a Computer Assisted Technique for Binary Systems Reliability Analysis*.

[40] M. Astolfi, S. Contini, Van Den Muyzenberg, G. Volta (1978) *Euratom*, JRC

Ispra, Italie EUR 6185e.

[41] M. Astolfi, C. A. Clarotti, S. Contini, P. Picchia (1980) *Euratom*, JCR Ispra, Italie, Eurocopi Report number 12.

[42] R. B. Worrell and D. W. Stack (1978) *A SETS User's Manual for the Fault Tree Analyst Sandia Laboratories*, Albuquerque, New Mexico, NUREG/CR-0465, SAND 77-2051, 1978 (program available from Argonne Software Center, Argonne, Illinois, 60439, USA).

[43] M. D. Olman (1981) *Quantitative Fault Tree Analysis Using the SEP Evaluation Program (SEP)*, USNRC Report NUREG/CR-1935.

[44] R. C. Erdmann and F. L. Leverenz (1978) *WAMCUT: A Computer Code for Fault Tree Evaluation*, NP.803, Electric Power Research Institute, Palo Alto, California, June 1978 (program available from EPRI).

[45] B. F. Putney and H. R. Kirch (1981) *WAMCUT II: a Fault Tree Evaluation Program*, SAI-SR-234-81 PA, Science Applications, Inc., Palo Alto, California (program available from EPRI).

[46] F. L. Leverenz (1981) *SPASM: a Computer Code for Monte Carlo System Evaluation*, EPRI NP-1685, Electric Power Research Institute, Palo Alto, California.

[47] A. Blin, A. Carnino and J. P. Signoret (1973) *PATREC, code de calcul pour les arbres de défaillances*, Rapport CEA-DSN 235f.

[48] B. V. Koen, A. Carnino (1974) Reliability Calculations with a List Processing Technique, *IEEE Transactions of Reliability*, **B-23**(1).

[49] B. V. Koen, A. Carnino et al. (1977) *The State of the Art of PATREC: a Computer Code for the Evaluation of Reliability and Availability of Complex Systems*, The National Reliability Conference, Nottingham, England.

[50] J. G. Waddington and A. Wild (1990) *The Fault Tree as a Tool in Safety Analysis in Nuclear Power Plants*, INFO, 0036, Atomic Encrgy Central Board, Ottawa, Canada (program available from Atomic Energy Control Board, Ottawa, Canada).

[51] F. L. Leverenz and H. Kirch (1976) *User's Guide for the WAM-BAM Computer Code*, Science Applications, Inc., EPRI 217-2-5 (programme available from the same organization).

[52] A. Dubreuil-Chambardel (1985) *ELSA: Un ensemble de logiciels quantificant les Séquences Accidentelles*, EDF-DER, HT/13/4/85.

[53] D. W. Stack (1984) *A SETS User's Manual for Accident Sequence Analysis*, NUREG/CR-3547, USNRC, Washington DC 20555.

[54] R. B. Worrell (1984) *SETS Reference Manual*, SANS 83 2675, Sandia National Laboratories, Albuquerque.

[55] I. L. Rondiris (1982) *Reliability Evaluation of Nuclear Power Plants*, Ph.D Thesis, UMIST, United Kingdom (program available from Nuclear Installation Inspectorate).

[56] M. Bouissou (1984) *Nouvelle méthode de calcul de la fiabilité par les probabilités de séquences dans un graphe d'états*, Le programme GRAFFITI, EDF-DER, HT/13/18/84.

[57] C. A. Clarotti and G. De Cola (1982) ENEA-NIRA, Italy, Report T-40 000 XX131-002.

[58] A. Blin (1978) *Notice de présentation et d'utilisation du programme MARK-GE*, CEA-SETSSR No. 236.

[59] *Manuel d'utilisation MARK-EXD* (1985) Contrôle et Prévention (CEP), 34 rue Rennequin 75017 Paris.

[60] A.G. Colombo, R. J. Jaarsma and C. Thonet (1977) *Euratom*, Joint Research Center, Ispra, Italie, EUR 5851e.

[61] E. J. Henley and H. Kumamoto (1981) *Reliability Engineering and Risk Assess-*

ment, Prentice Hall, Englewood Cliffs, N.J. (programme available from the authors).

[62] J. E. Doucet (1981) *SURF: Système d'évaluation de la sûreté de fonctionnement*, 2e partie: *Notice d'utilisation*, Laboratoire d'Automatique et d'Analyse des Systèmes (LAAS) du Centre National de la Recherche Scientifique, Toulouse, LASS-SIS No. 81052.

[63] J. E. Doucet (1982) *SURF Users' Guide*, Laboratoire d'Automatique et d'Analyse des Systèmes (LAAS) du Centre National de la Recherche Scientifique, Toulouse, LASS-SIS No. 82020.

[64] C. L. Cate and J. B. Fussel (1978) *BACKFIRE: a Computer Program for Common Cause Failure Analysis*, NERS 77-02 Nuclear Engineering Department, University of Tennessee, Knoxville (program available from the same organization).

[65] M. Modarres, N. C. Rasmussen and L. Wolf (1980) *Reliability Analysis of Complex Technical Systems Using the Modularization Technique*, MITNE-228, Department of Nuclear Engineering, Massachusetts, Institute of Technology, Cambridge, Massachusetts (program available from the same organization).

[66] J. J. Rooney and J. B. Fussel (1978) *BACKFIRE II: a Computer Program for Common Cause Failure Analysis of Complex System*, University of Tennessee, Knoxville.

[67] G. R. Burdick, N. H. Marshall and J. R. Wilson (1976) *COMCAN: a Computer Program for Common Cause Failure Analysis*, ERDA Report ANCR, 1314, Aerojet Nuclear Company (program available from Argonne Software Center, Argonne, Illinois, 60439, USA).

[68] D. M. Rasmuson et al. (1978) *COMCAN II: a Computer Program for Common Cause Failure Analysis*, USDOE Report TREE, 1289, EFF Idaho, Inc., Idaho Falls, Idaho.

[69] D. M. Rasmuson, N. H. Marshall, J. R. Wilson and G. R. Burdick (1979) *COMCAN II: a Computer Program for Automated Common Cause Failure Analysis*, USDOE Report TREE, 1361, EGG Idaho, Inc., Idaho Falls, Idaho, 1979 (program available from Argonne Software Center, Argonne, Illinois, 60439, USA).

[70] B. F. Putney (1981) *WAMCOM: Common Cause Methodologies Using Large Fault Trees*, NP-1851, Electric Power Research Institute, Palo Alto, California (program available from the same organization).

[71] M. Astolfi and J. Elbaz (1977) *Euratom*, JRC, Ispra, Italy, EUR 5804.

[72] U.S. Nuclear Regulatory Commission (1975) *Reactor Safety Study. An Assessment of Accident Risks in U.S. Commercial Nuclear Power Plants*, WASH 1400, NUREG 75/014 (program available from Argonne Software Center, Argonne, Illinois, 60439, USA).

[73] S. D. Matthews (1977) *MOCARS: a Monte Carlo Simulation Code for Determining the Distribution and Simulation Limits*, TREE-1138 (program available from Argonne Software Center, Argonne, Illinois, 60439, USA).

[74] A. G. Colombo and R. J. Jaarsma (1980) *A powerful numerical method to combine random variables, IEEE Trans. Rel.*

[75] A. G. Colombo and R. J. Jaarsma (1980) *SCORE: a Computer Program for the Systematic Combination of Random variables*, Euratom, JRC, Ispra, Italie, EUR 6819 EN.

[76] J. J. Cairns and K. N. Fleming (1977) *STADIC: a Computer Code for Combining Probability Distributions*, General Atomic Co, GA, A14055.

[77] D. D. Orvis and M. V. Frank et al. (1977) Guidebook for the Reliability and Maintainability Analysis of NWIS Repository Equipment, ONWI-334, Office of

Nuclear Waste Isolation, Battelle Memorial Institute, Colombus, Ohio (program available from General Atomic Company).

[78] C. F. Desieno and L. L. Stine (1964) A probability method for determining the reliability of electric power systems, *IEEE Transaction on Power Apparatus and Systems*, **83**, 174–81.

[79] J. Endrenyi (1980) *Reliability Modeling in Electric Power Systems*, John Wiley and Sons, New York.

[80] J. Batut (1986) *Fiabilité prévisionnelle du réseau à très haute tension d'EDF*, 5ᵉ colloque international de fiabilité et de maintenabilité, Biarritz, France.

[81] K. D. Tocher (1963) *The Art of Simulation*, The English Universities Press, London.

[82] M. C. Shooman (1968) *Probabilistic Reliability: an Engineering Approach*, McGraw-Hill, New York.

[83] C. Singh and R. Billinton (1977) *System Reliability Modelling and Evaluation*, Hutchinson, London.

[84] A. Pages and M. Gondran (1980) *Fiabilité des systèmes*, Collection de la Direction des Etudes et Recherches d'Electricité de France.

[85] J. P. Signoret, M. Gaboriaud and A. Leroy (1986) *Modélisation et évaluation probabiliste de la production d'hydrocarbures d'un ensemble de puits sousmarins*, 5ᵉ colloque international de fiabilité et de maintenabilité, Biarritz, France.

[86] S. Garribda, G. Reina and G. Volta (1974) *Repair Processes*, Euratom, JRC, Ispra, EUR 5232e.

[87] S. Garribda, G. Reina and G. Volta (1976) Availability of repairable units when failure and restoration rates age in real time, *IEEE Transactions on Reliability*, **R-225**(2).

[88] G. Reina (1982) *How to use AVACOM-ETARP*, Euratom, JRC, Ispra, T.N. No. I.06.01.82.05.

[89] A. Poucet (1983) *Computer Aided Fault Tree Synthesis*, Euratom Report EUR-8707, EN, Ispra, Italie.

[90] A. Poucet (1983) *Betrouwbaarheidsanalyse van complexe system*. Doctorat dissertation, Katholick Universiteit Leuven (Belgium), OKt.

[91] A. Poucet (1985) *CAFTS: Computer Aided Fault Tree Analysis*, International Topical Meeting on Probabilistic Safety Methods and Applications, San Francisco, USA.

[92] P. Lé (1985) *Les logiciels EBAI (Expressions Booléenne Automatiquement Interprétées) et AIDE: application à la Méthode des Combinaisons de Pannes*, EDF-DER, HT/13/44/85.

[93] M. Gondran and A. Pages (1976) *Calcul de la fiabilité des systèmes réparables: le code FIABC*, EDF-DER, HI 2199/02.

[94] M. Bouissou (1986) *Recherche et quantification automatiques de séquences accidentelles pour un système réparable*, 5ᵉ colloque international de fiabilité et de maintenabilité, Biarritz, France.

[95] J. K. Vaurio and D. Sciandone (1979) *Unavailability Modeling and Analysis of Redundant Safety Systems AND*, 79-87, Argonne National Laboratory, Argonne, Illinois (programme available from the same organization).

[96] H. E. Lambert and F. M. Gilman (1979) *The IMPORTANCE Computer Code*, ERDA Report UCRL-79269, Lawrence Livermore National Laboratory, Livermore University of California (program available from Argonne Software Center, Argonne, Illinois, 60439, USA).

[97] J. S. Arendt *et al.* (1978) *EXCON: a Computer Program for System Risk Assessment*, JBFA, 129-78, JBF Associated, Inc., Knoxville, Tennessee.

[98] C. Ancelin, M. Bouissou and S. de Saint-Quentin (1989) *Lesseps software: a*

living PSA for the French Paluel nuclear power plant, ANS/ENS-PSA-89, International Topical Meeting on Probability, Reliability and Safety Assessment, Pittsburg, USA, April 2–7.
[99] J. R. Taylor (1981) *Automatic Fault Tree Construction with RIKKE. A Compendium of Examples*, Vol. 1: *Basic Models*, vol. 2: *Control and Safety Loops*, RISO, M-2311.

21
AUTOMATIC ASSESSMENT OF DEPENDABILITY

21.1 INTRODUCTION

Is there a reliability engineer who has never dreamt of having a computer program for analysing system dependability! Analysis, especially qualitative analysis, is such a long and arduous task that automatic analysis is the dream of every specialist.

The first attempt at automation in dependability analysis dates back to the late 1960s with the development of the GO method. Then, in the mid-1970s, another method, CAT, was created. Yet another method, ESCAF, was developed in the late 1970s, and later, in 1984, an improved version, S.ESCAF, with more applications, was introduced. The first projects using artificial intelligence date back to the mid-1980s.

The automatic analysis of dependability necessarily includes the following steps:

- modelling of a system, its components as well as their interactions and functional relationships;

- performance of the system dependability analysis—from the modelling mentioned above—by a computer program: the analysis is both quantitative and qualitative.

The major difficulty of this type of automatic analysis is the choice of a system model which can be used for all types of systems and all dependability problems. Such programs have several potential and expected advantages:

- rapid assessment of the effects of modifications in a system project on dependability;

- easy comparison of several computing systems;

- cost reduction of dependability;

- improvement of the study quality and exhaustivity;

- development and use of such models for failure diagnosis (mainly by alarm analysis).

In this chapter, the four main methods of automatic dependability assessment which have been implemented will be presented. The first three use system and component modelling based respectively on:

- logical operators: GO program [1-6];
- decision tables: CAT program [7-19];
- electronic gates: ESCAF and S.ESCAF [20-27].

Note that some comments on this subject can be found in references [28-34].

The fourth approach presented is based on the use of artificial intelligence (expert system) [35-40]: this approach will be illustrated by the description of EXPRESS, a program developed by Electricité de France (EDF).

21.2 MODELLING BY LOGICAL OPERATORS AND INDUCTIVE ANALYSIS: THE GO PROGRAM

21.2.1 Introduction

GO was created in 1968 by Kaman Sciences Corporation (United States) [1]. It was essentially used in the defence industry. The development of GO was motivated by the need for a simple and precise method of performing reliability analysis of electromechanical systems. Previously, the reliability analysis of such systems was performed first with the success diagram method and after with an analytical reliability assessment. But with the growing complexity of systems, this method proved to be more and more cumbersome and difficult to perform. In the mid-1970s, GO gave rise to numerous applications and the program itself was modified and improved for the nuclear industry, especially at the request of the Electric Power Research Institute (EPRI) [2-4]. Some projects resulted in the development of this approach [5, 6].

21.2.2 Modelling by logic operators

The method consists in using a model of a system—and of the functional relationships between its components—based on a set of standardized logic operators. Logic operators describe the component operating logic as well as its interactions with other components. A GO model is an arrangement of logic operators; it can generally be constructed from the system drawing by replacing the component symbols with logic operators. These logic operators model most of the components commonly used in industrial systems. The analyst then models the interactions of the logic operators by connecting their inputs and outputs. These inputs and outputs are known as signals. There are currently 16 GO operators available for developing system models [2, 3].

Figure 21.1 shows two simple systems modelled by the GO program; each component of the system is represented by a logic operator. A logic operator normally has the form shown in Figure 21.2.

There are 16 different types of logic operators. Below is a list of some of the logic operators used in Figure 21.1:

- 1st type: a two-state component (operating state and failed state). Example of such a component: resistor.

- 3rd type: a generator producing normal or inadvertent signals. Example of such a component: relay coil.

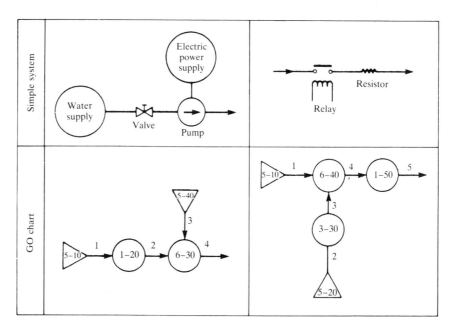

Figure 21.1 Go charts for two simple systems.

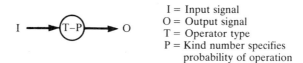

I = Input signal
O = Output signal
T = Operator type
P = Kind number specifies probability of operation

Figure 21.2 Go operator type.

The output signal ('close contact' command) is produced by an input signal (electric power) or due to inadvertent operation.

- 6th type: two-input component (signals 2 and 3) likely to be affected by a failure or premature operation.

The 5th type operator is rather particular: it is an 'initiating' operator which generates an input signal which may have two states (example: presence or absence of water supply).

Note that the GO method can be used to create 'super-operators' by combining a number of standardized logic operators.

21.2.3 The analysis principles

Analysis is inductive and performed by using a method of the consequence tree type; it consists in studying all the different combinations of signal states and logic operators. Figure 21.3 shows the consequence tree of the first simple system in Figure 21.1.

At each node of the tree, the values S of the different signals 1, 2, 3, 4 are given. The following notations are used:

$S_i = 0$: presence of signal ('sufficient water flow')

$S_i = 1$: absence of signal ('insufficient water flow')

The following probabilities are considered for the different branches of the consequence tree:

P_1 = no water supply probability

P_2 = probability of valve failure (plugging)

P_3 = probability of failure of the pump during operation

For simplicity's sake, the probability of electric supply was assumed to be 1. At each node of the consequence tree, the program calculates the probabilities of the different signal states by following the arrows in the GO model and by using the probability references given by the logic operators. The probabilities are calculated by multiplying the event probabilities: thus all these events are assumed to be statistically independent.

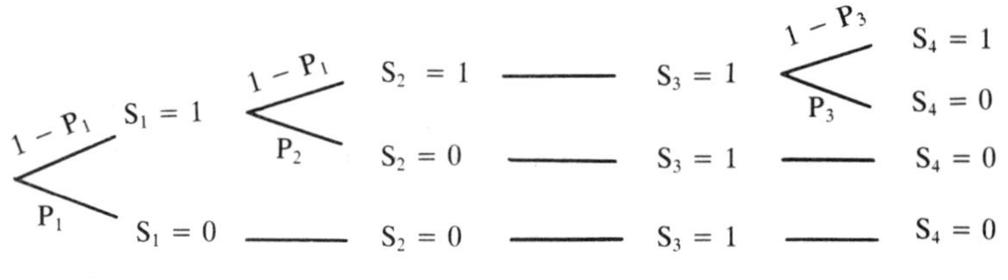

Figure 21.3 Consequence tree of the first simple system in Figure 21.1.

Obviously, the number of branches to be considered increases rapidly with the number of operators. The GO method uses two techniques to simplify the consequence tree:

- grouping of identical signal states when preceding signals are no longer useful for further calculations;

- setting a probability threshold and tree-pruning through elimination of those tree branches which may give probabilities lower than the fixed threshold during the calculation.

21.2.4 The GO program

The GO program can also be used to calculate the probability of an undesirable event modelled by a cause tree; AND and OR logic gates are then represented by logic operators. Note that, in such cases, GO examines the failure combinations up to order 4. The GO code is written in Fortran for CDC 7600 computers.

The GO method is well suited to the analysis of the reliability or availability of systems with clearly defined component failures and interactions. A GO model is simple enough to construct from engineering drawings of electromechanical or thermohydraulic systems by following the flow of the electric current, of the fluid, etc. Sensitivity studies can then be performed to determine the effect of modifications in reliability data or in system design.

The method can handle a large number of components with close physical and functional interactions. The possible creation of 'super-operators' simplifies the modelling of complex redundant systems. However, such a modelling technique takes account neither of the repairable nature of components nor of the possible time dependences of failures; in fact, the probability calculations performed by the GO method assume that events defined at the operator level are statistically independent.

21.3 MODELLING BY DECISION TABLES AND CAUSE TREE CONSTRUCTION: THE CAT PROGRAM

21.3.1 Introduction

CAT was developed by the Electric Power Research Institute (EPRI, US) [8–12] in conjunction with the University of California (School of Engineering and Applied Science) from the middle of the 1970s. System modelling is based on the decision table method [7]. CAT is used to develop the cause tree and, from this point, the engineer continues with the conventional method. To our knowledge, this code has not been widely used; nonetheless, the principles from which this program originated seem interesting. We would like to remind readers of the usefulness of decision tables for analysing systems with control loops [16–19].

21.3.2 Modelling with decision tables

Each component is represented by a decision table, which is defined as an extension of the truth table; any number of states can be used for each input of the table. A decision table groups

- the input states;
- the 'internal modes' of the component: they are the component operating and failure modes;
- the output states.

The connections between components are called nodes: at each node, component outputs are connected to the inputs of other components. The decision table describes the relations between input states, internal modes and output states. There are two methods of constructing a decision table:

- The first is inductive; it consists in listing all the possible combinations of input states and internal modes and in finding the output state of each combination. The states of the output parameters represent, in a way, component effects on the system whilst states of the input parameters represent that of the system on the component.

- The second is deductive and begins by identifying all possible output states and tracing back to all possible input states.

21.3.2.1 Inductive method for decision table construction

There are several phases:

- phase 1: finding all component input conditions and internal modes from data on the component physical and functional characteristics. For example, Table 21.1 shows the input parameters, the internal modes and the output parameter of a pump.

- phase 2: studying all combinations of input states and internal modes and defining the states of output parameters. The decision table is thus constructed. Twelve combinations can be studied for the pump mentioned above (Table 21.2).
    Row 1 of the decision table shows that, when there is a suction pressure and an electric power supply and the pump is in an operating state, the

Table 21.1 Pump model

Input 1	Main flow (pressure) input
	0 no pressure in or pressure too low
	1 normal pressure
Input 2	Power input
	0 No power in
	1 Power in
Internal modes	0 Pump in good condition
	2 Pump fails to start
	3 Pump fails to run normally
Output	Main flow (pressure) output
	0 No pressure out or pressure too low
	1 Pressure out

Table 21.2 Decision table for pump

Row	Input 1	Input 2	Internal modes	Output
1	0	0	0	0
2	0	0	2	0
3	0	0	3	0
4	0	1	0	0
5	0	1	2	0
6	0	1	3	0
7	1	0	0	0
8	1	0	2	0
9	1	0	3	0
10	1	1	0	1
11	1	1	2	0
12	1	1	3	0

discharge pressure is normal; row 10 indicates that, when there is no suction pressure and no electric power supply and the pump is in an operating state, there is an insufficient or inexistent discharge pressure.

Note that an FMEA can be profitably used during phases 1 and 2.

- phase 3: decision table reduction.

The method used to reduce the decision table is equivalent to a Boolean reduction. Let us look at rows 4, 5 and 6: the same output state (0) is obtained for the same input parameter states regardless of the internal mode state. The internal mode state does not then seem to have any influence on the output state for a given state of the input parameters; in this case, the state of the internal mode is considered to be a 'don't care' state. This 'don't care' state is characterized by '−1' in the corresponding row. Then rows 4, 5 and 6 are combined into a single row (0, 1, −1, 1). Below are the basic rules for reducing decision tables:

—decision table rows with identical output states are considered;

—if the input state combinations only differ in the state of an input parameter or an internal mode and if they include all the possible states of the chosen input parameter or all the internal modes, all the rows can be combined into a single row with the input parameter or the internal mode in a 'don't care' state.

The reduced decision table (Table 21.3) is obtained by the systematic application of the above rule to the previous decision table.

21.3.2.2 Deductive method for decision table construction

Cause trees are constructed for each component, and the decision table is derived from them. Let us take again the example of the pump; the information desired will be of the following type:

- what failure causes can produce insufficient or no discharge pressure?
- what events are required to obtain normal discharge pressure?

The different phases are the following:

- phase 1: identical to the first phase of the inductive method
- phase 2: determination of the output parameter states;
- phase 3: development of cause trees for the states of output parameters; the cause trees of the pump are shown in Figure 21.4;
- phase 4: construction of the decision table. Consider once again the first cause tree in Figure 21.4: insufficient or no suction pressure results in insufficient or no discharge pressure, whatever the states of the other input parameter or internal mode. As a result, the latter are 'don't care' states.

Table 21.3 Reduced decision table for pump by inductive method

Row	Input parameter		Internal modes	Output paramter
	1	2		
1	0	−1	−1	0
2	1	0	−1	0
3	1	1	0	1
4	−1	−1	2	0
5	−1	−1	3	0

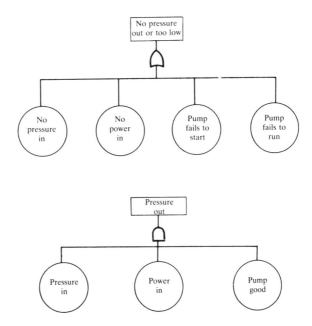

Figure 21.4 Cause trees for pump.

The decision table is shown by Table 21.4. This is clearly identical to the table obtained with the inductive method.

- phase 5: checking the completeness of the decision table to make sure that no relevant operating or failure modes have been overlooked in the construction of the table.

Decision tables for about 20 different components (amplifier, battery, fuse, pump, valve, pipes, ...) generally used in industrial systems have been developed in references [9, 11]. Note that component unavailabilities due to maintenance or testing are easily modelled by considering an input parameter known as repair or test parameter.

21.3.3 Cause tree construction

In the previous paragraphs we saw how a decision table is used to model each component. Components interact because the output parameters of some components are the input parameters of other components. The output parameter of a component may be connected to the input parameters of several components at a node. However, the opposite is not true for several output parameters cannot be connected to the input parameter of one component. If several outputs are connected in parallel (when there is a redundancy for instance), they will be modelled by a decision table.

Table 21.4 Reduced decision table for pump by deductive method

Row	Input parameter 1	Input parameter 2	Internal modes	Output paramter
1	0	−1	−1	0
2	−1	0	−1	0
3	−1	−1	2	0
4	−1	−1	3	0
5	1	1	0	1

The undesirable event has to be defined prior to constructing the cause tree. The undesirable event is one (or several) state(s) of an output parameter at a node of the model which is used as a starting point of the analysis. The combinations of the input parameter and the internal failure mode states resulting in the undesirable event can be found by examining the corresponding decision table. The search for fault combinations continues with the analysis of the states of the input parameters which are output parameters of other decision tables.

Let us consider the pump again. Let the undesirable event be 'insufficient or no discharge pressure'; then let us search for rows which have 1 as the output parameter state in the decision table (Tables 21.3 and 21.4). There are four of them: the corresponding combinations are thus connected to the undesirable event by an OR gate. Therefore, each row represents a combination of failures: these failures are connected to one another by an AND gate. In the previous example, each failure combination is reduced to one failure since the other states are 'don't care' states. The states of the pump input parameters are the output states of other components and the algorithm for cause tree construction goes on.

With this method, common-cause failures can be taken into account; thus, if a component output parameter is the input parameter of several components, the state of this parameter simultaneously affects the other components. Likewise, human errors made by operators on one or several components can be modelled: thus, if we consider a calibration error committed on one or several sensors, the decision table of a sensor will include calibration as an input parameter, and a decision table for the operator will be constructed with calibration as an output parameter.

References [8–11] give examples of the use of this methodology for the analysis of safety-related systems in nuclear power plants (thermohydraulic systems and control systems).

21.3.4 The CAT program

CAT, which works according to the previously explained principles, can handle a wide variety of components. The decision tables can be reduced by the program itself. Computing time is considerably shortened by the fact that already reduced decision tables are input.

The resulting cause tree can then be used in conjunction with a computing code such as PREP-KITT; the cause tree is then reduced, the minimal cut sets are determined and their probabilities are calculated. CAT is written in Fortran IV for IBM and CDC computers.

A version of CAT has been created by Electricité de France [13–15]; it generates a file which is compatible with the DEFAIL and FIABC programs (see Chapter 20) used at EDF for finding minimal cut sets and for quantitative analysis. Moreover, a module has been created to make data acquisition easier.

Clearly, CAT has the advantages that which go with automation but it has numerous limitations. The systems which can be modelled in this way are such that failure propagation follows the same direction as the flow of the fluid used. Thus, a component failure solely affects components downstream. Therefore, 'closed loop' systems, in which a component failure affects components upstream, are difficult to model. It should be stressed once again that CAT handles neither component repair nor the sequential occurrence of failures.

21.4 ELECTRONIC GATE-BASED MODELLING AND FAILURE SIMULATION: ESCAF AND S.ESCAF SYSTEMS

21.4.1 Introduction

ESCAF (Ensemble de Simulation et de Calcul de Fiabilité), designed by the Commissariat à l'Energie Atomique (CEA) in France [20–22] from 1979 onwards, is a unique system for simulating failures and computing reliability. It was produced on a wide scale by Télématique [25]. ESCAF is designed as a stand-alone unit with a console (monitor and keyboard). S.ESCAF, another version [23, 24], became available from 1984. It was developed to handle sequential systems, that is systems where the order in which the failures occur is essential.

ESCAF's principle of operation is very simple (Figure 21.5):

- The system whose reliability is to be studied is simply simulated by electronic gates on a board specially designed for this purpose; each gate corresponds to a component, and the gates simulate its operation or failure.

- A failure combination generator simulates the various likely failures in the circuit one by one (order 1), then two by two (order 2), etc, according to all the possible combinations. ESCAF determines if a combination triggers the system failure or not.

- A processing unit in ESCAF derives the possibility of the system failure

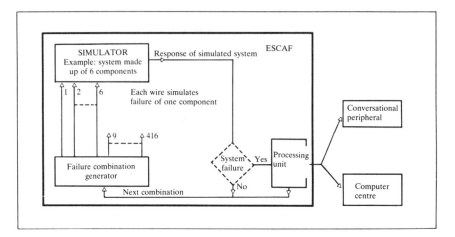

Figure 21.5 ESCAF architecture.

from data on component reliability. An external connection is available to send the data either on a storage medium (magnetic tapes or disks) or directly to a data processing centre for further processing if necessary. The diagram and the state of the system according to the various failures simulated can also be viewed on a screen. This idea of simulating failures using electronic circuits is also discussed in other publications [26, 27].

21.4.2 Modelling and analysis principles

ESCAF can directly model and analyse numerous systems, especially systems for which the logic data on the system operation or failures flow in only one direction. Each component is simulated by a gate with a first entry for the transmitted information and a second entry for the signal from the failure combination generator which defines the operating or failed state.

To illustrate this point, consider the fire-fighting equipment depicted in Figure 21.6; Figure 21.7 shows the same system modelled by AND and OR logic gates. When a fire is detected, an opening order is sent to the three valves. The diagram in Figure 21.7 is used to simulate the system operation and failures. Thus

- electronic gate AND_1 simulates valve No. 1
  $P_1 = 0$: fluid circulates when valve is opened
  $P_1 = 1$: fluid does not circulate due to valve failure;

- the electronic gates AND_2 and AND_3 simulate valves V_2 and V_3 respectively:

- electronic gate OR simulates the parallel operation of V_1 and V_2.

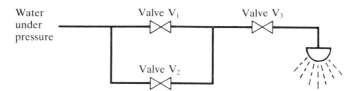

Figure 21.6 Fire extinguishing system.

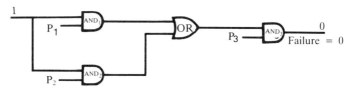

Figure 21.7 System simulation (electronic gates).

The system modelling on the board is connected to the failure combination generator. When ESCAF is ordered to start, a first failure is simulated by the combination ($P_1 = 1$, $P_2 = 0$, $P_3 = 0$); when no failure is indicated by output O ($O = 0$), ESCAF generates the next combination. For $P_3 = 1$, a system failure is found. ESCAF then simulates the failure combinations of order 2, then 3, ... up to order n; in this way, ESCAF identifies the failure combinations which cause the system failure and then deduces the minimal cut sets from these combinations.

21.4.3 ESCAF

The basic ESCAF system consists of 1-8 failure generating circuit boards: each circuit board can be used to simulate a maximum of 52 components. The whole system, with its 8 circuit boards, can handle up to 416 components. The electronic gates are quite conventional and they represent AND, OR, NOT and other types of logic gates. An index programming matrix can be inserted between the combination generator and the simulator. Certain parts of the system can then be tested, and common cause failures occurring simultaneously on several components (in addition to the independent random failure of the component) can be simulated.

ESCAF rapidly determines all the cut sets as well as the minimal cut sets; indeed, the failure combination generator is capable of generating a failure combination in a few microseconds and the simulator answers almost at once (roughly 1 μs). So the time required for ESCAF to determine the number of cut sets — minimal cut sets included — is roughly the number of combinations considered (C_n^r for order r combinations of n components) multiplied by 3 μs. For example

- 7 s are required to study 50 components for order 5 combinations,

- about 4 min are required to study 100 components for order 5 combinations,

- about 40 min are required to study 200 components for order 5 combinations.

Probabilities are computed by the truth table method using previously determined cut sets. This method assumes that the events considered are mutually independent; this implies, for example, that a component failure or repair does not affect the failure or repair of another component.

For non-repairable systems, input probabilities are failure probabilities at time t. For repairable systems, the inputs are unavailabilities at time t. In order to obtain the system asymptotic unavailability, (component) asymptotic unavailabilities can be entered when failure and repair rates are constant. It takes a longer time to compute probabilities; computing time is even more directly dependent on the number of cut sets. It is approximately 10 ms multiplied by the number of cut sets. Once the cut sets have been determined, it sometimes takes less time to perform this calculation on another computer. ESCAF handles non-coherent systems as well; an example can be found in reference [22]. Even though ESCAF cannot be used to obtain prime implicants, the method for calculating probabilities gives accurate results for this type of system.

Considering the method used, ESCAF can obviously analyse cause trees; each gate of the cause tree is then directly replaced by an electronic gate. ESCAF is thus capable of analysing cause trees with up to 416 logic gates whatever the way they are connected. The same sequence of processing operations as that mentioned above is then used.

Thus ESCAF can model the operation of a system with up to 416 components. It is well suited for assessing the reliability of complex systems; however, the modelling of each component should be simple. Precise values of the probability are computed, which is particularly useful when failure rates are high.

21.4.4 S.ESCAF

S.ESCAF is an improved version of ESCAF. It processes systems where the order in which failures occur is important; remember that such systems are known as 'sequential systems'. In S.ESCAF, the failure combination generator becomes a sequential event combination generator.

For studying sequential events, ESCAF needs to be supplemented with its S.ESCAF option consisting of 1–4 circuit boards, called S.ESCAF 20, 40, 60, 80, that can be inserted into the unit portion specially allocated to the user. All four circuit boards operating together make up a sequential event combination generator capable of handling up to 80 events and order 5 combinations. Unlike with ESCAF, the events are not output simultaneously, during the combination process, but in sequence, one after the other, and all possible sequences are gone through.

This sequential generator acts on one (or several) SEQ.SIM simulation circuit board(s) which the user utilizes for simulating almost directly the system to be studied. The events produced by the generator are connected to the 'failure' or 'repair' inputs of simulated components. The OUT output of these components changes to 0 when the corresponding failure is applied. An area is loaded with very conventional integrated electronic circuits containing AND and OR gates to simulate the interconnection between the system components. With S.ESCAF, the term event applies to all the phases in the life of a repairable component such as failure, repair, failure to start and interruptions.

S.ESCAF considers all the possible event combinations as well as all the allowable sequences for each combination. Some sequences are prohibited (for instance, a sequence with two successive failures of the same component). Two types of components are simulated: active components and standby components.

The next step consists in calculating the probabilities of each sequence of events; the method used is similar to that presented in Chapter 14. An approximation method is used; it is based on the assumption that the components have low failure rates compared to their repair rates. In other words, when a component is failed, in all likelihood it will be repaired before another failure occurs! Looped sequences are then negligible, and considering only direct event sequences is a good approximation. Besides, the same method can be used to assess the mean time to repair as well as the average event sequence duration.

S.ESCAF is particularly appropriate for repairable systems with a large number of components. Indeed, the application of the state space method to such systems is too unwieldy because of the large number of states and matrices involved!

21.5 USING EXPERT SYSTEMS FOR MODELLING: THE EXPRESS PROGRAM

21.5.1 Introduction

Research on Artificial Intelligence (AI) started in the early 1950s. The expression was coined in 1956 by MacCarthy, the author of the first programming language Lisp (LISt Processing, 1959) which is still the most commonly used language. AI is generally defined as the discipline dealing with the automatic processing of knowledge. AI tries to emulate the mental processes of the human brain. Ambitious goals, such as automatic translation, characterized most of the research work on AI. All projects to this end were however virtually abandoned in the 1960s. In the late 1960s, the first expert system (Dendral) was created by Djerassi, Feigenbaum and Lederberg, who was awarded the Nobel prize for his work in genetics. Dendral determines the formula of a molecule from its mass spectrometric analysis.

From the 1970s onwards, an ever-increasing number of expert systems were developed in all reknowned universities, then within a few pioneering companies; an example of such a system is Mycin which is designed to diagnose blood infection and meningitis and to recommend the appropriate treatment.

The Prolog language was developed from 1977 onwards and has had tremendous support since the Japanese chose it for their project of fifth generation computers.

Expert systems became fully operational in the 1980s, and in the mid 1980s, the potentialities of expert systems for dependability analysis started to be explored by Electricité de France (EDF). The Saga expert system [36] was born in 1985: it is used to generate, from cooling system faults, the accident sequences which may affect the core cooling function of the Creys–Malville breeder reactor plant. Two other expert systems have been developed since 1985. The first of them, EXPRESS (système EXPert pour la Réalisation d'Etudes de Sûreté des Systèmes), is used for tree construction [37, 38]; the second one, GSI, (Génération de Séquences par Inférence) is designed to generate event sequences and calculate their probabilities (see Section 20.8).

After discussing the operating principles of an expert system, we will show how EXPRESS is used to perform reliability analysis, then we will discuss the contribution of artificial intelligence to dependability analysis.

21.5.2 How does an expert system work?

An expert system is a software package which contains an expert's knowledge on a given subject and is programmed to use it as he would [39, 40]. Classical algorithmic computer programs have a major shortcoming: specific knowledge and the computations performed from this data are too closely linked, up to the point where they can no longer be differentiated. Even a trifling change in a large program can rapidly become a real headache. This drawback becomes insurmountable if the software has to process a large amount of specific knowledge which is likely to extend and require updating. As it is difficult to acquire knowledge and convert it into formal language for entry into a computer, a system in which knowledge can be separated from the program using this knowledge and expressed as much as possible in single units (rule by rule) is definitely preferable.

An expert system is made up of two independent parts:

- a knowledge base;
- an inference engine.

As a rule, the knowledge base is the memory in which the domain expert's knowledge and know-how are stored (description of objects and their relations, particular cases, solving strategies and their conditions of application). The inference engine is an algorithm capable of logical reasoning using the data fed into it and the knowledge contained in the knowledge base as well as an appropriate strategy to control the reasoning process; after all, an inference is a logic operation whereby a proposition is accepted by virtue of its link with other propositions which are already held to be true.

Two essential characteristics differentiate the expert system from an ordinary computer program; the system may at any time

- explain its behaviour to the expert

- receive new knowledge from the expert without requiring new programming

The knowledge base must therefore be legible and independent of the inference engine (which should be able to interpret it).

The knowledge base normally includes a data base and a set of production rules. A rule establishes a relation between the facts (objects, entities, events, ...); it is generally a relation of logical dependence which can be expressed as

if facts $A, B, C, \ldots$ are true,

then they result in facts $E, F, G, \ldots$

For example, let the following knowledge base be

initial facts base: A, B

rule set: R_1 $A \rightarrow C$

R_2 C and $D \rightarrow E$

R_3 $B \rightarrow D$

where $A \rightarrow C$ means 'if A is true, then C is true'.

The inference engine infers new facts from the knowledge base. Two inference methods can be used:

- *Forward chaining*. The inference engine starts from the initial facts base and fires all the rules whose **if** portion (left-hand side) are true; the inferred facts are placed in the facts base. The engine thus moves forward until all the possibilities are exhausted. Below are the data bases obtained by forward chaining:

$$\{A, B, C\}, \{A, B, C, D\}, \{A, B, C, D, E\}$$

- *Backward chaining*. The goal here is to prove that fact E is true. The inference engine then tries to find all the rules which contain this goal in the conclusion. Each of these rules is considered: if all the **if** parts are true in the initial facts base, then the goal is reached. Otherwise, the unknown **if** statements are recorded as new goals and the cycle has to start over again for each of them.

Note that there are two types of production rules:

- the first type is based on propositional calculus: rules are directly applied to facts; thus the rule premises and consequences are facts. Note that Boolean algebra is based on the same logic;

- the second type is based on predicate calculus (in first-order logic): rules can use variables; we can have rules like

if	x is a mmm then x in mmhd,
if	x is an element of y
and if	y is an element of z
then	x is an element of z.

Why do we need expert systems for dependability analysis?

Imagine that the facts are failures or faults; dependability engineers who construct an FMEA or any other model (for example, a cause tree) have to permanently reason in this way:

if such a failure were to occur **and if** another fault existed, etc, **then** the undesirable event would occur. The engineer normally uses both the inductive approach (forward chaining) and the deductive approach (backward chaining).

Thus, as a rule, the facts base will be the base of events likely to occur during the lifetime of the system and the rule set will contain the reasoning on these events whereby the relevant events will be identified.

Expert systems can therefore be used for qualitative analysis, as is shown in the example below.

21.5.3 Example of a system reliability analysis

In this chapter, we will describe the reliability analysis of a nuclear power plant (thermohydraulic) safety system using an expert system: the EXPRESS program was created for this purpose.

The system studied will be briefly described: the system in question is the auxiliary feedwater system (AFS) of the steam generators. Equipped with two trains (A and B), the system is designed to supply water (sucked from a tank) to steam generators. Each train is made up of a line incorporating an auxiliary feedwater motor-driven pump (Figure 21.8) (power supply) and a line incorporating a turbine-driven pump (steam extraction from the steam generators). During certain incidents, the AFS system starts and must function for a few hours: this mission is what we are particularly interested in. Then at least one steam generator must be supplied with water by a pump.

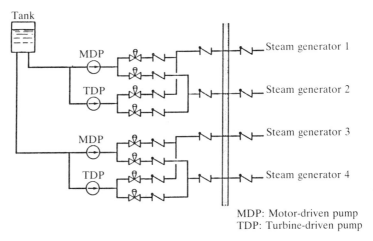

Figure 21.8 Auxiliary feedwater system (AFS).

How can an expert system be used to obtain the cause tree of the failure of the system mission?

We proceed from the system topology (description of components and their links) and the knowledge of the component failure modes; facts and rule sets are built and two inference engines are used:

- LRC: developed by EDF, LRC is based on propositional calculus [41],

- ALOUETTE: developed by EDF, ALOUETTE is based on predicate calculus [42].

Now the facts and rule sets (Figure 21.9) will be described:

1. Facts bases. Two types exist, the topological facts base and the functional facts base.

- *Topological facts base.* This uses two types of relationships to describe the topology and the main functions of the system:

  −(Y COMES AFTER X) means that component Y is located after X; for example, valve 5VP is downstream from 3VP: (5 VP COMES AFTER 3 VP),

  −(X, NATURE, Y), means that X is of Y nature; for example, component AFS 31 PO is a pump: (AFS 31 PO, NATURE, PUMP).

- *Functional facts base.* This specifies the causes and consequences of failures. It indicates, particularly, the possible paths between a source (the tank) and a destination (the steam generator) on the one hand, and the failure causes (blockage, leakage) on the other hand. In the examples considered, failures have two categories of effects: blockage (or loss of flow) or external leakage (partial water loss and flooding).

A difference has to be made between the initial facts base and the 'intermediate' facts base which also contains data created by the first inference through the use of the topological rule set.

2. **Rule sets**. There are three types of rule sets: the topological rule set, the generating rule set and the LRC rule set.

- *Topological rule set*. This set manages the topology and handles the facts base. Thus it builds all the concepts required for the identification of the 'blockage' or 'leakage' blocks (or macro-components) and of the path losses.

- *Generating rule set*. Two types of rules are considered:

    —topological rules which use the results of the first inference and translate the consequences of a component failure into 'path losses',

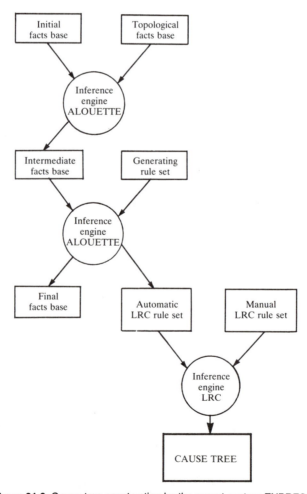

Figure 21.9 Cause tree construction by the expert system EXPRESS.

according to the 'leakage' and/or 'blockage' block(s) to which it belongs;
—functional rules which generally state the blockage or leakage causes precisely.

A second inference (still with ALOUETTE) can be made and used to deduce the automatic LRC rule set.

- *Automatic LRC rule set*. The rules which make up this set are very simply formulated; they indicate the paths lost for each component failure. Below is an example of a rule obtained in LRC language:

if
01BA – LEAKAGE – EXTERNAL = YES
then
PATH 1 – LOSS = YES

Some problems specific to the system were not automatically taken into account; that explains the presence of a manually constructed LRC rule set.

- *Manual LRC rule set*. This mainly contains rules which are related to
    —the undesirable event;
    —boundary conditions: these describe the AFS system interfaces with the external systems (electric power supplies, cooling systems, controls . . .);
    —specific problems: these describe, for the AFS system, particular operating conditions related to the minimal flow lines of the pumps.

For example, for the AFS system, we have obtained:

- automatic LRC rule set: 470 rules;
- manual LRC rule set: 83 rules that can be broken down as below:
    —description of the undesirable event: 29,
    —description of the boundary conditions: 50,
    —description of specific problems: 4.

The cause tree (several hundred events) is derived from two LRC rule sets using the LRC inference engine.

Now let us come back to the approach used: it is basically an inductive approach. The automatic LRC rule set contains the equivalent of the FMEA performed on all the system components in the form of rules; all the effects are either 'blockages' or 'leaks' and therefore ultimately result in path losses. The method has several points in common with the GFCM (see Section 11.3); the fault combinations whose effects add up and result in path losses are easily identified by the expert system. On the other hand, fault combinations whose effects do not add up must be translated into specific rules which can, for example, be introduced into the manual LRC rule set.

21.5.4 Prospects

As illustrated by the preceding example, artificial intelligence can be used to perform a large part of a system automatic reliability analysis; thus the work of a dependability engineer is made easier because he is exempt from having to analyse numerous failure combinations (an arduous task!). Thus freed from this time-consuming task, the engineer can spend more time on more difficult problems (common-cause failures, failures peculiar to a particular component or system, . . .).

Below are certain aspects which need to be particularly emphasized:

- Writing the rules which emulate the expert's reasoning requires much rigour in defining the knowledge about the system operation; this knowledge, which constitutes the basis of dependability analysis, is formalized, written and codified at this stage. Thus modifications (or errors) in the operating assumptions can be easily identified, then studying their effect on dependability analysis becomes possible.

- Revising and updating dependability assessment is greatly simplified; apparently, the engineer merely needs to modify facts and/or rules, to add them or to subtract them from the existing data.

Artificial intelligence seems to be the ideal assistant of a dependability engineer. In the long run, engineers should be able to make 'living' tools for assessing system dependability at the design stage by taking the multiple variants of a project into account. Moreover, such assessment tools can be easily updated and improved by using the knowledge obtained from the analysis of operating experience; they will thus be able to cover the entire life cycle of a system.

21.6 CONCLUSION

The automatic dependability assessment is a dream it is difficult to turn into reality! The difficulties involved have been clearly illustrated in this chapter. However, the tools already available are efficient for specific cases. Major methodological progress is still needed before the use of these tools can be extended to many fields. The future of artificial intelligence seems to be very bright.

REFERENCES

[1] W. Y. Gately, D. W. Stoddard and R. L. Williams (1968) *GO, A Computer Program for the Reliability Analysis of Complex Systems*, Kaman Sciences Corporation, Colorado Springs, Colarado, USA, KN-67-704(R).

[2] W. Y. Gately and R. L. Williams (1978) *GO Methodology*, Overview, EPRI

NP-765, Electric Power Research Institute, Palo Alto, Calif., USA.

[3] W. Y. Gately and R. L. Williams (1978) *Go Methodology*, System Reliability Assessment and Computer Code Manual, EPRI NP-766, Electric Power Research Institute, Palo Alto, Calif., USA.

[4] Kaman Sciences Corporation (1978) *Go Methodology*, Fault Sequence Identification and Computer Code Manual, EPRI NP-767, Electric Power Research Institute, Palo Alto, Calif., USA.

[5] J. Ancelin, E. Chaty and M. Carré (1981) *Notice d'utilisation du code GO d'analyse automatique de la fiabilité*, Electricité de France, HT/13/34/81.

[6] T. Matsuoka and M. Kobayashi (1985) *GO-FLOW: A Reliability Analysis Methodology Applicable to Piping Systems*, ANS/ENS International Meeting on Probabilistic Safety Methods and Applications, San Francisco, USA.

[7] S. L. Pollack (1971) *Decision Tables: Theory and Practice*, Wiley-Interscience Publishers, New York.

[8] S. L. Salem, G. E. Apostolakis and D. Okrent (1976) *A Computer Oriented Approach to Fault-Tree Construction*. EPRI NP-288. Electric Power Research Institute, Palo Alto, California, USA.

[9] J. S. Wu, S. L. Salem and G. E. Apostolakis (1977) Decision Table Development for use with the CAT Code for the Automated Fault-Tree Construction, UCLA ENG-7711, University of California, Los Angeles, USA.

[10] S. L. Salem, G. E. Apostolakis and D. Okrent (1977) A New Methodology for the Computer Aided Construction of Fault Trees, *Nuclear Energy*, **4**, 417–33.

[11] G. E. Apostolakis, S. L. Salem and J. S. Wu (1978) *CAT: A Computer Code for the Automated Construction of Fault Trees*, EPRI NP-705. Electric Power Research Institute, Palo Alto, California, USA.

[12] S. L. Salem, J. S. Wu and G. Apostolakis (1979) *Decision table development and application to the construction of fault trees,* Nuclear Technology, **42**, January.

[13] J. Ancelin, J. F. Barbet, P. Bonaventure and J. M. Reix (1980) *Méthode automatique de construction d'un arbre de défaillances, utilisation du code CAT*, EDF-HT/13/39/80, Electricité de France.

[14] J. Ancelin and J. M. Reix (1981) *Notice d'utilisation du code CAT de construction automatique d'arbres de défaillances dans sa version conversationnelle*, EDF-HT/13/6/81, Electricité de France.

[15] J. Ancelin (1981) *Le Code CAT d'analyse automatique de la fiabilité des systèmes*, Présentation des problèmes de modélisation, modifications du code, EDF-HT/13/31/81, Electricité de France.

[16] E. J. Henley and H. Kumamoto (1981) *Reliability Engineering and Risk Assessment*, Prentice-Hall, Englewood Cliffs, N.J.

[17] H. Kumamoto and E. J. Henley (1979) *Safety and reliability analysis of systems with control loops*, AICHE Journal, **25**(1), 108.

[18] E. Ogunbiyi (1979), Ph.D. Thesis, University of Houston, Dept of Chemical Engineering.

[19] E. J. Henley and H. Kumamoto (1977) Comments on computer aided synthesis of fault trees, *IEEE Trans. Relibility*, **R-26**(5), 316.

[20] A. Laviron, C. Berard and R. Quenee (1979) *ESCAF. Failure Simulation and Reliability Calculation Device—Second National Reliability Conference*, Birmingham (England), Vol 2.

[21] A. Laviron, A. Carnino and J. C. Manaranche (1982) ESCAF, A new and cheap system for complex reliability analysis and computation, *IEEE Trans. Reliability*, **R**-31, October.

[22] A. Laviron (1986) *ESCAF: MTBF, time evolution, sensitivity coefficients*. Cut sets importance and non-coherence of large systems, *IEEE Trans. Reliability*, **R-35**(2), 139–44.

[23] A. Laviron (1984) *Simulation et Analyse de la fiabilité des systèmes séquentiels complexes avec S.ESCAF*, 4e colloque international sur la fiabilité et la maintenabilité, Perros-Guirec, France.

[24] A. Laviron, A. Blin, J. M. Lanore and C. Rivet (1985) S.ESCAF. Sequential complex systems are analysed with ESCAF through an add-on option, *IEEE Trans Reliability*. **R-34**(3).

[25] Télématique S. A. ZIRST: 38240 Meylan, France.

[26] P. G. Bailey, G. B. Rothbart and R. R. Fullwood (1981) *Verification of Fault Tree Analysis*, Electric Power Research Institute, EPRI NP-1570, Vols. 1 and 2.

[27] N. D. Deans and D. P. Mann (1982) *High-speed Stochastic Reliability Simulator*, Reliability in Electrical and Electronic Components and Systems, Copenhagen, DK, June 14–18, 1982. Available from: Robert Gordon Institute of Technology, School Hill, Aberdeen, Great Britain.

[28] J. B. Fussell (1973) *Synthetic Tree Model, a Formal Methodology for Fault Tree Construction*, Report ANCR, 1098.

[29] E. Hollo and J. R. Taylor (1976) *Algorithms and Programs for Consequence Diagram and Fault Tree Construction*, Report RISO-M-1907.

[30] G. J. Powers and F. C. Tompkins (1974) Fault tree synthesis for chemical processes, *AICh.E Journal*. **20**(2), 376–87.

[31] S. A. Lapp (1978) *Computer Assisted Fault Tree Synthesis*, Thesis, Department of Chemical Engineering, Carnegie, Mellon University, USA.

[32] S. A. Lapp and G. J. Powers (1977) Computer-Aided Synthesis of Fault Trees, *IEEE Trans. Reliability*, 1-13, April.

[33] J. F. Barbet, E. Bourgade and M. Llory (1982). *Modélisation des systèmes et analyse automatique de la fiabilité et de la sécurité*, Second Séminaire Européen sur la sécurité des systèmes, La Baule, France.

[34] A. M. Amendola, E. Silvestri, S. Contini and G. Squellati (1983) *Algorithms for Automated Fault Tree Construction*, 4th Conference Euredata, Venice, Italy.

[35] F. Antomarchi, A. Castiel and E. Seyden (1986) *Pense... Machine. Pour comprendre l'intelligence artificielle*, Editions CESTA, Paris.

[36] P. Le (1985) *Centrale de Creys-Malville-Génération automatique des séquences accidentelles liées à la fonction évacuation de la puissance résiduelle*, EDF-DER-HT/13/96/85.

[37] M. Gondran, J. F. Hery and J. C. Laleuf (1986) *Une aide aux études de fiabilité par systèmes experts*, 5e colloque international de fiabilité et de maintenabilité, Biarritz, France.

[38] C. Ancelin, P. Le and S. de St Quentin (1987) *EXPRESS: an Expert System to Perform System Safety Study*, International Topical Conference on Probabilistic Safety Assessment and Risk Management, Zurich, Switzerland.

[39] M. Gondran (1984) *Introduction aux systèmes experts*, Eyrolles, Paris.

[40] H. Farreny (1985) *Les systémes experts: principes et exemples*, Cepadues Editions, Toulouse.

[41] J. F. Hery and J. C. Laleuf (1985) *Notice d'utilisation et analyse des logiciels LRC*, EDF-DER, HI/5037-02, HT 14/23/85.

[42] D. Mulet-Marquis and M. Gondran (1984) *Un langage pour les systèmes experts: ALOUETTE*, EDF-DER-IMA, HI 4773-02.

Part 5
CONCLUSIONS

22
DEPENDABILITY ASSESSMENT APPROACH

22.1 INTRODUCTION

Faced with the wide variety of methods and techniques presented in this book, the reader may feel at a loss and wonder whether he will be able to choose the method(s) best suited to his problem. In this chapter we will try to help the reader as well as specialists to select the appropriate methods.

First, we will list the merits and limitations of the various analysis methods and compare their main features. Second, we will define the dependability assessment procedure as well as its limitations. Then we will determine whether dependability measures are in agreement with measures provided by operating experience.

In addition, we will examine the problems raised by the organization of these studies during the lifetime of a system and within a company.

Again, because of the diversity of the methods and of the nature of industrial systems, our reflections in this chapter are provided as a guidance for the user of the methods though they are in no way to be regarded as universally valid.

22.2 ADVANTAGES AND DRAWBACKS OF THE VARIOUS ANALYSIS METHODS

We will rapidly survey the main advantages and drawbacks of the various methods.

- **Failure modes and effects analysis (FMEA)** (see Chapter 7)

Advantages
- systematically identifies the effects of the component failure modes;

- provides a preliminary inventory of the failure modes (and their causes) which affect dependability;

- can be complemented by an FMECA to classify failure modes according to the probability–severity criterion.

Drawbacks
- can be lengthy to little avail;

- identifies only single failures;

- does not directly produce a model for the quantitative assessment of dependability.

- **Success diagram method (SDM)** (see Chapter 8)

Advantages
- the success diagram can be constructed almost directly from the functional diagram;

- the failure combinations jeopardizing the system function can be easily identified;

- the reliability of a non-repairable system can be easily computed from the success diagram;

- the availability of a repairable system can be approximated when components are independent of one another.

Drawbacks
- is not adapted to the analysis of complex relationships between the failure effects and causes;

- is not adapted to repairable systems with complex maintenance strategies.

- **Cause tree method (CTM)** (see Chapter 9)

Advantages
- identifies all the failure causes (and their combinations) leading to an undesirable event;

- identifies some forms of dependencies between failures;

- provides a list of the minimal cut sets of an undesirable event;

- generally permits the calculation of the undesirable event probability;

- is applicable to a wide variety of systems.

Drawbacks
- can lead to very large cause trees which are then unmanageable even with a computer program;

- for a complex system (such as a nuclear power plant or an aircraft), it does not always identify failure sequences leading to an undesirable event;

- cannot deal with repairable systems with complex maintenance strategies.

- **Truth table method (TTM)** (see Chapter 10)

Advantages
- systematically examines all the component failure combinations;

- probabilities can be easily computed from the truth table;

- the decision table method derived from the TTM can take into account components with more than two states and allows an automatic analysis to be performed for some systems.

Drawbacks
- is not applicable to systems with numerous components;

- is not suited for the analysis of complex relations between the failure effects and causes;

- is not adapted to the analysis of a repairable system.

- **Gathered fault combination method (GFCM)** (see Chapter 11)

Advantages
- can identify all the malfunctions and failure modes of a system;

- provides a detailed analysis of the causes and effects of failures (and their combinations);

- can analyse a set of closely interacting elementary systems and identify all their interactions:

- can theoretically identify the initiating events for a system.

Drawbacks
- the analysis may be very unwieldy and lengthy;

- can hardly be used for systems that are 'too' sequential or require complex maintenance strategies;

- for the quantitative analysis, other methods (such as the CTM and the SSM) are sometimes needed.

- **Consequence tree method (CQTM)** (see Chapter 12)

Advantages
- identifies all the consequences of an initiating event;

- yields the probabilities of unacceptable event sequences; this calculation is easy when all the events are independent, but more complicated if they are not (for instance, use of semi-Markovian processes);

- shows the interactions between elementary systems.

Drawbacks
- there is no way of proving that all the initiating events have been considered;

- should generally be used in conjunction with other methods for the analysis of generic events;

- cannot allow for very complex maintenance strategies for the elementary systems provided to control the intitiating event.

- **Cause–consequence diagram method (CCDM)** (see Chapter 13)

Advantages
- simultaneously identifies the causes and consequences of an initiating event;

- brings to light the dependencies between the causes and consequences of an initiating event;

- theoretically, has the same advantages as the CTM and the CQTM.

Drawbacks
- the quantitative analysis combines the quantitative analyses of the CTM and CQTM methods and the computer programs are limited to specific cause–consequence diagrams;

- is ill-adapted to complex sets of elementary systems.

- **State-space method (SSM)** (see Chapter 14)

Advantages
- identifies all the operating and failed states of a repairable system as well as all their transitions;

- provides measures of the reliability, availability and maintainability of a repairable system;

- takes complex maintenance strategies into account.

Drawbacks
- the quantitative analysis can become very complex, if not impossible to perform, for a system with a large number of states;

- the quantitative analysis is fairly easy only for Markov and semi-Markov processes.

As regards the other methods, we shall refer the reader to the corresponding chapters. We should, however, again draw his attention to the considerable merits of the computerization and automation of analysis methods (see Chapters 20 and 21).

22.3 COMPARISON OF THE VARIOUS ANALYSIS METHODS

We will now compare the main characteristics of the predictive analysis methods discussed in the second part of the book. We will draw a distinction between:

- inherent features which, normally, are independent of the system that must be analysed; and

- features which depend upon those of the system to be analysed.

This time again the comparison will not encompass the preliminary hazard analysis method. Indeed, this method should rather be regarded as a preparatory analysis method. Moreover, because of their specificity, automatic analysis methods will not be examined in this section (see Chapter 21).

22.3.1 Inherent features

Table 22.1 illustrates several aspects of this comparison. Two columns should be commented upon:

- *Nature of the method*. Is the method deductive, inductive or both?

    The adopted classification corresponds to the principles of the method. In

Table 22.1 Comparison of the various analysis methods

Inherent features	Nature of the method			Events taken into account at the beginning of the analysis	Identification of failure combinations	Representation means associated with the method
Methods	Inductive	Deductive	Both deductive and inductive			
FMEA (FMECA)	Yes			Components failure modes	No	Tables
Success diagram method (SDM)		Yes		Functions	Yes	Sucess diagram
Truth table method (TTM)	Yes			Components failure modes	Yes	Truth table
Cause tree method (CTM)		Yes		Undesirable event	Yes	Cause tree
Consequence tree method (CQTM)	Yes			Initiating event	Yes	Consequence tree
Gathered fault combination method (GFCM)	Yes			Components failure modes	Yes	—
Cause–consequence diagram method (CCDM)			Yes	Initiating event	Yes	Cause–consequence diagram
State-space method (SSM)	Yes			Operating and failed states	Yes	State graph

fact, methods are both deductive and inductive. We will mention, for instance,

- the FMEA: the identification of the failure causes is essentially deductive;
- the CTM: the analyst cannot help examining the effects of the intermediate and basic events he has identified, and this is perfectly desirable.

• *Representation means associated with the method.* The GFCM is the only method without specific representation means. It is essential to draw a distinction between the method and its means of representation. Thus, for instance, the cause tree representation may well be used to represent a global gathered fault identified with the GFCM.

22.3.2 System-dependent features

Some properties of the system help the analyst to choose the analysis method he will use. Let us examine them.

- *Non-repairable system.* All the components of such a system are considered to be non-repairable for a given function. Note that a system can be regarded as non-repairable for a given function and repairable for another one. Take the following example: in the event of an accident A_1, a safety system, which can be repaired, must operate for one hour; but as its operation time is too short to permit maintenance, the system is considered to be non-repairable for this mission. However, the same system can be regarded as repairable if it is required to operate over 24 hours in the event of an accident A_2.

  These systems are the simplest to analyse with nearly all methods. Note that the state space method (SSM) can be used—with caution, however—for partially repairable systems, some states having then no repair transition.

- *Repairable system.* To various degrees, all the methods are applicable to repairable systems:
  - the FMEA (or FMECA): by identifying the failure modes requiring maintenance operations, this provides an indication—though partial—of the system maintainability;
  - the SDM: the availability of a repairable system can be calculated as long as the components are independent;
  - the CTM: this can deal with repairable systems even if they are complex. Component independence is assumed for the quantitative analysis. When complex repair strategies (e.g. presence of only one repairman to fix numerous components) must be taken into account, the SSM should be used;

- the CQTM: the initiating event and generic events can be associated with repairable systems. The independence of these events is however generally assumed;
- the GFCM and the CCDM: in the qualitative analysis, the component repairability can be allowed for;
- the SSM: this is the method best adapted to repairable systems; it is irreplaceable when complex maintenance strategies have to be taken into account.

- *Sequential system.* It should be remembered that the failure of a sequential system mission strongly depends on the order in which failures occur. Generally, inductive methods are better adapted to these systems than deductive methods, and this is all the more so as the system complexity increases. Indeed, it is difficult to identify failure causes, for instance with the CTM, when the occurrence sequence of these causes is complex. This is why analysts have long given up the idea of constructing the tree of the direct causes of an airplane crash and of the core melt in a nuclear power plant!

- *Phased-mission system.* The analysis of such a system is very tricky. Under certain conditions, the only methods which can deal with such systems are the CTM, the CQTM and the SSM.

22.4 CRITERIA FOR CHOOSING THE METHODS

The choice of one or several analysis methods depends on numerous criteria; they are imposed by compelling technical reasons or derive from the analysts' habits! We will classify those selection criteria which are often decisive:

- *Criteria connected with the goals of the predictive analysis*
    - is it a reliability, availability, maintainability or safety assessment?
    - is the undesirable event known? Can there be other undesirable events?
    - should a dependability measure be calculated? Would a qualitative analysis suffice?

As a rule, the list of possible methods can be shortened by precisely defining the goals of the analysis.

- *Criteria connected with the systems to be studied.* The nature of the system and its degree of complexity are important factors in the choice of the methods. Note that complexity is not necessarily directly related to the number of components. It rather depends on the nature of the component functional and structural interrelationships: a system with a large number of components that can be grouped into macro-components may well prove easy to analyse. The existence of redundancies, the possible intervention of man during operation, the component repairability, the existence of software, etc. all complicate the analyses.

Before choosing the methods, a preliminary functional and technical analysis should be peformed.

The system properties (non-repairable, repairable, sequential,...) which will determine the choice of the methods (as seen in Section 22.3) are then examined.

- *Criteria connected with existing studies*. Sometimes designers and plant operators have already predicted faults, accidents, etc, and taken them into account in the design and operation, before dependability assessments were made. For instance, as regards a nuclear power plant, the list of the incidents and accidents allowed for in the design is a preliminary inventory of initiating events!

As a rule, all the knowledge acquired on the prediction of the system behaviour plays a significant part in the choice of the methods.

- *Criteria related to the study means*. Where goals are concerned, these methods vary in efficiency and in the means required. The methods should be adapted from the start to the means available, in terms of specialists as well as dependability data. Thus, sophisticated quantitative methods should by no means be adopted if there is too much uncertainty in the data available!

22.5 DEPENDABILITY ASSESSMENT: APPROACH DEFINITION

The dependablity assessment, based on a predictive analysis, consists of the following steps:

- functional and technical analysis;
- qualitative analysis;
- quantitative analysis;
- synthesis and conclusions.

Generally, to achieve the qualitative as well as quantitative goals of the dependability assessment, the analysis methods described in the second and third parts should be combined.

By way of an example, we will mention the combination of methods used for the assessment of a safety-related system in a nuclear power plant projected by EDF. Consider a thermal-hydraulic redundant safety-related system. It is regarded as a non-repairable system because, after an accident, it operates over a time too short for the system to be repaired should it have a failure, even if partial. The analysis methods are as follows:

- *FMEA*. Performed for the main system components, this serves to check that no relevant failure mode results in the failure of a mission and helps to perform the following analyses.

- *CTM*. This is used to analyse the causes of failures resulting in the failure of the system missions. Quantification consists in integrating the failures identified with the following analyses.

- *Common-cause failure analysis*. Such failures may prevent the system from having a high reliability record. Two methods are used:
  - Predictive analysis according to generic causes (see Section 15.8);
  - Quantitative analysis of certain common-cause failures using the beta-factor or the shock methods (see Section 15.9).

  The identified common-cause failures are integrated into the cause trees.

- *Human error analysis*. Some human errors can be made during periodic tests, preventive and unscheduled maintenance operations. A probabilistic human reliability assessment (HRA) will therefore be performed for these operations. The identified human errors are integrated into the cause trees.

Cause trees may be modelled with expert systems such as the EXPRESS software (see Section 21.5). This will greatly facilitate the subsequent updating of the model.

The assessment will include two additional steps:

- The component dependability data are collected on similar operating systems.

- The operating experience, that is all the incidents recorded on similar systems, is analysed in detail. Even when no complete failure of the system has been observed, this analysis is worthwhile. Indeed, by analysing the partial failures, we can check whether the dependability model has anticipated such failures both quantitatively and qualitatively; parts of the cause tree can thus be validated.

Finally, the quantitative analysis which evaluates the failure probabilities of the system missions includes:

- an estimation of the importance measures (see Appendix 3);

- studies of sensitivity to significant and uncertain hypotheses or data;

- calculation of uncertainties associated with the final probability (see Appendix 4).

Thus the dependability specialist will have to define an actual approach to dependability assessment, including for instance:

- analysis methods used as well as their sequencing, sensitivity studies if any, uncertainty calculation, etc;

- supporting activities: collection of dependability data, reliability test performance to gather dependability data, etc;

- partial or global validation against operating experience analysis.

22.6 LIMITATIONS OF DEPENDABILITY ASSESSMENT

We shall thoroughly examine important points or problems in dependability assessment which usually bring its limitations to light.

22.6.1 Limitations of the qualitative assessment

- **Not everything can be anticipated!** Although this goes without saying, analysts should always be reminded of this fact because they tend to get carried away by the considerable potentialities of the methods capable of identifying an impressive number of minimal cut sets for an undesirable event!

Comparison of the results with the operating experience fortifies our confidence in these methods, but at the same time calls for a certain humility, especially as regards the identification of common-cause failures and human errors.

First, we should specify the required degree of accuracy in the prediction, and, to avoid any misunderstanding, we shall take the example of an operator who misses out a step in a test procedure because he wants to be finished with the test and to go home to solve his family problems!

The analyst has merely to anticipate the omission of a step in the procedure and to assess the probability order of magnitude precisely; in this case, the cause is not important so long as the analyst has anticipated the effect and its probability. Indeed, never mind the cause when the effect and its probability have been foreseen!

- **Conservative assessment** The analyst, feeling quite uncertain about the effects, will sometimes be tempted to assume that the failures will result in the undesirable event, for lack of a functional study to the contrary. Generally, dependability assessment demands a thorough knowledge of the system operation and of the effects of the most insignificant failures of its components; as a matter of fact the designer has not necessarily acquired this knowledge when the system was designed. Indeed the designer, in support of his design, needs only to demonstrate he has provided margins with respect to prescribed criteria, for instance to allow for accident conditions. However, the dependability specialist will wish to know and assess these margins more precisely. As a result, dependability models are often conservative, at least in the way identified failures are taken into account!

- **Failure interdependency** It is vital to determine this interdependency since it can deeply affect the conclusions of the dependability assessment. Common cause failures and human errors require special attention. Don't forget that, generally, it is assumed there are no design errors.

22.6.2 Limitations of the quantitative assessment

- **Dependability data** Quantitative assessment is very much conditioned by the quality of the dependability data. To make up for the bad quality of the data, a sensitivity study can be performed.

As a rule, the dependability data used concern similar components: when the latter are taken into account to assess a project under study, the new components are assumed to have at least an as good dependability as the older ones. This implies, among others, numerous hypotheses such as the following:

- quality assurance and preoperational tests are maintained at least at an equivalent quality level;

- the equipment will be operated (tests, preventive maintenance, scheduled maintenance, ...) under the same conditions.

Uncertainty as to these data results both from the randomness of the failures and from our limited knowledge of the environment and conditions the components will be operated in. These studies have the merit of highlighting and taking account of these uncertainties, thus compelling the analyst to specify the limitations of his knowledge and expertise.

Data, even when marred by a high level of uncertainty, are evidence of an improved knowledge and are preferable to the customary engineering judgement which usually amounts to nothing better than an opinion with little true foundation. Nevertheless, the analysts must be warned against excessive faith in figures!

- **Dependability measures** These measures must be assigned uncertainty levels, especially for all the decision-oriented studies.

As already underlined, sometimes predictive values are only approximated; this is not a real drawback when an order of magnitude is sufficient.

As a rule, the constant and detailed comparison of the predictive model and the operating experience helps to reduce these limitations.

22.7 VALIDATION OF THE DEPENDABILITY ASSESSMENT

Is there a good fit between calculated dependability measures and observed measures derived from operating experience? This fundamental question raises the problem of the dependability assessment validation we shall examine now.

Very many relevant failures or faults which had been identified in the course of a dependability assessment have actually occurred after that! (for instance the DC 10 air crash of 1974, see Section 15.3). For this reason, we shall essentially speak of the quantitative analysis and examine comparisons concerning, first, electronic, chemical and nuclear systems and, second, industrial facilities like nuclear power plants.

The first comparison concerns analogue and digital telecommunications equipment (printed circuit boards) used by the French Postal and Telecommunication System [1]; it covers 37 different components.

The predicted failure rate is determined using the Parts Count Method (see Section 8.5) applied to data from the CNET data base (see Section 5.4.1); the operating failure rate is deduced from the data collected during operation.

We shall use the term 'validation ratio' to mean the ratio of the observed value (V_0) to the calculated value (V_p). In 75% of the cases this ratio is below 2. In the remaining 25%, the high operating failure rates observed are due to component faulty design or manufacture.

The second comparison concerns 130 very different types of systems from the nuclear and chemical industries. It was perfomed by the National Centre of Systems Reliability (United Kingdom) [2]; 146 calculated reliability values were thus studied. Apparently:

- 63% of the calculated values are within the interval $[V_0/2; 2V_0]$,
- 93% of the calculated values are within the interval $[V_0/4; 4V_0]$.

The third comparison was effected on the elementary turbine protection system of the 900 MWe pressurized water reactor-type nuclear units, thirty of which are operating in France. This system is responsible for unscheduled nuclear power plant shutdowns, at a rate of 0.5 per year and per unit, due to failures of the system components. The predicted number of shutdowns is one per year and per unit [3]. The validation ratio is therefore approximately 0.5. The analysis method adopted was the cause tree method and the reliability data used were the habitual generic data used for this type of components.

Thus the studies confirm that the quantitative analysis can generally predict the reliability of a system to a reasonable degree (for instance, values differing by a factor of about 2–4). Such good results are certainly not possible for highly redundant systems where the share of common-mode failures or of human errors can be very important and even predominant. To the author's knowledge, today there are no extensive studies similar to the second mentioned above for such systems. However, in the nuclear field, experience shows that the reliability of elementary safety systems can be assessed with values differing by a factor below 10.

We shall now examine the assessment of major risks; due to the still limited number of such risks outside the nuclear field, we shall limit our comparison to this sector.

Note that the evaluated annual probability of a core melt is often in the range of 10^{-3}/year and 5×10^{-5}/year (see Section 19.4). Only a few PSAs give lower probabilities.

International experience with nuclear power plants amounted in 1986 to some 3800 reactor-years; note that over 350 power plants were operating in 1986. Two major core melt accidents have so far taken place: one at the American Three Mile Island nuclear power plant in March 1979 (see Section 15.3) and one at the Chernobyl nuclear power plant in the USSR in April 1986 [4]: no PSA had been performed for these facilities so that we cannot say what their predicted risk level was before the accident. Moreover, the risk levels, of course, vary considerably from one type of nuclear power plant to another. From the international operating experience, the annual core melt risk in a nuclear power plant is estimated at 5×10^{-4}/year; this risk does not contradict the results of the predictive analysis.

Let us now further compare the probabilistic risk assessment performed on the Surry plant in 1975 (Rasmussen Report) and the accident scenario of the Three Mile Island plant since the design of the latter is fairly similar to Surry's. In the Rasmussen report the initiating event of the accident sequence had been identified as a 'small break in the primary circuit' followed by a failure of the safety-related system (safety injection system); the probability of a human diagnosis error inducing an operator to stop the safety system had been deemed negligible. We have already stressed that precursor events had occurred (see Section 15.3). They informed the analyst that the operator's behaviour was not as predicted since the operator sometimes found reasons for stopping an important safety system. If this new piece of information had been introduced in the previous model, that is the Rasmussen report, the probability of the future accident scenario in the Three Mile Island power plant would have proved important. A similar conclusion can be drawn from the analysis of the DC 10 jumbo jet crash at Ermenonville in France, which is explained in Section 15.3.

This goes to show that *the validation of the dependability model must be a constant concern* and that the model should be updated to take the operating experience into account.

As already mentioned, the predictive dependability analyses to be performed and improved must be based on a number of hypotheses on the failure effects, the system boundaries, the component repairability, the human factor impact, the component failure rates, etc. These assumptions may vary in the course of the system operation and analyses must be regularly updated to take these changes and alterations into account.

Model validation against operating experience is the often indispensable guarantee of the model quality and of its consistency with the operating reality.

22.8 ORGANIZATION AND MANAGEMENT OF DEPENDABILITY ASSESSMENT

Assessment methods can bring the best and the worse. We shall therefore examine now some principles demonstrated to be important by experience and whose application increases the efficiency of the methods.

- **Predict dependability: a specialist's job** Only dependability specialists who deal every day with the occurrence of failures (and their combinations) can use the methods!

Specialists who design or compute the system operation often have difficulties in admitting the occurrence of failures ('it can't happen') and will tend to look upon failures as nonsensical! This is normal since their mind is set on making the system operate. It is hard to have both approaches coexist within the mind of the same specialist!

Dependability specialists must have an insatiable thirst for technological knowledge, must be capable of rapidly assimilating a new technology, have a real power of abstraction and a good morale to stand constantly dwelling in a world of failures and catastrophes. Experience is essential in this field to avoid two extreme attitudes consisting in asserting:

- quantitative assessment is useless or not credible!

- figures are indisputable! Therefore specialists hang on to them as though their life depended on them.

And last but not least, specialists must be intellectually honest in this field where engineering judgement is often resorted to and where data can be tampered with.

- **A multidisciplinary activity** Dependability assessment is aimed at covering as many factors as possible during the system life cycle. To reach this goal, specialists have to work at a great detail level and have to address problems as diverse as component behaviour under accident conditions, the format of a procedure, preoperational tests and their efficiency, the rate of periodic tests, etc. Dependability assessment is, therefore, intrinsically a multidisciplinary activity. Dependability specialists must work in close collaboration with the designers, the plant operators, the operation specialists, the ergonomists, etc.

Dependability assessment may structure the various related fields of activity: new working and communication networks may emerge within the company. First, it may help to instil a greater rigour in work and spread new dependability management methods and second, it can contribute to the birth of a true dependability culture.

• **A necessary organization** Considering what has been pointed out earlier, any extensive program supposes that a real organization of these assessment studies be implemented within the company. Due to the potentially important repercussions such studies may have, the selected organization, even if it is well adapted to the problem it is meant to solve, must take account of the knowledge acquired by the company on dependability and must be accepted by those who will have to use the study lessons. We must emphasize this last point: it is sometimes vital that the most relevant lessons reach the decision makers at the highest levels of the hierarchy; the space shuttle accident (in 28 January 1986) clearly demonstrated the importance of this organizational aspect [5].

• **Quality assurance of these studies** In view of the considerable bulk and potential repercussions of the assumptions used in the studies as well as of their great variety, a real quality assurance program must be devised. A large amount of information and many lessons are derived from the verification and validation of the quantitative analyses performed by the specialists. The detailed analysis of the quantitative results and their systematic verification against approximated calculation results are elementary precautions that must never be omitted! Note that, in the nuclear safety field, independent control tends to become general practice.

Whenever possible, specialists must strive to validate their predictive analysis, to check that the system component behaviour coresponds to what they have predicted. We shall discuss this point further in Chapter 23.

• **Dependability assessment management** Generally, the assessment is valid only at a given time in the system life cycle. The gradual change of the project during the design phase and, later, during the operating phase makes an updating of the dependability assessment necessary; this may have an impact on the methods adopted and encourage the use of computer programs to facilitate the assessment updating.

REFERENCES

[1] J. Y. Planchais and R. Prato (1986) *Etude, par un constructeur de matériels de télécommunications, de la comparaison du taux de défaillance en exploitation par rapport au prévisionnel, à l'aide du système SADE de l'Administration Française des PTT*, 5ᶜ Colloque International de fiabilité et de maintenabilité, Biarritz, France.

[2] E. R. Snaith (1981) *The Correlation Between the Predicted and the Observed Reliabilities of Components, Equipment and Systems*, NCSR R18, National Centre of Systems Reliability, Culcheth, Warrington.

[3] M. Bouissou and E. Bourgade (1985) *Etude de fiabilité du Système GSE de protection turbine des centrales 900 MWe*, EDF, HT/13/89/85.

[4] Société Française d'Energie Nucléaire (1987) *Actes de la journée 'L'Accident de Tchernobyl: les aspects de sûreté et d'environnement'* Paris.

[5] W. P. Rogers (1986) *Report of the Presidential Commission on the Space Shuttle Challenger Accident*, US Government printing office: 1986 0-157-336.

23
DEPENDABILITY ASSESSMENT APPLICATION

23.1 INTRODUCTION

Dependability is the science of failures, their study, assessment, mesurement and prediction. Since its inception in 1940, the aim has been to forecast the future of a system from the analysis and modelling of past and future systems.

From the 1960s onwards, the engineers' ambition—then a true Utopia—to quantify resulted in the considerable development of qualitative analysis methods. In 1982, Favez insightfully summed up the situation at the time in the following words: systems dependability ... 'has won recognition by overcoming the paradox whereby a major qualitative progress originated in what was initially thought of as a quantification requirement.' In general, dependability is 'then more a mental attitude towards failures than an accurate scientific tool. On this account, reliability has become a very tangible and precious industrial reality' [1].

Since the beginning of the 1980s, there has been an ever-increasing influence of quantitative assessments on the design and operation of engineering systems. The identified measurable quantities are used to make technical choices and, much as performance and cost, are regarded as essential parameters for the description of systems.

How can these qualitative and quantitative assessments be used? And to what end? This chapter provides a brief and synthetic answer to these questions since we will mention only general-purpose applications. Indeed, specific answers can be found throughout the book.

23.2 SYSTEM DESIGN

23.2.1 Dependability assessment in design

We shall examine the main uses of dependability assessments in the design of a system.

- *Comparative evaluation of different projects*. A designer having to make his choice among several different system designs performs a comparative dependability assessment. The quantified evaluation can be carried out in conjunction with an economic comparison of the project costs. The latter is very meaningful since it does not depend much on the dependability data taken into consideration. However, this is not always true, so comparative estimations which are based solely on one figure which may be sometimes uncertain should be used with care.

- *Assessment of a project dependability*. Naturally, methods used for such an evaluation may change with time as the project takes shape. The first assessments must be performed very early, mainly to check whether the major principles chosen are compatible with the dependability requirements. Thereafter the project must be examined in detail to make sure that these requirements are still complied with in the last development phases. We must always remember that, if a single failure with a high occurrence probability is overlooked, the most meticulously developed projects can be jeopardized, especially if safety is involved.

Usually predictions cover a system's useful life; their aim is not to forecast the probability of early failures. Predicted reliability should be compared only with operational reliability after early failures have been eliminated, that is after preoperational tests have been performed.

We must recall that component dependability data usually corresponds to the useful life period. In this case, the better the qualification and preoperational tests are designed and carried out, thereby eliminating defects difficult to quantify, the more accurate the predictions will be.

This brings us logically to the subject of assessment and quality assurance. The better the quality assurance, the better the predictions. Dependability assessment studies and good predicted dependability values should not be reasons for disregarding quality assurance but, on the contrary, should motivate its enhancement. The following example used to illustrate this point is an incident drawn from real life. A predictive reliability analysis performed on a handling crane showed that the crane was perfectly reliable, equipped as it was with devices such as redundant wire ropes and braking systems. But during the first test performed with a nominal load, the load fell. The surprised user blamed the failure on the predictive reliability analysis! But the incident analysis revealed that the reduction gear of the handling crane had been undersized due to an error in the manufacturing drawing while the design study had been correctly performed. In fact quality assurance had been defective since it had left an error in the manufacturing of a major mechanical part undetected. The predictive analysis was therefore not to blame since the analysis had been performed based on the assumption that design and execution were satisfactory. Moreover, the reduction gear was theoretically the most crucial component, so it should have been subject to stricter quality control! This goes to show that dependability assessment can help to rank priorities in quality assurance and to identify components which need particular attention.

There is a need for a real **dependability program** if dependability assessment is to be made the most of. This program must provide for the necessary organizational structure, responsibilities, procedures, activities, expertise and means which will insure that a system will satisfy the dependability requirements of a particular project.

Likewise a **policy of dependability assurance** must be implemented to guarantee that the required goals will be attained and, at the same time, that dependability will be what it is supposed to be. The dependability assurance program will include among others:

— dependability auditing;

— project or design review;

— creation of a dependability data collection system and later its use;

— continuous monitoring of dependability measures.

This assurance policy is enforced right from the design stage and must, of course, be maintained during operation.

However, during the design stage, there is a danger of the analyst getting carried away by the powerful methods he uses and the extremely high number of failures or accidents envisaged. The analyst must therefore keep in mind, first, that it is virtually impossible to foresee all the possible failures or accidents that reality and hazard may have in store and, second, that his dependability model may be based on numerous hypotheses which may not always be valid. The validation of the model against the operating experience will therefore provide an extra guarantee. This approach will be discussed in Section 23.3.

The role of the dependability assessments performed at the design stage raises the problem of the approach used during this phase. This problem will be examined below.

23.2.2 Deterministic design and probabilistic design

Any system design must take into account the uncertainties linked to the system manufacture and operation. For instance its internal failures or its operating conditions are random in essence. To solve this problem there are two types of design methods: the 'deterministic' design and the 'probabilistic' design.

- *Deterministic design*. Briefly, this method consists in choosing a reference or envelope scenario among many possible scenarios. Therefore, if the system is reliable and safe in the reference scenario, then it will be considered to be reliable and safe in 'most' scenarios regarded as less severe. Thus, in this approach the scope of possible scenarios is replaced by a deterministic scenario looked upon as equivalent. This design technique makes systematic use of practices developed over the years in the fields concerned.

- *Probabilistic design*. Briefly, it is based on dependability assessment. The method consists of identifying all possible scenarios and then in assigning a probability and consequences to each of them. Using a decision criterion combining both the probabilities and the consequences, the probabilistic method tries to define 'optimal' answers for all the scenarios chosen.

Any design is implicitly based on probabilistic reasoning. Take a car for instance: the fact that a car is equipped with only one spare tyre implicitly implies that an 'acceptable risk' will be the simultaneous puncture of two tyres. The label 'probabilistic design' will therefore be reserved for a design based, on the one hand, on dependability assessment, and, on the other hand, on the fulfilment of quantified dependability objectives.

The deterministic method is generally easier to handle, but the choice of a reference scenario is rather tricky. The probabilistic method *a priori* includes all scenarios but the difficulty lies in the choice of the criterion. Moreover this method is much more complicated.

Deterministic and probabilistic design methods are increasingly used in conjunction. In fact the two methods are complementary for the deterministic design method is relatively simple, so combining it with the probabilistic design guarantees that a minimal level of dependability will be sought, reached and preserved.

In Chapter 19, we saw that quantified safety goals were being increasingly used. Likewise, the introduction of quantified objectives becomes widespread in the fields of reliability, availability and maintainability. This can be explained, on the one hand, by the desire to help the designer to make rational choices in his job and, on the other hand, to guarantee the user a minimum level of dependability. Below are examples of such goals:

- *Telephone system switches*. The sum total of downtime hours due to equipment failures must not exceed 2 h for every 40 years of operation [2].

- *Off-shore oil-rigs*. The dynamic positioning systems are supposed to have an MTBF of at least 100 000 h [3].

An increasing number of contracts include clauses which contain such goals (example, reliability clause [4–6]); assessment and goal verification methods are normally included in the contracts together with penalties when objectives are not fulfilled. The fear of sometimes heavy penalties induces decision makers to pay more attention to the contents of these clauses.

23.3 SYSTEM OPERATION

We shall now examine the main applications of these methods to system operation.

- **Assessment of an operating system** Even though an analysis of the operating experience (incidents, accidents) of an industrial system can be used as a basis for appraisal, it is sometimes necessary to perform such predictive assessments on operating systems. For example, this approach is used to find out the occurrence probability of a newly identified hazard in high-risk plants. In this case, the analysis methods are practically the same as those used to evaluate a project: however, dependability data are specific to the plant and it is normally possible to take real operating conditions into account.

- **Analysis of the operating experience** Consider an industrial system with safety problems; two different cases will be examined:

—There is no model to predict accident risks: the occurrence of certain incidents indicates clearly that the plant has a safety problem although the margins between these incidents and a serious accident cannot be readily assessed. The incident is then regarded as an initiating event or as an event likely to jeopardize the smooth operation of the plant after the occurrence of another initiating event. The accident sequence may be predicted, for example by using the consequence tree method: from the calculated accident probabilities, the user can derive the margins left in the design and check their conservatism so that he can determine whether the plant design should be modified. Thanks to this approach, incidents can be ranked according to the severity of their consequences, and, in particular precursory events can be identified.

—There is a model to predict accident risks: the operating experience is integrated into the model. When the event introduced (e.g. a new failure mode, failure mode with a higher probability, ...) causes an important increase in the measured accident risk, the event is considered to be a precursory event. As it deeply affects the evaluated risk, such a precursory event requires that the plant be modified to prevent its reccurrence.

As a rule, thanks to these assessments, incidents can be ranked in order of importance so as to deal first with the most dangerous.

- **Definition of periodic tests** Normally, the safety systems of industrial plants are standby systems which need to be tested regularly. Appendix 2 discusses the availability assessment of such systems. There is an optimum test interval which can be determined from certain dependability parameters and which minimizes system unavailability.

- **Definition of operating rules** Complex industrial systems must be operated in conformity to certain operating rules also known as 'technical specifications' in nuclear plants. Some of these rules can be derived from dependability assessments. Consider a nuclear power plant where a safety-related component is unavailable. What should the operator decide to do? Due to component redundancy, power operation of the plant can continue; the operator, however, now runs an additional risk of accident, for if a precursory event occurred, his means to face the problem would be lessened.

After assessing the additional risk, the operating conditions will be limited in an effort to bring risks down to acceptable levels. As a result appropriate decision rules are derived: the methods and examples are explained in detail in Appendix 5.

REFERENCES

[1] B. Favez (1983) *Les études de fiabilité: utopie ou réalité industrielle?* International Reliability, Availability, Maintainability Conference, Toronto, Canada.
[2] R. Lavergne, P. Mahieu and J. P. Cremieux (1982) *System E12: Reiability of Hardware and Software Sophistication*, Third International Conference on Reliability and Maintainability, Toulouse, France.
[3] A. Leroy (1982) *Utilisation des méthodes probabilistes pour assurer le sécurité des installations et des opérations pétrolières en mer*, Revue de l'énergie, No. 346, August–September.
[4] J. P. Plantard (1978) *Les Clauses de fiabilité dans les contrats*, Colloque International sur la fiabilité et la maintenabilité, Paris.
[5] J. G. Ligeron and A. Delage (1982) *Les clauses de fiabilité dans les grands contrats à l'exportation, leurs contenus et leurs pièges*, 3ᶜ conférence international sur la fiabilité et la maintainabilté, Toulouse, France.
[6] J. N. Baismaison (1982) *Recommandations pour l'élaboration de clauses de fiabiité opérationnelle, de coût de maintenance, de disponibilité dans les contrats de fourniture d'équipments aéronautiques* (1982) Groupe de travail Fiabilité et Maintenabilité de la Commission Technique du Groupe des Equipements du GIFAS, 3ᶜ conférence internationale sur la fiabilité et la maintenabilité, Toulouse, France.

Part 6
CASE STUDIES

24

ANALYSIS OF A SET OF ELEMENTARY SYSTEMS BY DIFFERENT METHODS*

24.1 GOALS

Among the methods presented in the previous chapters, which should be chosen for a particular purpose? Should one construct a cause tree or a consequence tree or apply the gathered fault combination method (GFCM)?

Such are the questions an analyst may ask when he must assess the dependability of an actual system. Each system being specific, we will not in this chapter dictate a given approach. Using a deliberately simplified example, we will endeavour to:

- illustrate the major advantages and drawbacks of the methods described in the second part of the book: some difficulties will be mentioned, and the reader will most probably find others;

- show that the various methods yield the same result but in a very different way.

24.2 PRESENTATION OF THE SYSTEM

The industrial equipment considered is used to cool nitric acid before it enters a chemical reactor. It is similar to the equipment studied and reported on in *IEEE Transactions on Reliability* [1–4] apropos of the digraph method for the computer-aided synthesis of fault trees. This example has been largely used at Electricité de France for training purposes [5, 6].

---

* This chapter was written in collaboration with Mrs Ancelin (Electricité de France)

24.2.1 General description

The studied equipment will be regarded as a system made of elementary systems. It is shown in Figure 24.1. The nitric acid outlet temperature T_2 is controlled by a control valve in the cooling water system. In addition, a safety system is provided to cut off the nitric acid feed if the water flow rate is insufficient.

The system is therefore essentially provided to maintain the value of the nitric acid temperature T_2 within an acceptable range. Once the main system function has been clearly formulated, the degraded condition 'failure to maintain temperature T_2 within an acceptable range' can be defined.

Since a preliminary hazard risk analysis has shown that an excessive temperature in the reactor could result in an explosion, the obvious undesirable event is 'excessive rise in temperature T_2.'

In the case of more complex systems fulfilling several functions or interacting with other systems, inventorying the undesirable events is not always so easy. Deductive reasoning is not enough and, as seen above, the GFCM may be of help for the identification of these events.

To clarify the way the system operates and is designed we will further specify the following points:

- The system is supplied by a nitric acid feed system whose temperature and flow rate can be modified; thus the inlet temperature and flow rate can increase, resulting in a rise in the nitric acid temperature T_2. In addition to the small so-called 'normal' disturbances, two types of changes in the inlet conditions can occur (Figure 24.2):

  —moderate increase in the flow rate or temperature: the control system has enough time to maintain temperature T_2 within an acceptable range: this is called 'correctable rise in temperature T_2';

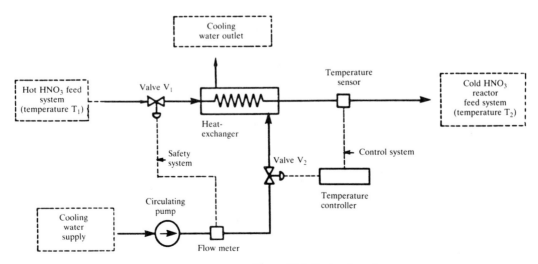

Figure 24.1 System layout.

PRESENTATION OF THE SYSTEM

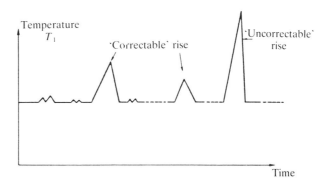

Figure 24.2 Variation of the upstream nitric acid temperature.

- significant increase in the flow rate or temperature: the control system does not have enough time to maintain temperature T_2 within an acceptable range even if valve V_2 opens completely. These conditions are considered to be exceptional and are referred to as 'uncorrectable' rise in T_2.

- Control valve V_2 opens or closes in compliance with the control system commands; under normal conditions, this valve is assumed to be half open,

- The system is supposed to operate continuously: it is assumed to be shut down after every detected fault, and transitions to no-risk conditions will not be taken into account. However, the system is assumed to continue operating after the loss of 380 V and 48 V electric power supplies (see Section 24.2.2),

- The operator's task is restricted to monitoring and control operations. In the event of a rise in the nitric acid temperature (whether this rise is correctable or uncorrectable), the operator has no time to take action. Under normal operating conditions, he periodically checks the safety and control systems. On this occasion, he may leave the temperature controller in manual operating mode so that this controller is unable to respond to the nitric acid temperature rise.

24.2.2 Detailed description of the elementary systems

In the present case, there are five elementary systems (Figure 24.3):

- *The nitric acid system (elementary system S_1).* S_1 comprises mainly:
  - on–off motor-operated valve V_1 (mechanical part and motor);
  - the heat-exchanger tubes. This is a counterflow double-wall tubular

632 ANALYSIS OF A SET OF ELEMENTARY SYSTEMS BY DIFFERENT METHODS

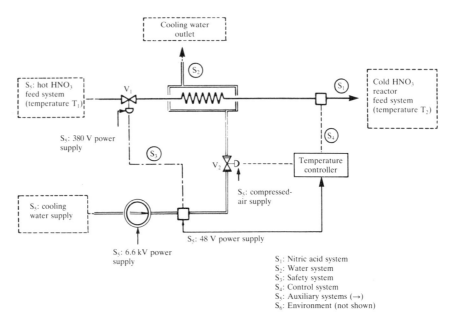

Figure 24.3 System breakdown into elementary systems.

heat exchanger. Nitric acid flows in the tubes at a pressure below that of the cooling water.

Note. For simplicity's sake, the pipes will not be analysed.

- *The water system (elementary system S_2).* S_2 mainly comprises:
    - the motor-operated circulating pump;
    - the air-operated control valve (mechanical part and control air);
    - the heat-exchanger shell.

- *The safety system (elementary system S_3).* S_3 comprises:
    - the flowmeter and associated instrumentation;
    - the relay circuitry closing valve V_1 on a low-water flow rate threshold. The 'low-water flow rate' threshold is defined as follows: as long as the water flow rate remains above this threshold, normal nitric acid temperature values cannot alone lead to an excessive temperature T_2.

- *The control system (elementary system S_4).* S_4 is made up of:
    - the temperature sensor recording temperature T_2 and sending an electric signal to the controller;
    - the controller which orders valve V_2 to open or close.

- *The auxiliary systems (elementary system S_5).* This system groups:
    - the electric power supplies including:

- the 6.6 kV system supplying the circulating pump motor;
- the 380 V system supplying valve V_1 motor;
- the 48 V system supplying the flow meter and associated instrumentation, the system safety relay circuitry and the control system;
- the compressed (control) air supply to the control valve V_2;
- the water supply which ensures an adequate suction pressure for the circulating motor-driven pump;
- the nitric acid feed.

- *The environment (elementary system S_6).* The system S_6 covers all the environmental conditions which, if they vary significantly, can affect the operation of the above-mentioned system (for instance, a significant change in the external temperature induced by a fire).

24.3 FAILURE MODES AND EFFECTS ANALYSIS

It should be borne in mind that the failure modes and effects analysis (FMEA) (see Chapter 7) is an inductive method used to thoroughly analyse a system. For each component, it identifies all the possible failure modes and causes as well as the failure effects on the studied system and on other systems. The detection means and possible operator actions are also examined.

The FMEA results are presented in Table 24.1. For simplicity, we showed only the main failure modes of the elementary system components. This analysis yields a preliminary model of the elementary system failures, for which the assumptions described below must be defined. As a matter of fact, the FMEA requires a thorough knowledge of the normal and upset operation of the studied system.

The inventory of the failure modes and their possible causes is generally based on the operating experience acquired with similar equipment. However, major difficulties are encountered in the following respects:

- defining the 'effects' of these failures: parameter variations (flow rate, temperature, heat removed, ...), component state changes, automatic or manual action, etc;

- assessing these effects.

Indeed, if the normal operating states of a system are generally well described in the documents at the specialist's disposal, this is not always true of the upset operating conditions, which cannot all be dealt with in these documents.

While performing the FMEA, the dependability specialists and systems specialists must therefore constantly remain in contact. Concerning failure combinations which have not been allowed for during system design, this dialogue allows the analysts to make hypotheses as realistically as possible.

Table 24.1 FMEA of elementary systems
ELEMENTARY SYSTEM: S_1 (nitric acid system)

Component identification (identification code, name, type, location)	Functions, states	Failure modes	Possible failure causes (Internal, external causes)	Effects on elementary system	Effects on other elementary systems	Observations
Name Valve V_1 *Type* On–off motor operated valve	*Function* Isolation of nitric acid feed line when water flow rate is insufficient *States* Open under normal operating conditions	• Inadvertent closing	• Mechanical defect • Electrical defect • Control defect • S_3 induced defect	• Nitric acid flow interruption		
	Closed upon low-water flow rate signal	• Failure to close	• Mechanical defect • Electrical defect • Control defect • S_3 induced defect • Loss of 48 V supply (S_5) • Loss of 380 V supply (S_5)	• failure of V_1 to close when water flow rate is insufficient, rise in T_2 uncorrectable by control system	• Control system (S_4) required to operate, opening of valve V_2	• When 48 V supply lost: loss of safety system and control system (V_1 and V_2 remain as they are)

Table 24.1 (cont.)
ELEMENTARY SYSTEM: S_1 (nitric acid system)

Component identification (identification code, name, type, location)	Functions, states	Failure modes	Possible failure causes (Internal, external causes)	Effects on elementary system	Effects on the other elementary systems	Observations
Name Heat-exchanger (heat-exchanger tubes) *Type* Double-wall	*Functions* Circulation and cooling of the nitric acid before it enters chemical reactor	• Internal leak	• Leaktightness defect (corrosion, vibrations, defective weld, . . .)	• Water ingress in tubes, violent chemical reaction, rise in T_2 uncorrectable by control system	• Control system (S_4) required to operate, opening of valve V_2	
		• Plugging	• Foreign body, fowling	• Reduction of nitric acid flow rate	• Control system (S_4) required to operate	
		• Loss of cooling function	• Fire in heat-exchanger room			

Table 24.1 (*cont.*)
ELEMENTARY SYSTEM: S_2 (water system)

Component identification (identification code, name, type, location)	Functions, states	Failure modes	Possible failure causes (Internal, external causes)	Effects on elementary system	Effects on the other elementary systems	Observations
Name Pump *Type* Motor-operated pump	*Function* Cooling water circulation	• Pump shutdown	• Mechanical defect • Electrical defect (motor) • 6.6 V supply loss (S_5)	• Water flow interruption	• Safety system required to operate (S_3), nitric acid flow interruption due to valve V_1 closing (S_1)	• When S_3 does not operate: uncorrectable rise in T_2
Name Valve V_2 *Type* Air-operated control valve	*Function* Water flow regulation according to the nitric acid temperature T_2 *States* The valve position is set by S_4; when 48 V supply (S_5) is lost valve V_2 remains as it is	• Valve remains stuck as it is • Inadvertent closing	• Mechanical defect • Control defect • S_4 induced defect • 48 V supply loss (S_5) • Mechanical defect • Control defect • S_4 induced defect • Air pressure control system loss (S_5)	• Water flow rate remains constant • Water flow interruption	• No correction of the nitric acid temperature T_2 fluctuations (S_1) (in particular: no correction of temperature T_2 rise) • Safety system required to operate (S_3), nitric acid flow interruption due to valve V_1 closing (S_1)	• When S_3 does not operate: uncorrectable rise in T_2

Table 24.1 (cont.)
ELEMENTARY SYSTEM: S_2 (water system)

Component identification (identification code, name, type, location)	Functions, states	Failure modes	Possible failure causes (Internal, external causes)	Effects on elementary system	Effects on the other elementary systems	Observations
		• Inadvertent opening	• Fire in the instrumentation room (S_6) • Control defect S_4 induced defect	• Inadvertent increase in water flow rate	• Drop in T_2 (S_1)	
		• External leak	• Leaktightness defect (defective weld …)	• Water flow rate reduction undetectable by flow meter	• Rise in T_2 (S_1), control system required to operate (S_4), opening of valve V_2 (S_2), uncorrectable rise in T_2	• Only a significant out leakage is taken into account
Name Heat-exchanger (shell part)	*Function* Cooling water circulation around the tubes	• External leak	• Leaktightness defect (corrosion, vibrations, defective weld …)	• Water flow rate reduction undetectable by flow meter	• Rise in T_2 (S_1), control system required to operate (S_4), opening of valve V_2 (S_2), uncorrectable rise in T_2	
		• Plugging	• Foreign body • Fowling	• Reduction of cooling water flow rate	• Safety system (S_3) required to operate, nitric acid flow interruption due to value V_1 closing (S_1)	• A partial plugging is not taken into account as it would have little effect
		• Loss of cooling function	• Fire in the heat-exchanger room (S_6)		• Control system (S_4) required to operate, complete opening of valve V_2 (S_2)	• A large fire likely to result in uncontrollable rise in T_2 is considered

Table 24.1 (cont.)
ELEMENTARY SYSTEM: S_3 (safety system)

Component identification (identification code, name, type, location)	Functions, states	Failure modes	Possible failure causes (Internal, external causes)	Effects on elementary system	Effects on the other elementary systems	Observations
Name Flowmeter *Type* Orifice plate	*Function* Water flow rate measurement	• Failure upon demand (when low flow rate threshold is reached) • Inadvertent operation	• Sensor primary failure • Loss of 48 V supply (S_5) • Sensor drift	• Failure to detect a low water flow rate • Spurious detection of low water flow rate, safety system inadvertently required to operate	• Valve V_1 (S_1) does not close when water flow rate is insufficient; as a result uncorrectable rise in T_2 • Closing of valve V_1 (S_1), inadvertent nitric acid flow interruption	
Name Relay circuitry *Type* Electromechanical	*Function* Signal emission of order to close valve V_1 when low water flow rate	• Failure upon demand • Inadvertent operation	• Primary failure (relay defect) • Loss of 48 V supply (S_5) • Fire in instrumentation room (S_6) • Relay circuitry defect	• Low water flow rate signal not transmitted • Signal to close valve V_1 is inadvertently emitted	• Valve V_1 (S_1) failure to close when water flow rate is insufficient; as a result uncorrectable rise in T_2 • Inadvertent closing of V_1 (S_1), inadvertent interruption of nitric acid flow	• When fire in instrumentation room: valve V_2 closes completely, valve V_1 remains open, this resulting in an uncorrectable rise in T_2

Table 24.1 (cont.)
ELEMENTARY SYSTEM: S_4 (control system)

Component identification (identification code, name, type, location)	Functions, states	Failure modes	Possible failure causes (Internal, external causes)	Effects on elementary system	Effects on the other elementary systems	Observations
Name Temperature sensor	*Function* Measures the nitric acid temperature at the heat-exchanger outlet	• Inadvertent high-temperature signal	• Sensor drift	• Signal from control system to open valve V_2	• Opening of valve V_2 (S_2), increase in cooling water (S_2) flow rate, nitric acid temperature T_2 (S_1) stabilizes at a value below the normal value	• In case of correctable increase in nitric acid conditions, non correction risk: in the rest of the analysis, only those drifts sufficiently large to initiate the safety system are taken into account, the other drifts are assumed to result in acceptable temperatures T_2
		• Inadvertent low-temperature signal	• Sensor drift	• Signal from control system to close valve V_2	• Closing of valve V_2 (S_2), lower cooling water S_2 flow rate, T_2 (S_1) stabilizes at a value above the normal value	

Table 24.1 (*cont.*)
ELEMENTARY SYSTEM: S_4 (control system)

Component identification (identification code, name, type, location)	Functions, states	Failure modes	Possible failure causes (Internal, external causes)	Effects on elementary system	Effects on other elementary systems	Observations
Name Controller	*Function* Adjustment of water flow rate (using valve V_2) to the nitric acid temperature T_2	• Failure to operate upon demand	• Sensor defect • Loss of 48 V (S_5) supply	• No signal for control system	• When the low-water flow rate threshold is reached, safety system (S_2) is required to operate, nitric acid flow (S_1) stops • Water flow rate remains constant; no correction of temperature T_2 (S_1) fluctuation, in particular no correction of rise in T_2	• When 48 V supply is lost: safety system and control system are lost (V_1 and V_2 remain as they are)
		• Inadvertent signal to open valve V_2	• Drift of set point (operator wrongly sets set point)	• Controller signal to open valve V_2	• Opening of valve V_2 (S_2), increase in cooling water (S_2) flow rate; nitric acid temperature T_2 (S_1) stabilizes at a value below the normal value	
		• Inadvertent signal to close valve V_2	• Drift in set point (operator wrongly sets set point)	• Controller signal to close valve V_2	• Closing of valve V_2 (S_2), lower cooling water (S_2) flow rate	• In case of correctable increase in nitric acid conditions, non-correction risk

Table 24.1 (cont.)
ELEMENTARY SYSTEM: S_4 (control system)

Component identification (identification code, name, type, location)	Functions, states	Failure modes	Possible failure causes (Internal, external causes)	Effects on elementary system	Effects on the other elementary systems	Observations
		• Failure to operate upon demand	• Primary failure of controller • Operators' error (controller left on manual position) • Loss of 48 V (S_5) supply • Fire in the instrumentation room (S_6)	• No signal sent to valve V_2	• T_2 (S_1) stabilizes at a value above normal value when low water flow rate threshold is reached, safety system (S_3) is required to operate, nitric acid flow (S_1) interruption • Water flow rate remains constant; no correction of temperature T_2 (S_1) fluctuation	• Only drifts sufficiently important to initiate the safety system operation are taken into account; the other drifts are assumed to result in acceptable temperatures T_2 • When 48 V supply (S_5) is lost: safety system and control systems are lost (V_1 and V_2 remain as they are)
		• Destruction of control system		• Destruction of S_4, signal from controller to close valve V_2	• Cooling water (S_2) flow loss, valves V_1 (S_1) fails to close resulting in uncorrectable rise in T_2	

Table 24.1 (cont.)
ELEMENTARY SYSTEM: S_5 (auxiliary systems)

Component identification (identification code, name, type, location)	Functions, states	Failure modes	Possible failure causes (Internal, external causes)	Effects on elementary system	Effects on other elementary systems	Observations
Name Nitric acid feed system	*Function* Supply of nitric acid to chemical reactor	• Moderate increase in upstream nitric acid flow rate			• Rise in nitric acid temperature T_2 (S_1) correctable by control system (S_4); control system (S_4) required to operate; opening of valve V_2 (S_2)	
		• Moderate rise in upstream nitric acid temperature • Significant increase in upstream nitric acid flow rate			• Same as above • Rise in nitric acid temperature T_2 (S_1) not correctable by control system	

Table 24.1 (*cont.*)
ELEMENTARY SYSTEM: S_5 (auxiliary systems)

Component identification (identification code, name, type, location)	Functions, states	Failure modes	Possible failure causes (Internal, external causes)	Effects on elementary system	Effects on the other elementary systems	Observations
Name Feed water system	*Function* Supply of the nitric acid cooling system with water	• Significant rise in downstream nitric acid temperature • Insufficient pressure			• Control system (S_4) required to operate, complete opening of valve V_2 (S_2) • Same as above • Significant reduction of water flow rate (S_2), safety system (S_3) requested to operate, nitric acid flow (S_1) interruption	
Name Control air system	*Function* Supply of compressed air to valve V_2	• Pressure loss			• V_2 (S_2) closes completely, safety system (S_3) is required to operate, nitric acid flow (S_1) stops	

Table 24.1 (cont.)
ELEMENTARY SYSTEM: S_5 (auxiliary systems) (continued)

Component identification (identification code, name, type, location)	Functions, states	Failure modes	Possible failure causes (Internal, external causes)	Effects on elementary system	Effects on the other elementary systems	Observations
Name 6.6 kV power supply system	*Function* Power supply to pump motor	• Loss of 6.6 kV system			• Pump shutdown, water flow (S_2) stops, safety system (S_3) required to operate, nitric acid flow (S_1) interruption	• If the safety system fails to operate, there will be an uncorrectable rise in T_2
Name 380 V power supply system	*Function* Power supply to motor of valve V_1	• Loss of 380 V supply			• Valve V_1 remains as it is	• Valve V_1 (S_1) fails to close when water flow rate is insufficient: as a result uncorrectable rise in T_2 • There is an alarm 'loss of 380 V supply'
Name 48 V power supply system	*Function* Power supply of the safety and control systems	• Loss of 48 V power supply			• Loss of signal from control system (S_4), valve V_2 (S_2) remains as it is • Loss of safety system (S_3), valve V_1 (S_1) remains as it is	• There is an alarm 'loss of 48 V supply'

In the present case, the following assumptions will be made:

- upon a loss of compressed air, control valve V_2 completely closes;

- upon the loss of the 48 V supply, the loss of the control system S_4 leaves control valve V_2 as it is and the loss of the safety system (S_3) leaves valve V_1 as it is;

- upon the loss of the 380 V supply, valve V_1 remains as it is;

- a fire in the heat exchanger room induces a significant rise in T_2 which cannot be corrected by the control system;

- in the event of a fire in the instrumentation room where the safety system relay circuitry and the control system are located, the most serious consequences are assumed:
  - the destruction of the control system (S_4) results in the closing of valve V_2 (S_2);
  - the destruction of the safety system relay circuitry (S_3) leaves valve V_1 (S_1) open.

For simplicity's sake, no FMEA has been performed for system S_6 because of its elementary nature.

24.4 CAUSE TREE METHOD

The cause tree method (CTM) is a deductive method in which the undesirable event(s) to be studied is (are) assumed to be known (see Chapter 9).

24.4.1 Beginning of the construction

When beginning the construction of a cause tree, the 'component' approach adopted for the FMEA should be given up for a general representation of system operation (at least at the very beginning!). Here, the relationships between the main parameters involved are summarized as shown in (Figure 24.4).

Remember that the undesirable event is

'Excessive rise in temperature T_2'

What are the causes of this event? To find them by deduction, we must use again the state equations of elementary system 1, in particular equations concerning the heat exchanger. The causes could be as follows:

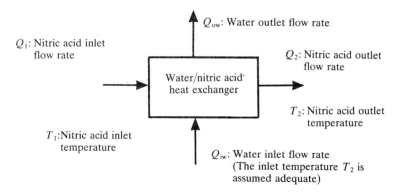

Figure 24.4 Schematic representation of the main parameters of S_1.

- *Increases in the upstream nitric acid flow rate or temperature.* As before, a distinction is drawn between:

  —increases that cannot be corrected by the control system;
  —increases that can be corrected by the control system; when they are combined with a defect of the control system, these increases result in unacceptable consequences (UC);

- *Inadequate flow rate.* That is

  —inadequate flow rate that cannot be detected by the safety system: is this possible? The analyst who has not previously performed an FMEA cannot easily answer this question;
  —inadequate flow rate detectable by the safety system: combined with a failure of the safety system to operate, this incident leads to the unacceptable consequences (UC);

- *Deterioration of the heat exchangers.* The following causes are considered:

  —external heat source like a fire in the heat exchanger room;
  —internal heat source like one resulting from an internal leakage in the heat exchanger and from the resulting exothermic reaction. However, these two causes are equivalent to an 'uncorrectable' increase in the nitric acid conditions.

Therefore, the cause tree should necessarily begin as shown in Figure 24.5.

This purely deductive analysis shows the difficulties raised by this approach! Indeed, the analyst would most certainly have overlooked the inadequate flow rate undetectable by the safety system. However, assuming he would have thought of this event, he would probably have tried to identify its causes using an inductive reasoning of the FMEA type!

Suppose now that an FMEA has been performed before constructing the cause tree. An inductive approach can be used.

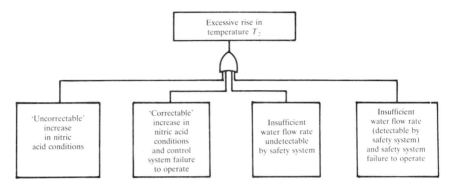

Figure 24.5 Beginning of the cause tree construction.

The nitric acid inlet conditions (Q_1, T_1) can be

- normal;
- 'correctable' by the control system;
- 'uncorrectable' by the control system.

Likewise, there are three conditions of flow rate Q_{iw}:

- adequate flow rate Q_{iw};
- inadequate flow rate Q_{iw} that can be detected by the safety system;
- or inadequate flow rate Q_{iw} undetectable by the safety system.

With the data presented in Table 24.2, the first level of the cause tree can be built. The eight framed configurations lead to the undesirable event. The four intermediate events at the top of the cause trees are found again (Figure 24.5) under the first OR logic gate.

Table 24.2 Conditions for the undesirable event occurrence

	Normal conditions	Conditions 'correctable' by the control system	Conditions 'uncorrectable' by the control system
Insufficient water flow rate		Control system operation; otherwise: UC	UC
Insufficient water flow detectable by safety system	Safety system operation; otherwise: UC	Safety system operation; otherwise: UC	UC
Insufficient water flow rate undetectable by safety system	UC	UC	UC

24.4.2 Cause tree

The identification of these main intermediate events is the most tricky step in the construction of the cause tree. As a matter of fact, it requires a deep knowledge of the operation of the studied system and determines the thoroughness of the identification of the failure combinations leading to the undesirable event. The cause tree construction then proceeds until the events introduced match the failure modes studied with the FMEA.

The cause tree presented in Figure 24.6 is not reduced.

The construction of a cause tree raises some problems which are listed below:

- All the causes of the 'control system failure to operate' have not been indicated. What does the 'control system' stand for here? It covers the control function fulfilled by systems S_4 and S_2, and is therefore not restricted to system S_4. This highlights the importance of the wording of the intermediate events and of the way the analyst understands them. Let us carry on with the identification of the failure causes of the control function fulfilled by S_2: in addition to the sticking of valve V_2, we could consider all the causes of insufficient flow rate. In fact, these causes result either in the actuation of the safety system—and should not therefore be taken into account—or directly in the undesirable event if the inadequate flow rate cannot be detected—so that they are otherwise taken into account.

- The loss of the 48 V power supply prevents the control and safety systems from operating. Therefore, the basic event 'loss of 48 V power supply' is incompatible with the basic event 'inadvertent closing signal of valve V_2 from controller'. For simplicity's sake and given the very low probability of the loss of the 48 V power supply (see Section 24.8) this event was dropped from consideration.

This cause tree is easy to reduce (Figure 24.7). Order 1 minimal cut sets are of the type:

- significant rise in the upstream nitric acid temperature;
- large increase in the upstream nitric acid flow rate;
- fire in the heat exchanger room;
- heat exchanger internal leakage;
- heat exchanger external leakage;
- valve V_2 external leakage;
- fire in the instrumentation room.

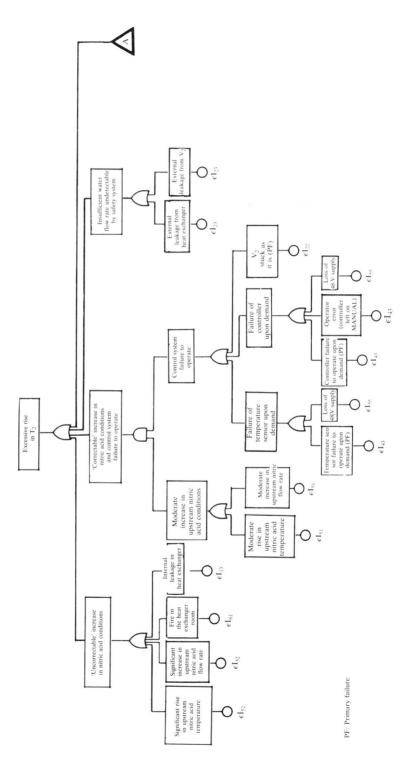

PF: Primary failure

Figure 24.6 Cause tree.

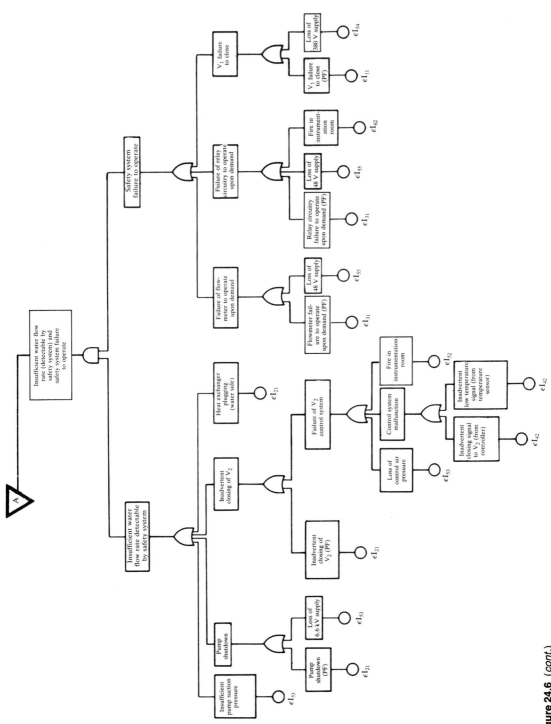

Figure 24.6 (cont.)

CAUSE TREE METHOD 651

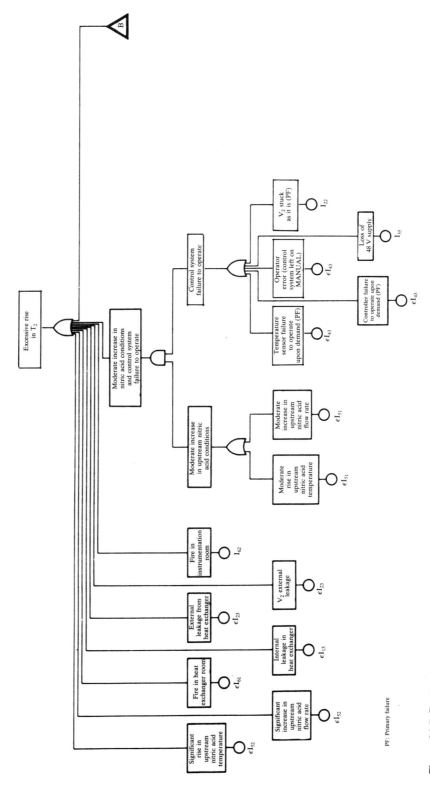

Figure 24.7 Reduced cause tree.

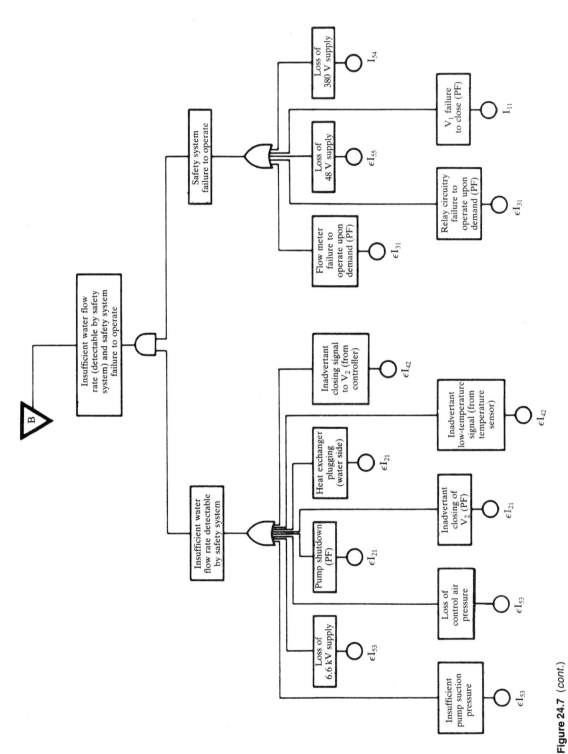

Figure 24.7 (cont.)

Order 2 minimal cut sets are more numerous (50) and of the type:

- moderate increase in the upstream nitric acid conditions and control system failure to operate;

- inadequate flow rate (detectable by the safety system) and safety system failure to operate.

24.5 GATHERED FAULT COMBINATION METHOD

It should be remembered that the GFCM is an inductive method consisting in grouping and combining failure modes (see Chapter 11). This method therefore entirely relies on the FMEA and especially on the two columns concerning the failure mode effects (effects on the elementary system, on the one hand, and effects on the other elementary systems, on the other). Let us follow the various steps of the method.

24.5.1 Identification of internal gathered faults

The failure modes with identical effects are grouped. The IGFs of the 6 elementary systems considered are presented in Table 24.3. All the failure modes belong to at least one gathered fault. Remember that only the causes internal to the failure mode mentioned are considered barring human errors since we have not regarded the operator as an elementary system. Sometimes, IGFs must be combined to bring to light new effects. Thus, in S_5, I_{53} ('loss of auxiliary systems resulting in flow interruption') and I_{54} ('valve V_1 inability to close') have been combined. The resulting fault I_{56} ('flow interruption with safety system failing to operate') directly leads to an excessive rise in T_2. Note that the analyst could also have grouped faults I_{52}, I_{56} and I_{57} into the same IGF since they all induce the excessive rise in T_2. This would not have changed the results of the identification of the causes of the undesirable event! In this exercise, we have tried to perform a refined analysis of the effects for illustration purposes, and therefore we have not always sought simplification.

24.5.2 Identification of external and global gathered faults

Among the IGFs, and their combinations, of a given elementary system we try to find those which have effects on the studied elementary system: they are called external gathered faults (EGFs). Given the large number of combinations which must be studied, we will try to simplify this identification. The question which arises then is: is there an order for the construction of EGFs which can facilitate their identification, while at the same time guaranteeing the optimum exhaustivity?

Table 24.3 Internal, external and global gathered faults

Elementary system	Internal gathered faults		External gathered faults		Global gathered faults	
	No.	Name	No.	Name	No.	Name
S_6	I_{61}	'Fire in the heat exchanger room'			G_{61}	'Fire in heat exchanger room' (excessive rise in T_2)
	I_{62}	'Fire in the instrumentation room'			G_{62}	'Fire in instrumentation room' (excessive rise in T_2)
S_5	I_{51}	'Moderate increase in upstream conditions': moderate increase in the upstream nitric acid flow rate or temperature			G_{51}	'Moderate increase in upstream conditions' ('correctable' rise in T_2)
	I_{52}	'Significant increase in upstream conditions': significant increase in upstream nitric acid flow rate or temperature			G_{52}	'Significant increase in upstream conditions' ('uncorrectable' rise in T_2)
	I_{53}	'Loss of auxiliary systems resulting in water flow interruption': insufficient pump section pressure, loss of control air pressure, loss of 6.6 kV supply system			G_{53}	'Loss of auxiliary systems resulting in water flow interruption' (insufficient water flow rate detectable by safety system)
	I_{54}	'Impossibility to close V_1': loss of 380 V supply			G_{54}	'Impossibility to close V_1'
	I_{55}	'Loss of 48 V supply'			G_{55}	'Loss of 48 V supply' (loss of safety and control systems)
	I_{56}	'Water flow interruption with safety system failure to operate' $I_{53}.I_{54} + I_{53}.I_{55}$			G_{56}	'Water flow interruption with safety system failure to operate' (excessive rise in T_2)
	I_{57}	'Moderate increase in upstream conditions and control system failure to operate': $I_{51}.I_{55}$			G_{57}	'Moderate increase in upstream conditions and control system failure to operate' (excessive rise in T_2)

Table 24.3 (cont.)

Elementary system	Internal gathered faults		External gathered faults		Global gathered faults	
	No.	Name	No.	Name	No.	Name
S_4	I_{41}	'Opening signal to V_2 from temperature controller': inadvertent high-temperature signal from temperature sensor, inadvertent opening signal to V_2 from temperature controller			G_{41}	'Opening signal to V_2 from temperature controller' (maximum water flow rate)
	I_{42}	'Closing signal to V_2 from temperature controller': inadvertent low-temperature signal from temperature sensor, inadvertent closing signal to V_2 from temperature controller			G_{42}	'Closing signal to V_2 from temperature controller' (nitric acid flow interruption)
	I_{43}	'Control system failure to operate upon demand': temperature sensor failure to operate upon demand, temperature controller failure to operate upon demand, temperature controller left on manual			G_{43}	'Control system failure to operate upon demand'
			G_{62}	'Fire in instrumentation room'	G_{44}	'Fire in instrumentation room' (excessive rise in T_2)
			G_{55}	'Loss of 48 V supply' (loss of safety and control systems)	G_{45}	'Loss of 48 V supply' (loss of safety and control systems)
			G_{51}	'Moderate increase in upstream conditions'	G_{46}	'Moderate increase in upstream conditions' ('correctable' rise in T_2)
					G_{47}	'Moderate increase in upstream conditions and control system failure to operate upon demand' (excessive rise in T_2): $G_{46}.G_{43}$
					G_{48}	'Moderate increase in upstream conditions and loss of 48 V supply' (excessive rise in T_2): $G_{46}.G_{45}$

Table 24.3 (cont.)

Elementary system	Internal gathered faults		External gathered faults		Global gathered faults	
	No.	Name	No.	Name	No.	Name
S_3	I_{31}	'Safety system failure to operate upon demand': flow meter failure to operate upon demand, relay circuitry failure to operate upon demand			G_{31}	'Safety system failure to operate upon demand'
	I_{32}	'Inadvertent operation of safety system': inadvertent operation of flow meter or relay circuitry			G_{32}	'Inadvertent operation of safety system'
			G_{62}	'Fire in instrumentation room'	G_{33}	'Fire in instrumentation room' (excessive rise in T_2)
			G_{55}	'Loss of 48 V supply' (loss of safety and control systems)	G_{34}	'Loss of 48 V supply' (loss of safety and control systems)
			G_{21}	'Water flow interruption'	G_{35}	'Water flow interruption and safety system required to operate': $G_{35} = G_{21}$
					G_{36}	'Water flow interruption and safety system failure to operate' (excessive rise in T_2) $G_{36} = G_{21}.G_{31} + G_{21}.G_{34}$

Table 24.3 (cont.)

Elementary system	Internal gathered faults		External gathered faults		Global gathered faults	
	No.	Name	No.	Name	No.	Name
S_2	I_{21}	'Water flow interruption': pump shutdown, inadvertent closing of V_2, heat exchanger plugging			G_{21}	'Water flow interruption' $G_{21} = I_{21} + G_{53} + G_{42}$
	I_{22}	'Constant water flow rate': V_2 stuck as it is			G_{22}	'Constant water flow rate' $G_{22} = I_{22} + G_{43}$
	I_{23}	'Insufficient water flow rate undetectable by safety system': external leakage from V_2, external leakage from heat exchanger			G_{23}	'Insufficient water flow rate undetectable by safety system' (excessive rise in T_2) $G_{23} = I_{23}$
	I_{24}	'Inadvertent increase in water flow rate': inadvertent opening of V_2			G_{24}	'Inadvertent increase in water flow rate' $G_{24} = I_{24} + G_{41}$
			G_{53}	'Loss of auxiliary systems resulting in water flow interruption'		
			G_{41}	'Opening signal to V_2 from temperature controller'		
			G_{42}	'Closing signal to V_2 from temperature controller'		
			G_{43}	'Control system failure to operate upon demand'		
			G_{44}	'Fire in instrumentation room'	G_{25}	'Fire in instrumentation room' (excessive rise in T_2) $G_{25} = G_{44}$
			G_{45}	'Loss of 48 V supply'	G_{26}	'Loss of 48 V supply' (loss of safety and control systems) $G_{26} = G_{45}$
			G_{46}	'Moderate increase in upstream conditions'	G_{27}	'Moderate increase in upstream conditions' $G_{27} = G_{46}$

Table 24.3 (cont.)

Elementary system	Internal gathered faults		External gathered faults		Global gathered faults	
	No.	Name	No.	Name	No.	Name
			G_{47}	'Moderate increase in upstream conditions and control system failure to operate upon demand'	G_{28}	'Moderate increase in upstream conditions and constant water flow rate' (excessive rise in T_2) $G_{28} = G_{47} + G_{27}.G_{22} + G_{27}.G_{26} + G_{48}$
			G_{48}	'Moderate increase in upstream conditions and loss of 48 V supply'	G_{29}	'Water flow interruption and loss of 48 V supply' (excessive rise in T_2) $G_{29} = G_{21}.G_{26}$

Table 24.3 (cont.)

Elementary system	Internal gathered faults		External gathered faults		Global gathered faults	
	No.	Name	No.	Name	No.	Name
S_1	I_{11}	'No nitric acid flow interruption when safety system is actuated': V_1 failure to close			G_{11}	'No nitric acid flow interruption on low water flow rate signal' $G_{11} = I_{11} + G_{54} + G_{31}$
	I_{12}	'Nitric acid flow interruption': inadvertent closing of V_1, heat exchanger plugging			G_{12}	'Nitric acid flow interruption' $G_{12} = I_{12} + G_{32}$
	I_{13}	'Nitric acid water reaction': internal leakage in heat exchanger			G_{13}	'Excessive rise in T_2' $G_{13} = I_{13} + G_{61} + G_{52} + G_{56} + G_{57} + G_{33} + G_{36} + G_{23} + G_{28} + G_{29} + G_{16} \cdot G_{11} + G_{14} \cdot G_{15} + G_{14} \cdot G_{17} + G_{15} \cdot G_{16}$
			G_{61}	'Fire in heat-exchanger room' (excessive rise in T_2)		
			G_{51}	'Moderate increase in upstream conditions' ($G_{51} = G_{27}$)	G_{14}	'Moderate increase in upstream conditions': $G_{14} = G_{51}$
			G_{52}	'Significant increase in upstream conditions' ('uncorrectable' rise in T_2)		
			G_{54}	'Impossibility to close V_1' (loss of 380 V supply)		
			G_{56}	'Water flow interruption with safety system failure to operate (excessive rise in T_2)		
			G_{57}	'Moderate increase in upstream conditions and control system failure to operate' (excessive rise in T_2)		

Table 24.3 (cont.)

Elementary system	Internal gathered faults		External gathered faults		Global gathered faults	
	No.	Name	No.	Name	No.	Name
S_1 (continued)			G_{31}	'Failure of safety system to operate upon demand'		
			G_{32}	'Inadvertent operation of safety system'		
			G_{33}	'Fire in instrumentation room' (excessive rise in T_2) ($G_{33} = G_{25}$)		
			G_{34}	'Loss of 48 V supply' ($G_{34} = G_{26}$)	G_{15}	'Loss of safety and control systems': $G_{15} = G_{34}$
			G_{35}	'Water flow interruption and safety system required to operate'	G_{16}	'Water flow interruption and safety system required to operate': $G_{16} = G_{35}$
			G_{36}	'Water flow interruption and failure of safety system to operate' (excessive rise in T_2)		
			G_{22}	'Constant water flow rate'	G_{17}	'Constant flow rate of cooling water in heat exchanger': $G_{17} = G_{22}$
			G_{23}	'Insufficient water flow rate indetectable by safety system' (excessive rise in T_2)		
			G_{24}	'Inadvertent increase in water flow rate'	G_{18}	'Drop of T_2': $G_{18} = G_{24}$
			G_{28}	'Moderate increase in upstream conditions and constant water flow rate' (excessive rise in T_2)		
			G_{29}	'Water flow interruption and loss of 48 V supply' (excessive rise in T_2)		

The answer to this question can vary widely from one study to another. However, we can notice in real life that there is at least one system whose operation is not affected by the faults of other systems. In other words, this system has practically no EGF. Thus it is often recommended to start the analysis of a set of interconnected systems by studying the electric power supplies: indeed, the reliability of this system conditions the operation of all the dependent systems but, conversely, few are the systems that can affect its reliability.

In the present case, it is also desirable to start with the analysis of systems S_6 (environment) and S_5 (auxiliary systems). Obviously no IGFs of systems S_1, S_2, S_3 or S_4 have effects on S_5 or S_6. The construction of the GGFs of these two systems is then easy since it is directly performed from the IGFs. The main consequences for the GGFs thus constructed are then used to identify the EGFs of the systems whose operation is directly related to that of systems S_6 and S_5.

Thus G_{61} ('fire in heat exchanger room') has direct effects on the nitric acid system and is therefore an EGF for S_1. Similarly, G_{53} ('loss of auxiliary systems inducing flow interruption') is an EGF for S_2 since it directly leads to the water flow interruption.

Figure 24.8 contains a model of the interactions between elementary systems; it represents the effects of the faults of each system on the other systems and, thus, helps to build all the different GGFs. Of course, this figure is obtained iteratively; however, the major characteristics of the interactions it models can be summarized by the analyst using, among others, reasonings such as those described above for selecting the order in which the systems should be preferably analysed.

A first list of the GGFs is obtained for each elementary system by proceeding as follows:

- Each IGF, possibly associated with the EGFs having the same effects, is integrated into a GGF. Thus, for S_4, faults G_{41} to G_{43} are created.

- Each of the EGFs that has not been grouped with an IGF is then integrated into a new GGF.
  As faults G_{62}, G_{55} and G_{57} don't have the same effects as faults I_{41} to I_{43}, each produces a GGF: G_{44}, G_{45} and G_{46} are thus created.

- The GGFs thus created are paired or, when necessary, combined by threes.

Consider again the case of system S_4: the combination of faults G_{46} ('moderate increase in upstream conditions') and G_{43} ('control system failure to operate upon demand') has different effects than each of the two faults considered by itself since the combination results in an undesirable event; a new fault G_{47} is therefore created. Following the same reasoning, G_{48} is also created.

When the complete gathered fault combination method is applied, all the GGFs of S_1 are worked out.

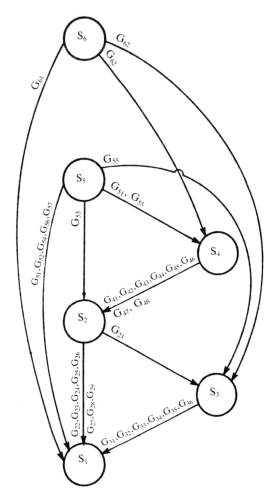

Figure 24.8 Model of interactions between elementary systems.

24.5.3 Undesirable event

Fault G_{13} is the undesirable event; it consists of the following:

- IGFs of system S_1 or GGFs of the other systems which have effects on system S_1: this contribution can be written:

$$I_{13} + G_{61} + G_{52} + G_{56} + G_{57} + G_{33} + G_{36} + G_{23} + G_{28} + G_{29}$$

- the combinations of the GGFs of system S_1 (beside G_{13}); this contribution can be written:

$$G_{16} \cdot G_{11} + G_{14} \cdot G_{15} + G_{14} \cdot G_{17} + G_{15} \cdot G_{16}$$

We must now determine the expression for G_{13} according to the IGFs we have

$$G_{13} = I_{62} + I_{61} + I_{52} + I_{23} + I_{13} + I_{51} \cdot (I_{55} + I_{43} + I_{22})$$
$$+ (I_{53} + I_{42} + I_{21}) \cdot (I_{55} + I_{54} + I_{31} + I_{11})$$

Note that the reduction is fairly long; this is due to the fact that, through caution we have maintained many faults which could have been grouped.

These results are identical to what was obtained in the form of reduced cause trees (Figure 24.7). We have indicated the number of the IGF associated with each basic event on this figure. It is easy thus to see the results of the method as well as how the faults are grouped.

24.5.4 Discussion

As the faults of system S_1 were being sought, several problems appeared. We examine here the major ones:

• **The first problem** This concerns the way interactions between elementary systems should be modelled and therefore which definitions of the effects should be chosen. Take fault G_{51}, 'moderate increase in upstream conditions'. Should we take into account the effects of this fault on system S_1 (owing to the nitric acid temperature rise at the heat exchanger outlet), on system S_4 (due to the required operation of control system) or on system S_2 (resulting from the required operation of valve V_2), etc?

In the framework of the GFCM, it is advised to stick to the direct effects on the system. Here we should therefore consider the effects of G_{51} on S_1, and then the effects of a new fault of S_1 (including G_{51}) on S_4, ... In this way, after constructing the faults of S_1, we would construct again the faults of S_4 (Figure 24.8) with the inescapable consequence of making the process even more complex.

Therefore we have directly introduced the effects of fault G_{51} on system S_4 (Table 24.3, system S_1); the reader can see without difficulty that the faults of S_1 which affect S_4 do not introduce new effects.

The analyst could have chosen another procedure; remember that the main parameter we are interested in is the nitric acid temperature at the exchanger outlet. Therefore it is sensible to define the effects of a fault in terms of the resulting variations in this parameter. Thus:

— When a fault affects the nitric acid system S_1,

— either the fault considered directly results in the undesirable event and then it need not be combined with other events: for instance, fault G_{61} (fire in the heat exchanger room) is directly related to system S_1;
— or the fault considered induces moderate variations in the nitric acid temperature at the exchanger outlet; then it must be combined with other faults to result in the undesirable event.

For this fault, only those effects it has on the system(s), thereby 'directly' modifying its operation, will be taken into account. Fault G_{51} described above has 'direct' effects on the control system; this is why it will be introduced in Figure 24.8.

—Faults with no direct effects on system S_1 (like fault G_{55}: loss of the 48 V power sypply) will also have to be combined with other faults to result in the undesirable event. These faults are therefore connected to the system(s) whose operation they directly condition. As an example, G_{55} affects the control system S_4 and the safety system S_3.

This approach would undoubtedly have yielded the same results. As a conclusion we shall add that nothing *a priori* goes against considering all the cascade effects of a given fault; modelling would only be more unwieldy.

- **The second problem.** This concerns the conversion of the EGFs into GGFs. Consider again the example of fault G_{51} 'moderate increase in upstream conditions'. Faults G_{64} and G_{51}, which are regarded as equivalent, are not in fact strictly identical; indeed, G_{46} is rather 'the moderate increase in upstream conditions' combined with 'the proper control system operation upon demand'. When the second condition is not met, we have fault G_{47}. The analyst should not forget this second condition, which is not systematically fully formulated for simplicity's sake, since it can prove quite important especially for the quantitative assessment.

24.6 CONSEQUENCE TREE METHOD

After identifying the initiating event, the consequence trees are built (see Chapter 12); the links with the GFCM are then discussed.

24.6.1 Initiating events

The analyst, once he has performed a preliminary analysis of the system, naturally thinks of a number of events such as:

- significant rise in the upstream nitric acid temperature;
- significant increase in the upstream nitric acid flow rate;
- fire in the heat exchanger room;
- fire in the instrumentation room;
- moderate rise in the upstream nitric acid temperature;
- moderate increase in the upstream nitric acid flow rate;
- inadequate water flow rate.

The first four events directly result in an excessive temperature rise T_2; only the last three ones should therefore be analysed. Is this list complete? The analyst, if he has not performed an FMEA, will probably have difficulties in thinking of other initiating events such as:

- internal leakage in the heat exchanger;
- external leakage from the heat excanger;
- external leakage from valve V_2.

Is the list complete this time? There is nothing to prove it. To answer this question, we shall use an analysis of the GFCM type.

24.6.2 Consequence tree

We have determined that three initiating events should be analysed:

- moderate rise in the upstream nitric acid temperature;
- moderate increase in the upstream nitric acid flow rate;
- inadequate water flow rate.

As the first two events result in identical functional responses of the system, we shall group them under an initiating event termed 'moderate increase in upstream conditions'. Here, we shall merely analyse this initiating event, leaving the reader to examine the one called 'insufficient water flow rate'.

The two functions which limit the consequences of this initiating event are the control and the heat exchanger cooling; they involve system S_4 and S_2. The consequence tree obtained is represented in Figure 24.9.

Now, we seek the two generic event causes with the help of the CTM:

- 'control system (S_4) fails to operate' (after the occurrence of the initiating event):

  —'failure of control system to operate upon demand': I_{43};

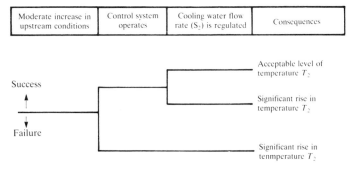

Figure 24.9 Consequence tree.

—'loss of 48 V power supply': I_{55};
—'fire in instrumentation room': I_{62} (already allowed for as the initiating event);

- 'the cooling water flow rate is not regulated' (after the initiating event occurrence and the control system operation):

  —'water flow interruption': I_{21};
  —'constant value of the water flow rate': I_{22};
  —'insufficient water flow rate undetectable by safety system': I_{23};
  —'loss of auxiliary systems resulting in water flow interruption': I_{53}.

Given the consequences of these faults, only I_{22} needs to be considered.

For simplicity's sake, we have used the same notation as in the GFCM. Finally, we obtain (F: undesirable event):

$$F = I_{51} \cdot (I_{43} + I_{55}) + I_{51} \cdot I_{22}$$
$$F = I_{51} \cdot (I_{43} + I_{55} + I_{22})$$

Thus we find all the fault combinations linked to the initiating event 'moderate increase in upstream conditions'. The reader, if he has determined the fault combinations associated with the initiating event 'insufficient water flow rate', finds:

$$F = (I_{53} + I_{42} + I_{21}) \cdot (I_{55} + I_{54} + I_{31} + I_{11})$$

Thus, all the minimal cut sets are found again, as long as all the initiating events have been determined!

24.6.3 Links with the gathered fault combination method

We know (Chapter 12) the GFCM enables us to inventory all the initiating events; indeed, the latter are among the first-order GGFs of S_1 which result in a variation of the parameters of S_1—a variation which is not unlikely to lead to the undesirable event—that is: I_{13}, G_{61}, G_{51}, G_{52}, G_{56}, G_{57}, G_{33}, G_{35}, G_{36}, G_{23}, G_{28}, G_{29}.

This list contains the following:

- The initiating events directly resulting in a significant rise in T_2: I_{13}, G_{61}, G_{52}, G_{33}, G_{23}. We obtain here all the order 1 minimal cut sets of the undesirable event.

- The initiating event related to 'the moderate increase in the upstream conditions':

  —'moderate increase in the upstream conditions': G_{51};
  —'moderate increase in the upstream conditions and control system failure to operate': G_{57};

—'moderate increase in the upstream conditions and constant water flow rate': G_{28}.

The last two events produce the undesirable event; therefore only G_{51} should be analysed as an initiating event:

- Initiating events related to the insufficient water flow rate:

  —'water flow interruption and failure of safety system to operate': G_{56};
  —'water flow interruption and required operation of safety system': G_{35};
  —'water flow interruption and safety system failure to operate': G_{36};
  —'water flow interruption and loss of 48 V power supply': G_{29}.
  Only the second initiating event does not directly result in the undesirable event and should therefore be examined.

We thus make sure that the list of the initiating events considered in Section 24.6.1 is complete.

24.7 CAUSE–CONSEQUENCE DIAGRAM METHOD

This method (see Chapter 13) provides a means of representing the links between the causes of an initiating event and its likely consequences when one or several system(s) did not operate. It combines the deductive character of 'cause identification' using the cause tree and the inductive character of the 'consequence identification'.

As for the CQTM, the main difficulties stem from the identification and choice of the initiating event to be studied.

The initiating events naturally coming to the analyst's mind are already listed in Section 6.1; we could here associate a cause–consequence diagram with each of them. As this method is both deductive and inductive, initiating events can be deduced; for instance, if we consider the first four intermediate events in the cause tree (see Section 24.4) which directly lead to the undesirable event or which imply the operation (or failure) of a safety system we have the following initiating events:

IE_1: 'uncorrectable' increase in the nitric acid conditions;

IE_2: moderate increase in the upstream nitric acid conditions;

IE_3: insufficient water flow rate undetectable by safety system;

IE_4: insufficient water flow rate detectable by safety system.

The cause–consequence diagrams corresponding to the initiating events previously inventoried are represented in Figures 24.10, 24.11, 24.12 and 24.13; we have indicated which cause trees should be developed. The reader will be left to do it and he will thus find trees very similar to the subtrees of the cause tree built in Section 24.4. The method is then continued so that the minimal cut sets described in Section 24.4 can be identified.

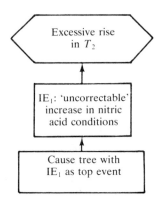

Figure 24.10 Cause–consequence diagram for initiating event IE_1.

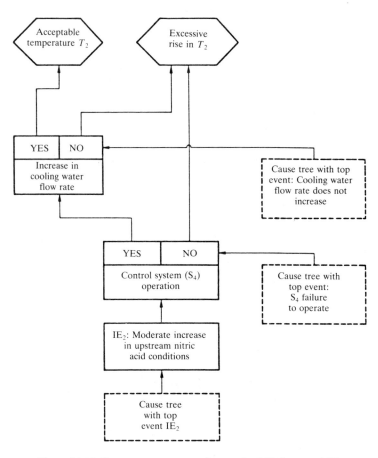

Figure 24.11 Cause–consequence diagram for initiating event IE_2.

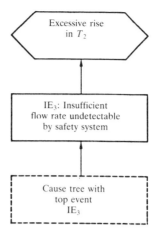

Figure 24.12 Cause–consequence diagram for initiating event IE$_3$.

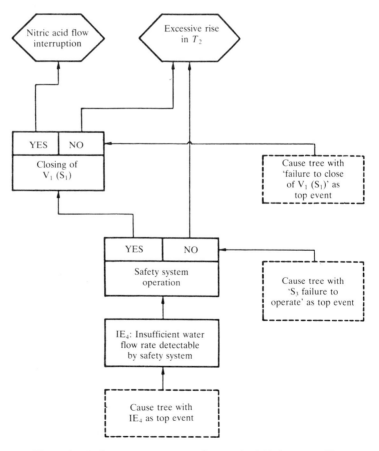

Figure 24.13 Cause–consequence diagram for initiating event IE$_4$.

24.8 QUANTITATIVE ANALYSIS

This analysis is chiefly performed with the CTM. The reader can train himself to find the same results with other methods. We also discuss the merits of using semi-Markovian processes.

24.8.1 Occurrence probability of the undesirable event

Dependability data are presented in Table 24.4 solely for the exercise purposes and will not therefore be justified.

We try to calculate the annual probability of an 'excessive rise in temperature T_2'. The reduced cause tree is represented in Figure 24.7. The probability value is approximated from the minimal cut sets obtained in Section 24.4.

- Order 1 minimal cut sets:

  —Significant rise in the upstream nitric acid temperature: $\lambda = 5 \times 10^{-8}$/h;
  —Significant increase in the uptream nitric acid flow rate: $\lambda = 5 \times 10^{-8}$/h;
  —Fire in heat exchanger room: $\lambda = 10^{-7}$/h;
  —Internal leakage in heat exchanger: $\lambda = 2 \times 10^{-7}$/h;
  —External leakage from the exchanger: $\lambda = 10^{-8}$/h;
  —External leakage from valve V_2: $\lambda = 10^{-8}$/h;
  —Fire in instrumentation room: $\lambda = 10^{-7}$/h.

  The equivalent failure rate is then deduced:

  $$\Lambda_{C_1} \simeq 5.2 \times 10^{-7}/h \simeq 4.6 \times 10^{-3}/\text{year}$$

- Order 2 minimal cut sets:

  —Moderate increase in the upstream nitric acid conditions and control system failure to operate.

  The occurrence rate of 'moderate increase in the nitric acid conditions' is $\lambda = 2 \times 10^{-4}$/h.

  We shall now calculate the probability of the 'control system failure to operate'. The control system (temperature sensor and controller) is assumed to be tested every month; the failure probability of the temperature sensor (or controller) to operate upon demand is equal to the equivalent unavailability rate $\lambda T/2$ (T being the time interval between tests). Moreover the equivalent unavailability rate of the 48 V power supply is $\lambda \tau$, where τ is the mean time to repair the 48 V power supply.

  As a result, the probability of the control system failure to operate is 4.7×10^{-3}/d; human error is here the main contributor ($\simeq 60\%$).

  The equivalent failure rate is then deduced:

  $$\Lambda_{C_2}^1 \simeq 9.4 \times 10^{-7}/h \simeq 8.2 \times 10^{-3}/\text{year}$$

Table 24.4 Dependability data

Elementary system	Component	Failure mode	Failure rate (error factor)
S_1	On–off motor-operated valve V_1	• Failure to close	$10^{-3}/d$ (3)
	Heat-exchanger (tubes)	• Internal leakage	$2 \times 10^{-7}/h$ (10)
S_2	Pump	• Shutdown	$2 \times 10^{-4}/h$ (2)
	Air-operated control valve V_2	• Stuck as it is	$10^{-3}/d$ (3)
		• Inadvertent closing	$2 \times 10^{-6}/h$ (10)
		• External leakage	$10^{-8}/h$ (10)
	Heat-exchanger (shell)	• External leakage	$10^{-7}/h$ (10)
		• Plugging	$10^{-7}/h$ (10)
S_3	Flow meter	• Failure of flow meter (constant output signal)	$10^{-6}/h$ (10)
	Relay circuitry	• Failure upon demand	$10^{-3}/d$ (10)
S_4	Temperature sensor	• Inadvertent low-temperature signal	$10^{-6}/h$ (10)
		• Constant output signal	$10^{-6}/h$ (10)
	Temperature controller	• Inadvertent closing signal	$10^{-6}/h$ (10)
		• Failure upon demand: operator error (controller left on Manual)	$3 \times 10^{-3}/d$ (10)
S_5	Nitric acid feed system	• Moderate increase in upstream nitric acid flow rate	$10^{-4}/h$ (10)
		• Moderate rise in upstream nitric acid temperature	$10^{-4}/h$ (10)
		• Significant increase in upstream nitric acid flow rate	$5 \times 10^{-8}/h$ (10)
		• Significant rise in upstream nitric acid temperature	$5 \times 10^{-8}/h$ (10)
	Water-supply system	• Insufficient pump section pressure	$10^{-5}/h$ (10)
	Control air system	• Loss of control air pressure	$10^{-5}/h$ (10)
	6.6 kV electric power supply	• Loss of 6.6 kV power supply	$6 \times 10^{-6}/h$ (3)
	380 V electric power supply	• Loss of 380 V power supply	$6 \times 10^{-6}/h$ (3)
			(MTTR: 10 h)
	48 V electric power supply	• Loss of 48 V power supply	$6 \times 10^{-6}/h$ (3)
			(MTTR: 10 h)
S_6	Heat-exchanger room	• fire	$10^{-7}/h$ (10)
	Instrumentation room	• fire	$10^{-7}/h$ (10)

— Insufficient water flow rate (detectable by the safety system) and safety system failure to operate.

The occurrence rate of the 'insufficient water flow rate (detectable by the safety system)' is $\lambda \simeq 2.3 \times 10^{-4}/h$.

Let us now calculate the probability of the 'safety system failure to operate'. The safety system is assumed to be tested every six months. This probability amounts to $4.2 \times 10^{-3}/d$, the failure of the flowmeter to operate upon demand (with some 50%), the relay circuitry failure to operate upon demand (with some 25%) and the failure to close of valve V_1 (with some 25%) being the main contributors.

The equivalent failure rate is deduced:

$$\Lambda^2_{C_2} \simeq 9.7 \times 10^{-7}/h \simeq 8.5 \times 10^{-3}/year$$

As a result, the probability of the undesirable event is

$$P \simeq 4.6 \times 10^{-3}/year + 8.2 \times 10^{-3} + 8.5 \times 10^{-3}/year$$
$$P \simeq 2.1 \times 10^{-2}/year$$

When each dependability data item is assumed to follow a lognormal distribution — whose mean is the value assigned to this data item — a Monte Carlo-type simulation gives for the probability distribution of the undesirable event:

$$\text{median} \quad \simeq 10^{-2}/year$$
$$\text{error factor} \quad \simeq 3.4$$

24.8.2 Conclusions; lessons

We give here a list of the minimal cut sets according to their contribution to the final probability:

- Moderate increase in the upstream nitric acid conditions and operator error (controller left on manual operation)   25%

- Pump shutdown and flowmeter failure to operate upon demand   20%

- Internal leakage in the heat exchanger   8%

- Moderate increase in the upstream nitric acid conditions and valve V_2 stuck as it is   8%

- Pump shutdown and relay circuitry failure to operate upon demand   8%

- Pump shutdown and valve V_1 failure to close upon demand   8%

- Fire in the instrumentation room   4%

- Fire in the heat exchanger room   4%

It can thus be observed that two minimal cut sets contribute to 45% of the probability, while numerous other cut sets account for the remaining probability.

It becomes also quite rapidly apparent that small modifications of the system cannot contribute to lower significantly the probability of the undesirable event. We shall nevertheless examine the possible alterations:

- The probability of an operator leaving the controller on manual operation after a test can be reduced by fitting in an alarm: the human error probability could be thus reduced by a factor of 10.

- The probability that the flowmeter fails to operate upon demand can be reduced thanks to monthly tests.

The probability comes down to $P \simeq 1.3 \times 10^{-2}$/year; the alterations improve the probability by a factor of approximately 2.

24.8.3 Quantitative analysis method: discussion

The quantitative analysis method used, that is the CTM, proved quite easy to implement since most failures considered were not repairable.

We must draw the reader's attention to the calculation of order 2 minimal cut sets; the equivalent failure rate is the product of the initiating event (IE_2 or IE_4) occurrence rate and the failure probability of the studied safety system. This can be readily demonstrated by considering a three-state semi-Markovian process and by assuming that the system can be rapidly repaired after the initiating event occurrence and the safety system operation.

Suppose we assume that new hypotheses exist, such as the following:

- after the occurrence of an order 2 minimal cut set, there is some time left ($d_{c_2}^1$ or $d_{c_2}^2$) before unacceptable consequences take place;

- this time interval can be profitably used to repair the safety systems concerned;

- after initiating events IE_2 or IE_4 have taken place, followed in turn by the operation of the safety systems concerned, the system is put back to service within a (constant) time interval τ.

The CTM is no longer the best-suited quantification method. A semi-Markovian process should then be used or the CQTM with simplifying assumptions.

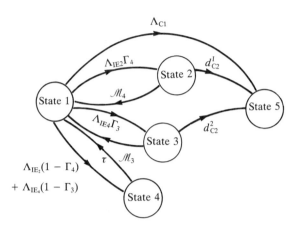

Figure 24.14 Modelling by a semi-Markovian process.

As an example, the equivalent semi-Markovian process is represented in Figure 24.14:

- state 1 is the fully operational state;

- state 2 is the state produced by the occurrence of initiating event IE_2 (moderate increase in the upstream nitric acid conditions) followed by the control system failure to operate;

- state 3 is the state produced by the occurrence of initiating event IE_4 (insufficient water flow rate detectable by safety system) followed by the failure of the safety system to operate;

- state 4 is the state produced by the occurrence of initiating events IE_2 and IE_4, followed by the operation of the safety systems concerned;

- state 5 is the state with unacceptable consequences.

The transitions between state 2 and 5, 3 and 5, 4 and 1 are constant-time transitions. The reader will be left to assess the probability of the undesirable event using the semi-Markovian solving method (see Section 14.9 and Appendix 6).

REFERENCES

[1] S. A. Lapp and G. J. Powers (1977) Computer-aided synthesis of fault trees, *IEEE Trans. Reliability*, **R-26**, 2–13.
[2] M. O. Locks (1974) Synthesis of fault trees: an example of non-coherence, *IEEE Trans. Reliability*, **R-28**, 2–5.
[3] H. E. Lambert (1979) Comments on the Lapp–Powers computer aided synthesis of fault trees, *IEEE Trans. Reliability*, **R-28**, 6–8.

[4] S. A. Lapp and G. J. Powers (1979) Update of Lapp Powers fault-tree synthesis algorithm, *IEEE Trans. Reliability*, **R-28**, 12–15.
[5] M. Llory and A. Villemeur (1980) *La méthode des combinaisons de pannes: présentation, application à un ensemble simple de systèmes*, EDF-HT/13/68/80.
[6] Analyse de la fiabilité et de la sûreté des systèmes industriels (1982) Centre d'Actualisation Scientifique et Technique. Document de stage. Coordinateurs: MM. Villemeur (EDF), Ralko (CAST-CEFIMA). INSA, Bât. 705, 20, avenue Albert Einstein, 69621 Villeurbanne, France.

25

HUMAN RELIABILITY ASSESSMENT*

25.1 PRESENTATION OF THE EXAMPLE

The purpose of this example is to illustrate the HRA approach presented in the chapter concerning human factors (see Chapter 16). We have thus chosen a simple example, on a fictional installation. Nonetheless, the results found are representative of those that might come out of the assessment of a real system. In particular, the probabilities fall within a likely range.

The system studied is a chemical unit making up product C out of two substances, A and B. The reaction is exothermic and so the unit must be continually cooled. Without coolant, there is a risk of explosion. In the case of cooling system loss, the inlet of reagents is therefore automatically cut off.

We will study the aftermath of a loss of coolant, followed by a failure of the automatic reagent inlet isolation (RII). The 'unacceptable consequences' are explosion of the unit. We will go through the phases of HRA as laid out in Section 16.5.

25.2 IDENTIFICATION OF POTENTIAL HUMAN ERRORS

- **Examination of procedures** Following the 'loss of coolant' alarm, the operator must verify that the RII is working. If this automatic mechanism fails, he must use procedure P_1. This asks him to send an operator to close the valves manually, 'on site'. In case of a new failure, he must trigger, from the control room, the injection of product S into the unit, which will stifle the reaction. This solution is only to be used as a last recourse, because it has the disadvantage of seriously damaging the installation.

The examination of procedure P_1 does not reveal any notable potential error in carrying out secondary actions. We can foresee the following errors in performing principal actions:

- the operator does not carry out (or carries out too late) manual RII, or injection of S;

---

*This chapter was written with the collaboration of M. Mosneron-Dupin (Electricité de France).

- the operator chooses the wrong valve during the manual RII on site (selection error), or else opens the valves wider instead of closing them;

- the operator reverses the operations (he injects product S before attempting RII).

• **Simulator test** A few trials on the simulator show that some operators are very reluctant to inject product S, which they know seriously damages the installation. This decision is not easy to make. The tests do not reveal any potential non-required action.

• **Feedback from experience and prior studies** Loss of coolant happens on the average once a year across operating installations, but the automatic RII has always functioned.

• **System studies** Some errors emerged in the course of a predictive systems safety analysis, in particular in the cause tree regarding product S injection failure (Figure 25.1).

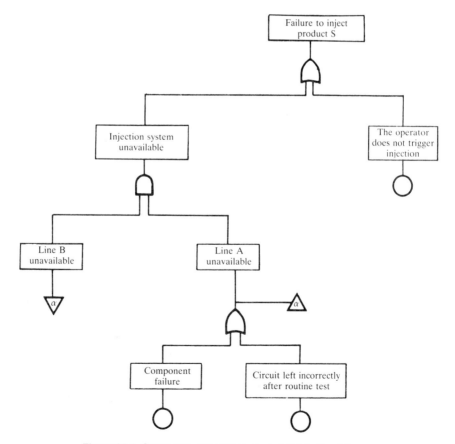

Figure 25.1 Cause tree concerning product S injection failure.

25.3 SELECTING SIGNIFICANT ERRORS

Among the errors brought out, we can eliminate the reversal of injection and RII, which does not lead to unacceptable consequences, even in combination with other errors.

Otherwise, the errors in the tree shown in Figure 25.1 belong to cut sets of the first or second order. There is not enough distance between these orders for us to make a selection *a priori*.

25.4 DETAILED ANALYSIS OF SIGNIFICANT ERRORS

Here we present the detailed analysis of errors related to manual RII on site.

- **Action characteristics** Functional studies show that RII must be performed 10 minutes at the most after coolant loss. Three minutes are necessary to go from the control room to the valve site and one minute to close each of the two valves. It is indispensable to close both valves; thus five minutes are left over for diagnosis, consultation of the procedure and decision making.

- **Man–machine interface characteristics** A visit to an existing installation allows us to gather some interesting information.

The principal means of accident detection and identification are the following:

- a Category 1 alarm (horn and flashing red light) for loss of coolant and another alarm for subsequent automatic RII failure;

- a unit temperature register and a gradient indicator—these parameters are amongst those most watched during normal operation;

- coolant flow indicators;

- state signals indicating RII valve position.

The procedure is clear and well presented. It has been checked by quality control and is available in the control room. It indicates valve coordinates and location.

Access to the valves is easy and they are clearly distinguishable. They are tagged and labelled. They are equipped with an unremovable twist handle, easily turned in the customary direction. They are marked with an index which allows their position to be checked. There is no similar valve nearby.

Communications between the control room and other rooms are ensured by loudspeakers and telephone.

- **Environmental characteristics** Average characteristics, nothing to report.

- **Organizational characteristics** Two technical agents are available for on-site manoeuvres. The operator has an assistant. In case of difficulty, the latter can be sent on-site. In the situation under consideration, the utilization of the designated procedure is obligatory.

- **Operator characteristics** Loss of coolant happens approximately once a year and thus is well known by the operator. Subsequent failure of automatic RII has never happened. Practice for accident situations on the simulator does not systematically include this accident, but it is mentioned in training. Interviews with operators confirm their reluctance regarding S injection.

- **Means of recovery** The operator's assistant provides some redundancy. The indicators described above are available to the operator in the control room to check that RII has been performed on-site.

- **Error-consequences** The non-performance of RII within the time limit will lead to the explosion of the unit, unless the operator proceeds to inject product S. If the agent chooses the wrong valve, the consequences are identical.

25.5 INTEGRATION WITH SYSTEM MODELLING

A first consequence tree is constructed and is presented in Figure 25.2.

Analysis shows that the failure of manual RII can schematically have two causes: human error and mechanical jamming of both valves.

The probability that the operator does not then proceed to S injection is not the same in the two cases. Indeed, if RII has failed through human error, it may be because the operator has not made the right diagnosis of the accident. He is unlikely then to go to the next step of injecting S. On the other hand, there is no notable dependence between the mechanical blocking of the valves and injection failure.

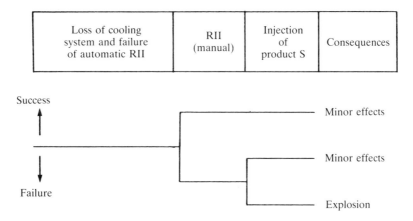

Figure 25.2 First consequence tree.

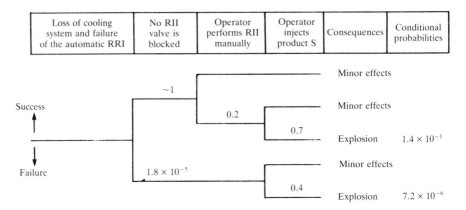

Figure 25.3 Final consequence tree.

The final event tree (Figure 25.3) allows us to take these dependencies into account, because it distinguishes the two causes of RII failure. Note that, in order to simplify, we have taken into account only the absence of valve blocking or simultaneous blocking of both.

25.6 QUANTIFICATION

RII by an operator is an operation which unfolds according to the pattern corresponding to Swain's models: detection, diagnosis, choice of procedure, task execution according to procedure.

We shall first use Swain's diagnostic model. Detailed analysis (Section 25.4) showed satisfactory conditions, and so we choose the nominal curve (see detailed selection criteria in Swain [26] pp. 12–23). Available time is approximately 5 minutes. The probability of failure to diagnose is thus about 2×10^{-1} (Figure 16.3). The diagnosis here is not very difficult. However, errors can be made in choosing the procedure, so we will consider that this number applies to the entire unit diagnosis-choice of procedure.

Now for the execution of the action. Analysis has shown that conditions are satisfactory, thus detailed quantification is unnecessary: the result would not modify the range found for diagnosis.

Thus we will retain the probability of 2×10^{-1} for the non-performance of manual RII by the operator.

To illustrate the use of Swain's model for task execution, we will detail the quantification of one of the errors included in the event 'circuit left incorrectly after routine test' (Figure 25.1). Figure 25.4 shows the circuit diagram.

In order to test the pump, B is opened and A is closed. At the end of the test, A is opened and B is closed. We will look at the error consisting of leaving A closed.

Qualitative analysis shows that we can foresee either a complete oversight (omission), or the opening of the close-by and similar valve X instead of

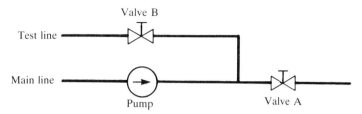

Figure 25.4.

valve A (selection error). Moreover, these valves are not well labelled. However, A is subject to administrative follow-up: it carries a lock and each manipulation must be logged in writing.

We can thus construct the tree seen in Figure 25.5 and use Table 25.1 in order to go about quantifying. The tree calculation gives us the probability of error:

$$P \simeq 0.95 \times (3 \times 10^{-3} + 3 \times 10^{-3}) + 5 \times 10^{-2} \times (10^{-2} + 8 \times 10^{-3})$$
$$= 6.6 \times 10^{-3}$$

The tree shown in Figure 25.5 is presented in the form suggested by Swain. It is possible, though, to adopt the usual consequence tree presentation (Figure 25.6).

We will not go into the quantification of other events. Note, however, that:

- The probability of both RII valves being blocked is outweighed by the

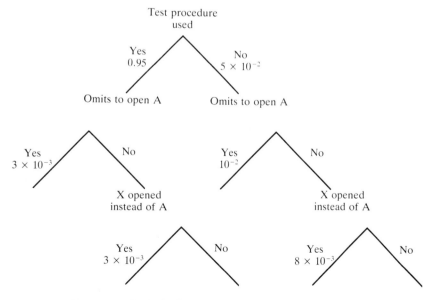

Figure 25.5 Analysis of the human error 'leaving valve A closed'.

Table 25.1 Elements for quantifying the human error 'leaving valve A closed'

Human error	Probability	Reference table†
• Failure to use written test procedure	5×10^{-2}	20.6/6
• Omission of an action when procedures without check-off provisions are used (more than 10 actions)	$3 \times 10^{-3}*$	20.7/4
• Omission of an action when written procedure is not used	$10^{-2}*$	20.7/5
• Selection error for locally operated valves (unclearly or ambiguously labelled, part of a group of two or more valves that are similar)		
— with procedure	$3 \times 10^{-3}*$	20.13/4
— without procedure	8×10^{-3}	20.13/4

*Lower limit and not nominal value given by table, in order to take account of better administrative surveillance of A (from Table 20.15/1 [26]).
†See Swain [26]. X – Y/Z: X – Y: Table No.; Z = rubric no. within table. Tables 20.7 and 20.13 can also be consulted in this book at 16.6.2.2 (Tables 16.2 and 16.3).

rate of common-cause failure of both valves (single failure rate: $1.8 \times 10^{-4}/d$; β factor $= 0.1$ (see Chapter 15)).

- The probability of S injection failure is outweighed by the probability of non-release by the operator. This value is found with Swain's diagnostic model. The basic value (0.2) is doubled in order to take into account operator's reluctance. In the case where RII has already failed through error, the probability is corrected thanks to Swain's dependence model, which gives 0.7 (strong dependence in relation to RII failure due to error).

The results are entered in the final consequence tree (Figure 25.6).

25.7 RESULTS

The probability of explosion of the unit in the case of loss of coolant followed by failure of automatic RII is approximately 1.4×10^{-1}. This conditional probability is estimated with an error factor of about 10. It can be attributed to the risk of non-performance of manual actions: RII and product S injection. Hardware failures have a negligible rate.

The main difficulty is diagnosis and choice of procedure. The quality of man–machine interface is good. Thus, improvement of training and a system modification are most apt to reduce the risk. A sensitivity study brings out the following elements:

- Systematic and periodic practice on the simulator with this accident could reduce risk by a maximum factor of about 10 (opening the way to using

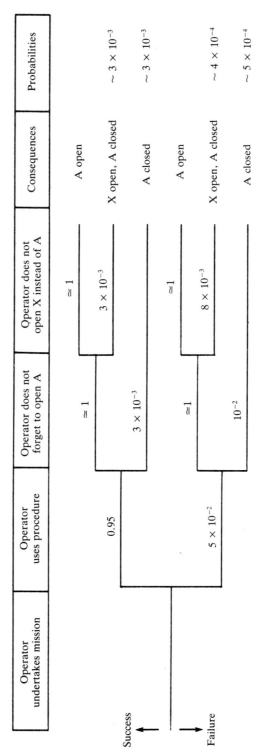

Figure 25.6 Consequence tree tracing the human error 'leaving valve A closed'.

the 'favourable conditions' curve in Swain's diagnosis model; disappearance of operator's reluctance).

- Modification of the system allowing five more minutes for intervention would divide the risk by about 2.

- Making the injection of S automatic would reduce the risk by a factor certainly higher than 10, or even 100, since the probability of hardware failure here is much less than the probability of the outstanding human errors.

The range of probabilities found and their relative values should be regarded as more important than the absolute values quoted.

APPENDICES

APPENDIX 1

MAIN DEFINITIONS

This appendix includes terms from both volumes, and is reproduced in both.

Accelerated test
A test in which the applied stress level is chosen to exceed that stated in the reference conditions in order to shorten the time duration required to observe the stress response of the entity, or to magnify the response in a given time duration.
Note: To be valid, an accelerated test shall not alter the basic fault modes and failure mechanisms, or their relative prevalence.

Acceptable
Qualifies an event judged acceptable with respect to dependability objectives.

Accident
Event having or liable to have catastrophic consequences.
Note: More precise and more technical definitions are sometimes used: in the nuclear field, an accident is defined as an event which can damage one or several barriers, thus leading to the release of radioactive products and calling for engineered safety systems to be put into service.

Accident scenario
Succession of (real or potential) events leading to an accident.

Accident sequence
Sequence of events leading to an accident.

Active component
Component with movable parts whose position is modified or whose configuration or properties must be changed with the help of an external energy source so that it can carry out its function.
Note: This definition is used in the nuclear field.

Active corrective maintenance time
That part of the active maintenance time during which actions of corrective maintenance are performed on an entity (IEC-50).

Active maintenance time
That part of the maintenance time during which a maintenance action is performed on an entity, either automatically or manually, excluding logistic delays (IEC-50).

Note: A maintenance action may be carried out while the entity is performing a required function (IEC-50).

Active preventive maintenance time
That part of the active maintenance time, during which actions of preventive maintenance are performed on an entity (IEC-50).

Active redundancy
That redundancy wherein all means for performing a required function are intended to operate simultaneously (IEC-50).

Administrative delay (for corrective maintenance)
The accumulated time during which an action of corrective maintenance on a faulty item is not performed due to administrative reasons (IEC-50).

Ageing failure
See wearout failure.

Asymptotic availability
For modelling purposes, the limit, if any, of the instantaneous availability when the time tends to infinity.
Note: It is denoted by $A(\infty)$.

Asymptotic failure rate
For modelling purposes, the limit, if any, of the instantaneous failure rate when the time tends to infinity.
Note: It is denoted by $\Lambda(\infty)$.

Asymptotic repair rate
For modelling purposes, the limit, if any, of the instantaneous repair rate when the time tends to infinity.
Note: It is denoted by $\mathcal{M}(\infty)$.

Asymptotic unavailability
For modelling purposes, the limit, if any, of the instantaneous unavailability when the time tends to infinity.
Note: It is denoted by $\bar{A}(\infty)$.

Auto maintainability
Probability that the human operator will return to appropriate behaviour before time t, when he has left it at time $t = 0$.

Availability (measure)
The probability that an entity is in a state to perform a required function under given conditions at a given instant of time.
Notes: 1. It is denoted by $A(t)$.
        2. This availability is also called 'instantaneous availability'.

Availability (performance)
The ability of an entity to be in a state to perform a required function under given conditions at a given instant of time.

Basic Event
In dependability assessment, a pertinent event serving as a basic (or elementary) unit in a model.
Note: Generally, basic events are component failures, operator errors, adverse environmental conditions etc. However, they can also relate to operation, maintenance, etc.

Bug
Defect introduced in a software by a human error in programming.

Burn-in (for a non-repairable entity)
A type of screening test employing the functional operation of an entity (IEC-50).

Burn-in (for repairable hardware)
A process of increasing the reliability performance of hardware employing functional operation of every entity in a prescribed environment with successive corrective maintenance at every failure during the early failure period (IEC-50).

Capability
The ability of an entity to meet a service demand with given quantitative characteristics under given internal conditions (IEC-50).
Notes: 1. Internal conditions refer, for example, to any combination of faulty and not faulty sub-entities (IEC-50).
2. For telecommunication services this is called trafficability performance (IEC-50).

Cascade failures
Dependent failures which can be classified in chronological order $A_1, \ldots, A_i, \ldots, A_n$ so that each failure $A_i (i \neq n)$ is the direct cause of the following failure A_{i+1}.

Cataleptic failure
A sudden failure which results in a complete inability to perform all required functions of an entity (IEC-50).
Note: This failure is also called catastrophic failure.

Catastrophic event
Any event which could potentially cause the loss of primary system function(s) resulting in significant damage to the system or its environment, and or cause the loss of life or limb.

Catastrophic failure
See cataleptic failure.

Cause–consequence diagram
A logic diagram showing the causes and the consequences of an initiating event.

Cause–consequence diagram method
Analysis method for identifying the causes and consequences of an initiating event; these causes and consequences can then be quantified.
Note: The causes and consequences of the initiating event are represented in a tree form called a cause–consequence diagram. The method is abbreviated to CCDM.

Cause tree
A logic diagram showing which event causes or combinations thereof result in a given undesirable event.
Note: The event causes can be failure modes, faults, external events, human errors, etc.

Cause tree method
Analysis method for identifying the causes of an undesirable event deductively; these causes can then be quantified.
Note: The causes of an undesirable event are represented in a tree form called a cause tree. The method is abbreviated to CTM.

Coherent structure function
Structure function possessing the following property:
$$x \geq y \Rightarrow \psi(x) \geq \psi(y)$$
Note: The term 'monotonic structure function' is also used.

Coherent system
System whose structure function is monotonic.
Note: A system is coherent if:

- the fault of all the components results in the system fault,
- the operation of all the components results in the system operation,
- when the system is failed, no additional failure will put the system back into operation,
- when the system is operating, no repair will result in the system fault.

Command failure
Failure of an entity whose direct or indirect cause is the failure of another entity and for which this entity was qualified and designed.

Common-cause failures
Dependent failures originating from the same direct cause.

Common-mode failures;
Common-cause failures which are made apparent by the same failure mode of the entities.
Note: The entities are generally assumed to be identical.

Complete failure
A failure which results in the complete inability of an entity to perform all required functions (IEC-50).

Compliance test
A test used to show whether or not a characteristic or a property of an entity complies with the stated requirements (IEC-50).

Component
The smallest part of a system which it is necessary and sufficient to consider for system analysis.

Confidence interval
Interval limited by the values of two statistics and such that the probability that the parameter to be estimated will belong to this interval is equal to a given value $(1 - \alpha)$.
Note: The upper and lower bounds of the confidence interval are called 'confidence limits'; the quantity $1 - \alpha$ is the confidence level.

Consequences (of an event)
Set of events which appear to be logical continuations of an event.
Note: There are catastrophic, critical, major, minor or significant consequences (see event).

Consequence tree
A logic diagram showing the consequences of an initiating event.

Consequence tree method
Analysis method for identifying all the consequences of an initiating event, in particular, those sequences of events resulting in undesirable events; these sequences can then be quantified.
Note: The sequences of events are represented in a tree form called a consequence tree. The method is abbreviated to CQTM.

Constant failure rate period
That possible period during which the failures occur at an approximately uniform rate (IEC-271, 1974).

Corrective maintenance
The maintenance carried out after fault recognition and intended to put an entity into a state in which it can perform a required function (IEC-50).

Corrective maintenance time
That part of the maintenance time during which corrective maintenance is performed on an entity, including technical delays and logistic delays inherent in corrective maintenance (IEC-50).

Covariance
The covariance of two random variables X and Y is the real number $\text{Cov}(X, Y)$, if any, defined by:

$$\text{Cov}(X, Y) = E[(X - E[X])(Y - E[Y])]$$

Note: $E[X]$ is the mean of the random variable X.

Critical component
Component whose failure, in a given operating state of the system, results in the system failure.

Critical event
Any event which could potentially cause the loss of the primary system function(s) resulting in significant damage to the said system or its environment and negligible hazard to life or limb.

Criticality analysis
Analysis for evaluating the probability–severity of the failure.

Cumulative distribution function
Function F giving, for any value x, the probability that the random variable X will be less than or equal to x.
Note: $F(x) = P[X \leq x]$.

Cut set
A combination of basic events resulting in the undesirable event.

Decision table method
Method for systematic analysis of all the combinations of events (operating states, failed states) of the components of a system; each component can have more than two states. The combinations can then be quantified.
Note: The results obtained using this method are presented in the form of a decision table. The method is abbreviated to DTM.

Deductive approach
When using this approach, the line of reasoning goes from the most general to the most specific.

Defect
Departure of a characteristic of an entity from the desired characteristic when this departure exceeds acceptable limits.

Deferred maintenance
Corrective maintenance which is not immediately initiated after a fault recognition but is delayed in accordance with given maintenance rules (IEC-50).

Degradation failure
A failure which is both a gradual failure and a partial failure (IEC-50).
Note: In time, such a failure may develop into a complete failure.

Dependability
The ability of an entity to perform one or several required functions under given conditions.
Note: This concept can encompass reliability, availability, maintainability, safety, durability, etc, or combinations of these abilities. Generally speaking, Dependability is considered to be the science of failures and faults.

Dependability assessment
Study to predict a dependability measure from an analysis leading to the construction of a model.

Dependability assurance
The implementation of adequate planned and systematic actions necessary to provide confidence that an entity will satisfy given dependability performance requirements.

Dependability audit
A systematic and independent examination to determine whether the activities and results comply with planned arrangements and whether these arrangements are effectively implemented and are suitable to achieve dependability performance objectives.

Dependability control
The operational techniques and activities that are employed to satisfy given dependability performance requirements of an entity.

Dependability growth
A condition characterized by a progressive improvement of a dependability performance measure of an entity with time.

Dependability improvement
A process undertaken with the deliberate intention of improving the dependability performance by eliminating failures and/or by reducing the occurrence probability of failures and/or by other means.

Dependability management
The administration of the functions and activities necessary to determine and satisfy the dependability performance requirements of an entity.

Dependability model
A model used for the prediction or estimation of one or several dependability measures of an entity.
Note: The model can be a mathematical one.

Dependability plan
A documented setting out the specific practices, resources and activities necessary to ensure that an entity will satisfy given dependability performance requirements relevant to a given contract or project.

Dependability programme
A documented set of time-scheduled activities, resources and events serving to implement the organization structure, responsibilities, procedures, activities, capabilities, and resources that together ensure that an entity will satisfy given dependability performance requirements relevant to a given contract or project.

Dependability surveillance
The continuing observation of the status of procedures, methods, conditions,

products, processes and services and analysis of records to assure that dependability performance requirements will be met.
Note: Dependability surveillance is often carried out by the customer or third parties to ensure that the contractual requirements are being met.

Dependability verification
A procedure applied for the purpose of determining whether the requirements for dependability performance measures for an entity have been met or not.
Note: The procedures may range from analysis of appropriate data to a dependability demonstration.

Dependent failures
Interdependent simultaneous or concomitant failures of multiple entities.
Note: We then have $P[E_1.E_2] = P[E_1].P[E_2|E_1] \neq P[E_1].P[E_2]$.

Determinate fault
For an entity which produces a response as a result of an action, a fault for which the response is the same for all actions (IEC-50).

Determination test
A test used to establish the value of a characteristic or a property of an entity (IEC-50).

Down time
The time interval during which an entity is in a down state (IEC-50).

Durability
The ability of an entity to perform a required function under given conditions of use and maintenance, until a limiting state is reached (IEC-50).
Note: A limiting state of an entity may be characterized by the end of the useful life, unsuitability for any economic or technological reasons or other relevant factors (IEC-50).

Duration
The difference between the end points of a time interval (IEC-50).

Early failure
Failure occurring at the beginning of the life of an entity and whose rate decreases rapidly with time.
Note: The life of an entity begins at a specific moment: upon leaving the assembly line or the factory, upon delivery, etc.

Early failure period
That possible early period beginning at a stated time and during which the failure rate decreases rapidly in comparison with that of the subsequent period (IEC-271, 1974).

Effect (of a failure, of a fault)
The set of manifestations of all types which result from the sole failure (or fault).

Effectiveness (performance)
The ability of an entity to meet a service demand of given quantitative characteristics (IEC-50).
Note: This ability depends on the combined aspects of the capability and the availability performance of the entity (IEC-50).

Endurance test
A test carried out over a time interval to investigate how the properties of an entity are affected by the application of stated stresses and by their time duration or repeated application (IEC-50).

Entity
Any part, component, device, subsystem, functional unit, equipment or system that can be individually considered.
Notes: 1. An entity may consist of hardware, software, or both, and may also in particular cases, include people.
2. An entity can also be called an 'item'.

Error mode
Effect by which a human error is observed.

Estimated
Qualifies a value obtained as the result of the operation made for the purpose of assigning, from the observed values in a sample, numerical values to the parameters of the distribution chosen as the statistical model of the population from which this sample is taken (IEC-50).
Note (IEC-50): The result may be expressed either as a single numerical value, a point estimate, or as a confidence interval.

Estimated value
Value obtained as the result of an operation aimed at assigning, from the values observed in a sample, numerical values to the parameters of the distribution chosen as the statistical model of the population from which this sample is taken.
Note: An estimated value can also be called an estimate. The result can be given either as a single numerical value, in the case of a point estimation, or as a confidence interval.

Expectation (of a random variable)
The quantity (written $E[X]$), if any, which is defined as follows:

- for a discrete random variable X taking values x_i, with probabilities p_i, by the sum

$$E[X] = \sum_i p_i x_i$$

  extended to all the values of x_i likely to be taken by X;
- for a continuous random variable X with probability density function $f(x)$, by the integral:

$$E[X] = \int xf(x)\,dx$$

extended to the entire domain of X.

Note: The terms expected value, mean value or mean of a random variable and mean of a distribution are also used.

External gathered fault
The set of internal gathered faults (or combinations thereof) pertaining to other elementary systems and likely to affect the operation of the elementary system being studied.

Extrapolated
Qualifies a predicted value based on observed or estimated values for one or a set of conditions, intended to apply to other conditions such as time, maintenance and environmental conditions (IEC-50).

Extrapolated value
Value predicted from values observed under or estimated for certain time, operating, maintenance or environmental conditions when this value is to be applied to other conditions.

Failed state
State of an entity characterized by the inability to perform a required function.

Fail safe
A design property of an entity which prevents its failures from resulting in catastrophic faults.

Failure
The termination of the ability of an entity to perform a required function (IEC-50).
Note: After failure, the entity has a fault.

Failure analysis
See fault analysis.

Failure cause
The circumstances during design, manufacture, or use which have led to a failure (IEC-50).

Failure density
The limit, if any, of the ratio of the conditional probability that the instant of time, T, of the first failure of an entity falls within a given time interval $[t, t + \Delta t]$, to the length of this interval, Δt, when Δt tends to zero, given that the entity is in an up state at time $t = 0$.
Note: It is denoted by $U(t)$.

Failure intensity acceleration factor
In a time interval of given duration, whose beginning is specified by a fixed

age of a repaired entity, the ratio of the number of failures obtained under two different sets of stress conditions.

Failure mechanism
The physical, chemical or other process which has led to a failure (IEC-50).

Failure mode
The effect by which a failure is observed (IEC-271).
Note: The terms 'failure mode' and 'fault mode' can be regarded as equivalent.

Failure modes and effects analysis (FMEA)
A qualitative method of system analysis which involves the study of the failure modes which can exist in every component of the system and the determination of the causes and of the effects of each failure mode.

Failure modes, effects and criticality analysis (FMECA)
A qualitive method of system analysis which involves a failure modes and effects analysis together with a criticality analysis.

Failure rate
The limit, if any, of the ratio of the conditional probability that the instant of time, T, of a failure of an entity falls within a given time interval, $[t, t + \triangle t]$, to the length of this interval, $\triangle t$, when $\triangle t$ tends to zero, given that the entity has not failed over $[0, t]$.
Notes: 1. It is also called 'instantaneous failure rate'.
2. It is denoted by $\Lambda(t)$.

Failure rate acceleration factor
The ratio of the failure rate under accelerated testing conditions to the failure rate under stated reference test conditions (IEC-50).
Note: Both failure rates refer to the same time period in the life of the tested entities (IEC-50).

Fault
The state of an entity characterized by its inability to perform a required function.
Note: A fault is often the result of a failure of the entity itself, but may exist without prior failure.

Fault analysis
The logical, systematic examination of an entity to identify and analyse the probability, causes and consequences of potential faults.
Note: It is also called 'failure analysis'.

Fault correction
Actions taken after fault localization to restore the ability of the faulty entity to perform a required function (IEC-50).

Fault coverage
The proportion of faults of an entity that can be successfully corrected (IEC-50).

Fault mode
Effect by which a fault is observed.
Note: The concepts of 'failure mode' and 'fault mode' can be regarded as equivalent.

Fault tolerance
The attribute of an entity that makes it able to perform a required function in the presence of certain given sub-entity faults (IEC-50).

Faulty
Pertaining to an entity which has a fault (IEC-50).

Field data
Observed data obtained during field operation (IEC-50).

Field test
A compliance test or determination test made in the field where operating, environmental, maintenance and measurement conditions present at the time of the test are recorded (IEC-50).

Fleeting fault
Intermittent fault of an entity which is hard to detect.

Free time
The time interval during which an entity is in a free state.
Note: It is also called 'idle time'.

Functional mode
The effect by which an entity is observed to operate.

Gathered fault combination method
Method for analysing combinations of failures (and/or faults) of system components and which is applied to obtain the failure modes and/or the undesirable events of the system by gathering the failures (and/or faults) which have the same effects; these failure modes and/or undesirable events can then be quantified.
Note: The method is abbreviated to GFCM.

Generic event
Event of a consequence tree occurring after the initiating event in a sequence.
Note: The terms 'event heading' and 'enabling event' can also be used.

Global gathered fault
The set of internal gathered faults or external gathered faults (or combinations thereof) of an elementary system having the same effects or consequences on the elementary system being studied as on other elementary systems.

Gradual failure
A failure due to a gradual change with time of given characteristics of an entity (IEC-50).

Notes: 1. A gradual failure may be anticipated by prior examination or monitoring and can sometimes be avoided by preventive maintenance (IEC-50).
2. A gradual failure is also called a 'drift failure'.

Hazard
Situation which is potentially dangerous for man, society or the environment.
Note: The term 'danger' can also be used.

Human availability
The ability of a human operator to be in a state to perform a required mission under given conditions at a given instant of time.

Human error
The departure of a human operator's behaviour from what it should be, this departure exceeding acceptable limits under given conditions.

Human failure
The termination of the ability of a human operator to perform a required mission.

Human failure mode
The effect by which a human failure is observed.

Human fault
The state of a human operator characterized by prolonged inability to perform a required mission.

Human incapability
The departure of a human operator's behaviour from what it should be because the conditions needed to carry out actions do not exist.

Human maintainability
The ability of a human operator to stay in or go back to a state in which he can fulfil a required mission under given conditions.

Human operator
Any individual, team or human organization having an action or a mission to carry out during the different phases of the life of a system.

Human reliability
The ability of human operator to perform a required mission under given conditions in a given time interval.

Human reliability assessment;
Dependability assessment concentrating on the human errors liable to be committed by the human operator having a mission to fulfil on a system.
Note: It is abbreviated to HRA.

Idle time
See free time.

Implicant
Combination of basic events and/or their complementary events whose occurrence leads to the undesirable event.

Incapability mode
Effect by which a human incapability is observed.

Incident
Event having or likely to have critical effects or consequences.

Inductive approach
When using this approach, the line of reasoning goes from the most specific to the most general.

Initiating event
Event in the life of an entity which has—or which can have—unacceptable consequences.
Note: The term initiator is also sometimes used. An initiating event can lead to an undesirable event.

Instant of time
A single point on a time scale (IEC-50).

Intermittent fault
A fault of an entity which persists for a limited time duration following which the entity recovers the ability to perform a required function without being subjected to any action of corrective maintenance (IEC-50).
Notes: 1. Such a fault is often recurrent (IEC-50).
2. It is also called a 'volatile fault' or 'transient fault'.

Internal gathered fault
Set of failure modes or combinations of failure modes which are internal to an elementary system and which have the same effects or consequences on the elementary system under consideration as on the other elementary systems.

Irrecoverability (of a human error)
The probability that the human error will not have been corrected before time t, given that it was committed at time $t = 0$.

Laboratory test
A compliance test or a determination test made under prescribed and controlled conditions which may or may not simulate field conditions (IEC-50).

Latent fault
An existing fault that has not yet been recognized (IEC-50).

Logistic delay
That accumulated time during which a maintenance action cannot be performed due to the necessity to acquire maintenance resources, excluding any administrative delay (IEC-50).

Note: Logistic delays can be due to, for example, travelling to unattended installations, pending arrival of spare parts, specialists, test equipment, information and suitable environmental conditions (IEC-50).

Maintainability (measure)
The probability that a given active maintenance action, for an entity under given conditions of use, can be carried out within a stated time interval, when the maintenance is performed under given conditions and using stated procedures and resources, given that the entity is faulty at time $t = 0$.
Note: It is denoted by $M(t)$.

Maintainability (performance)
The ability of an entity under given conditions of use, to be retained in, or restored to, a state in which it can perform a required function, when maintenance is performed under given conditions and using stated procedures and resources.
Note: The term 'maintainability' is also used as a measure of maintainability performance.

Maintenance
The combination of all technical and administrative actions, including supervision actions, intended to retain an entity in, or restore it to, a state in which it can perform a required function (IEC-50).

Maintenance support performance
The ability of a maintenance organization, under given conditions, to provide upon demand, the resources required to maintain an entity, under a given maintenance policy (IEC-50).
Note: The given conditions are related to the entity itself and to the conditions under which the entity is used and maintained (IEC-50).

Maintenance time
The time interval during which a maintenance action is performed on an entity either manually or automatically, including technical delays and logistic delays (IEC-50).
Note: Maintenance may be carried out while the entity is performing a required function (IEC-50).

Major event
Any critical or significant event.

MDT (mean down time)
The expectation of the down time (IEC-50).

Mean (of a randon variable)
The value obtained as the expectation of a random variable.

Mean availability
The mean of the instantaneous availability over a given time interval $[t_1, t_2]$.
Note: It is denoted by $A^m(t_1, t_2)$.

Mean unavailability
The mean of the instantaneous unavailability over a given time interval $[t_1, t_2]$.
Note: It is denoted by $\bar{A}^m(t_1, t_2)$.

Measure (in the probabilistic treatment of dependability)
A function or a quantity used to describe a random variable or a random process (IEC-50).
Note: For a random variable, examples of measures are the distribution function and the mean.

Minimal cut set
Cut set such that, if one of the events in a minimal cut set does not occur, then the undesirable event will not happen.

Minimal cut state
Failed state of a system where the repair of any one failed component leads to an operating state.

Minimal failed state
Failed state of a system where the repair of one system component leads to an operating state.
Note: This type of state has at least one transition towards operating states. Also called minimal faulty state.

Minimal operating state
Operating state of a system where the failure of one system component leads to a failed state.
Note: This type of state has at least one transition towards failed states.

Minimal success path
In a success diagram, a success path in which there is no other success path.

Minimal tie set
Tie set such that, if any entity does not operate, the combined operation of the remaining entities does not make up a new tie set.
Note: A minimal tie set is a tie set containing no other tie set.

Minimal unacceptable sequence
Unacceptable sequence of events such that, if any of the basic events of the sequence does not occur, a new unacceptable sequence cannot take place.
Note: It is important to keep in mind the order in which the basic events occur.

Minor event
Any event which could cause degradation of system performance function(s) resulting in negligible damage to either the system or its environment, and in no damage to life or limb.

MTBF (mean time between failures)
The expected operating time between two failures.

MTTF (mean time to failure)
The expected operating time to first failure.
Note: The MTTF is also called MTTFF (mean time to first failure).

MTTR (mean time to repair)
The expectation of the time to restoration (or to repair).

MUT (mean up time)
The expected operating time (or up time)
Note: It is also the expectation of the operating time after a repair.

Non-coherent system
System whose structure function is not monotonic.

Non-operating state
The state when an entity is not performing a required function (IEC-50).

Non-relevant failure
A failure that should be excluded in interpreting or assessing a dependability measure.

Non-repaired entity
An entity which is not repaired after a failure (IEC-50).
Note: A non-repaired entity may be repairable or not.

Non-required time
The time interval during which the user does not require the entity to be in a condition to perform a required function (IEC-50).

Non-sequential system
System for which the dependability model does not take into account the order in which the basic events occur.

Observed value
Values related to an entity or a process, obtained by direct observation.

Operating state
The state when an entity is performing a required function (IEC-50).

Operating time
The time interval during which an entity is in an operating state (IEC-50).

Operation
The combination of all technical and administrative actions intended to enable an entity to perform a required function, recognizing necessary adaptation to changes in external conditions (IEC-50).
Note: By external conditions are understood, for example, service demand and environmental conditions (IEC-50).

Operational value
Value determined under real and given operating conditions.

Partial failure
A failure which results in the inability of an entity to perform some, but not all, required functions (IEC-50).
Note: A partial failure can also be defined as 'a failure resulting from deviations in characteristic(s) beyond specified limits, but not such as to cause complete lack of the required function' (IEC-271).

Permanent fault
A fault of an entity that persists until an action of corrective maintenance is performed.
Note: It is also called 'persistent fault' or 'solid fault'.

Persistent fault
See permanent fault.

Precursor event
Event whose occurrence makes it likely that another event having a probability and/or consequences larger than expected will exist.

Predicted
Qualifies a value assigned to a quantity, before the quantity is actually observable, computed on the basis of earlier observed or estimated values of the same quantity or of other quantities using a mathematical model (IEC-50).
Note: A predicted value may also be designated by the term 'prediction' (IEC-50).

Predicted value
Numerical value assigned to a quantity, before the quantity is actually observable, computed from earlier observed or estimated values of the same quantity or of other quantities using a mathematical model.

Prediction
The process of computation used to obtain the predicted value(s) of a quantity (IEC-50).
Note: The term 'prediction' may also be used to denote the predicted value(s) of a quantity.

Preliminary hazard analysis
Analysis for identifying and assessing the (economic, human, etc) hazards inherent in using a system and which is carried out before using other more precise methods of analysis.
Note: It is abbreviated to PHA. It is called preliminary hazard and risk analysis when it is used together with a risk assessment.

Preliminary hazard and risk analysis
An analysis which involves a preliminary hazard analysis together with an assessment of risk.

Preventive maintenance
The maintenance carried out at predetermined intervals or according to

prescribed criteria and intended to reduce the probability of failure or the degradation of the functioning of an entity (IEC-50).

Preventive maintenance time
That part of the maintenance time during which preventive maintenance is performed on an entity, including technical delays and logistic delays inherent in preventive maintenance (IEC-50).

Primary failure
A failure of an entity, not caused either directly or indirectly by a failure or a fault of another entity (IEC-50).

Prime implicant
Implicant containing no other implicant.

Probabilistic risk assessment
Study aimed at evaluating the risks of a system using a probabilistic method.
Note: It is abbreviated to PRA.

Probabilistic safety assessment
Study aimed at evaluating the safety of a system using a probabilistic method.
Note: It is abbreviated to PSA.

Probability density function
The derivative, if any, of the cumulative distribution function of a random variable.
Note: It is denoted by $f(x)$; $f(x) = \frac{dF}{dx}(x)$

Quality
The totality of features and characteristics of a product or service that bear on its ability to satisfy stated or implied needs (ISO-8402)
Notes (ISO-8402-1986):
  1. In a contractual environment, needs are specified, whereas in other environments, implied needs should be identified and defined.
  2. In many instances, needs can change with time; this implies periodic revisions of specifications.
  3. Needs are usually translated into features and characteristics with specified criteria. Needs may include aspects of usability, safety, availability, maintainability, economics and environment.
  4. The term 'quality' is not used to express a degree of excellence in a comparative sense nor is it used in a quantitative sense for technical evaluations. In these cases a qualifying adjective shall be used. For example, use can be made of the following terms:
     (a) 'relative quality' where products or services are ranked on a relative basis in the 'degree of excellence' or 'comparative' sense;
     (b) 'quality level' and 'quality measure' where precise technical evaluations are carried out in a 'quantitative sense'.
  5. Product or service quality is influenced by many stages of interactive

activities, such as design, production or service operation and maintenance.
6. The economic achievement of satisfactory quality involves all stages of the quality loop (quality spiral) as a whole. The contributions to quality of the various stages within the quality loop (quality spiral) are sometimes identified separately for emphasis. Two examples: 'quality attributable to design', 'quality attributable to implementation'.
7. In some reference sources, quality is referred to as 'fitness for use' or 'fitness for purpose' or 'customer satisfaction' or 'conformance to the requirements'. Since these represent only certain facets of quality, fuller explanations are usually required that eventually lead to the concept defined above.

Quality assurance
All those planned and systematic actions necessary to provide adequate confidence that a product or service will satisfy given requirements for quality (ISO-8042-1986).
Notes (ISO-8042-1986):
1. Unless given requirements fully reflect the needs of the user, quality assurance will not be complete.
2. For effectiveness, quality assurance usually requires a continuing evaluation of factors that affect the adequacy of the design or specification for intended applications as well as verifications and audits of production, installation and inspection operations. Providing confidence may involve producing evidence.
3. Within an organization, quality assurance serves as a management tool. In contractual situations, quality assurance also serves to provide confidence in the supplier.

Random process
Set of time-dependent random variables whose values are governed by a given set of multidimensional distributions which correspond to all the combinations of the random variables.

Random variable
Variable which can take any one of a given set of values, each with an associated distribution.
Note: A random variable which can only take distinct and separate values is said to be 'discrete'. A random variable which can take all the values of a finite or infinite interval is said to be 'continuous'.

Recoverability (of a human error)
The probability that the human error will have been corrected before time t, given that it was committed at time $t = 0$.

Reduction
Process consisting in identifying and obtaining, from the set of failure or fault combinations considered, those most relevant to the dependability measure sought.

Redundancy
In an entity, the existence of more than one means for performing a required function (IEC-50).

Relevant failure
A failure that should be included in interpreting or assessing a dependability measure.

Reliability (measure)
The probability that an entity can perform a required function under given conditions for a given time interval, $[t_1, t_2]$ (IEC-50).
Notes: 1. It is generally assumed that the entity is in a state to perform this required function at the beginning of the time interval (IEC-50).
2. The term 'reliability' is also used to denote the reliability performance quantified by this probability (IEC-50).

Reliability (performance)
The ability of an entity to perform a required function under given conditions for a given time interval (IEC-50).
Notes: 1. It is generally assumed that the entity is in a state to perform this required function at the beginning of the time interval (IEC-50).
2. The term 'reliability' is also used as a measure of reliability performance (IEC-50).

Repair
The part of corrective maintenance in which manual actions are performed on the entity (IEC-50).

Repair density
The limit, if any, of the ratio of the conditional probability that the maintenance action terminates in a time interval, $[t, t + \Delta t]$, to the length of this time interval, Δt, when Δt tends to zero, given that the entity is faulty at time $t = 0$.
Note: It is denoted by $G(t)$.

Repaired entity
A repairable entity which is in fact repaired after a failure (IEC-50).

Repairman
Human operator whose mission is to perform maintenance on an entity.

Repair rate
The limit, if any, of the ratio of the conditional probability that the corrective maintenance action terminates in a time interval, $[t, t + \Delta t]$, to the length of this interval, Δt, when Δt tends to zero, given that the action had not terminated at the beginning of the time interval (IEC-50).
Note: 1. It is also called 'instantaneous repair rate'.
2. It is denoted by $\mathcal{M}(t)$.

Repair time
That part of active corrective maintenance time during which repair actions are performed on an entity (IEC-50).

Required function
A function, or a combination of functions, of an entity, which is considered necessary to provide a given service (IEC-50).

Required time
The time interval during which the user requires the entity to be in a condition to perform a required function (IEC-50).

Risk
Hazard measure combining a measure of the occurrence of an undesirable event and a measure of its effects or consequences.

Safety (measure)
Probability that an entity will not, under given conditions cause critical or catastrophic events.
Note: The term safety also refers to the safety (performance).

Safety (performance)
Ability of an entity not to cause, under given conditions, critical or catastrophic events.

Scenario (of events)
Succession of real or potential events.

Scheduled maintenance
The preventive maintenance carried out in accordance with an established time schedule (IEC-50).

Screening test
A test or a set of tests intended to remove or detect defective entities or those likely to exhibit early failures (IEC-50).

Secondary failure
A failure of an entity, caused either directly or indirectly by a failure or a fault of another entity and for which that entity has not been qualified or designed.

Sequence (of events)
Succession of basic events.
Note: A sequence is the modelling of a scenario.

Sequential system
System for which the dependability model must take into account the order in which the basic events occur.

Serveability performance
The ability of a service to be obtained—within specified tolerances and other given conditions—when requested by the user and to continue to be provided for a requested duration (IEC-50).
   Notes: 1. Serveability performance may be subdivided into the service accessibility performance and the service retainability performance (IEC-50).

2. Other word: serviceability performance.

Service
A set of functions offered to a user by an organization (IEC-50).

Serviceability performance
See serveability performance.

Service accessibility performance
The ability of a service to be obtained, within specified tolerances and other given conditions, when requested by the user (IEC-50).
Note: For example, in telecommunications, this takes into account the transmission tolerance and the combined aspects of propagation performance, trafficability performance and availability performance of the related systems (IEC-50).

Service retainability performance
The ability of a service, once obtained, to continue to be provided under given conditions for a requested duration (IEC-50).
Note: For example, in telecommunications this takes into account the transmission tolerance and the combined aspects of propagation performance, trafficability performance and availability performance of the related systems (IEC-50).

Service support performance
The ability of an organization to provide a service and assist in its utilization (IEC-50).

Significant event
Any event which degrades system performance function(s) without appreciable damage to either system or life or limb.

Solid fault
See permanent fault.

Standard deviation
The square root of the variance of a random variable.

Standby redundancy
That redundancy wherein a part of the means for performing a required function is intended to operate, while the remaining part(s) of the means are inoperative until needed (IEC-50).

Standby state
A non-operating up state during the required time (IEC-50).

Standby time
The time interval during which an entity is in a standby state (IEC-50).

State graph
A diagram showing the set of possible states of an entity and the possible one-step transitions between these states.
Note: It is also called 'state-transition graph' or 'state-transition diagram'.

State sequence
Succession of (real or potential) operating and/or failed states of entities.

State space method
Analysis method for identifying and assessing the operating and failed states of a repairable system.
Note: The operating and failed states, together with their transitions, are represented by state graphs. The method is abbreviated to SSM.

State variable
State variable x associated with an event and such that:

$$x = 1 \text{ if the event takes place}$$

$$x = 0 \text{ if the event does not take place.}$$

Note: The term 'binary indicator variable' is also used.

Steady-state
Qualifies a value determined for conditions of an entity when characteristic parameters of the entity remain constant (IEC-50).

Steady-state availability
The mean of the instantaneous availability under steady-state conditions over a given time interval (IEC-50).

Steady-state unavailability
The mean of the instantaneous unavailability under steady-state conditions over a given time interval (IEC-50).

Step stress test
A test consisting of several increasing stress levels applied sequentially for periods of equal time duration to an entity (IEC-50).

Stress analysis
Determination of the physical, chemical or other stresses an entity experiences under given conditions (IEC-50).

Stress model
A mathematical model used to describe the influence of relevant applied stresses on a reliability performance measure or on any other property of an entity (IEC-50).

Structure function
Function expressing the state variable associated with an undesirable event based on the state variables of the basic events.
Note: The structure function is denoted by $\psi(x)$.

Success diagram
Logic diagram in which each system component is represented by a block or a combination of blocks showing how component operation leads to system operation.
Note: The term 'reliability block diagram' is also used.

Success diagram method
Method for analysing system reliability which begins with the analysis of the system operation and whose objective is to obtain a reliability or availability model from the failures of the system components.
Note: The graphical representation of the model is a success diagram. The method is abbreviated to SDM.

Success path
The path in a success diagram representing a combination of entity operating states which ensure that the required function is performed.

Sudden failure
A failure that could not be anticipated by prior examination or monitoring (IEC-50).

System
Given set of discrete elements (or components) which are interconnected or are interacting.
Note: The following classification is proposed: (part) $\subset$ (component) $\subset$ (subsystem) $\subset$ (elementary system) $\subset$ (system).

System analysis
Process oriented towards the acquisition and orderly investigation and processing of information specific to the system and relevant to a decision or a given goal.
Note: The end product of this process is a model of the system.

Test
An experiment carried out in order to measure, quantify or classify a characteristic or a property of an entity (IEC-50).

Test Data
Observed data obtained during tests (IEC-50).

Tie set
Combination of entity operating states which ensure that the required function is performed.

Time acceleration factor
The ratio between the time durations necessary to obtain the same stated number of failures or degradations in two equal size samples under two different sets of stress conditions involving the same failure mechanisms and fault modes and their relative prevalence (IEC-50).
Note: One of the two sets of stress conditions should be a reference set (IEC-50).

Time interval
The part of a time scale between, and described by, two given instants of time (IEC-50).

Transient fault
See intermittent fault.

Transition density
The derivative, if this exists, of the transition probability.

Transition probability
Probability that a system will leave a state during the time interval [0, t] to enter another state in one transition, given that the first state was entered at $t = 0$.
Note: The term 'transition distribution' is also used.

Transition rate
Limit, if any, of the quotient of the probability of leaving one system state for another system state during the given time interval [t, $t + \triangle t$], by the duration of the time interval, as $\triangle t$ tends to zero.

Truth table method
Method for systematically analysing all the combinations of events (operating states, failed states) of the system components, each component having only two states. These combinations can then be quantified.
Note: The results of this method are presented in the form of a truth table. The method is abbreviated to TTM.

Unacceptable
Qualifies an event deemed intolerable considering the dependability objectives.

Unavailability (measure)
The probability that an entity is not in a state to perform a required function, under given conditions at a given instant of time.
Notes: 1. The term 'unavailability' is also used to denote the unavailability performance quantified by this probability.
2. The unavailability is denoted by $\bar{A}(t)$.

Unavailability (performance)
The inability of an entity to be in a state to perform a required function under given conditions at a given instant of time.
Notes: 1. This inability depends on the combined aspects of the reliability performance, the maintainability performance and the maintenance support performance.
2. The term 'unavailability' is also used as a measure of unavailability performance.

Unconditional failure intensity
The limit, if any, of the ratio of the conditional probability that the instant of time, T, of a failure of an entity falls within a given time interval [t, $t + \triangle t$], to the length of this interval, $\triangle t$, when $\triangle t$ tends to zero, given that the entity is in an up state at time $t = 0$.
Note: It is denoted by $W(t)$.

Unconditional repair intensity
The limit, if any, of the ratio of the conditional probability that the corrective maintenance action terminates in a time interval, $[t, t + \triangle t]$, to the length of this time interval, $\triangle t$, when $\triangle t$ tends to zero, given that the entity is in an up state at time $t = 0$.
Note: It is denoted by $V(t)$.

Undesirable event
Event (in the life of an entity) which should not occur or which should occur with a lower probability considering dependability objectives.

Undetected fault time
The time interval between failure and recognition of the resulting fault (IEC-50).

Unmaintainability (measure)
The probability that a given active maintenance action for an entity under given conditions of use cannot be carried out within a stated time interval, when the maintenance is performed under given stated conditions and using stated procedures and resources.
Notes: 1. The term 'unmaintainability' is also used to denote the unmaintainability performance quantified by the probability.
2. The unmaintainability is denoted by $\overline{M}(t)$.

Unmaintainability (performance)
The inability of an entity under given conditions of use, to be retained in, or restored to, a state in which it can perform a required function, when maintenance is performed under given conditions and using stated procedures and resources.
Note: The term 'unmaintainability' is also used as a measure of maintainability performance.

Unreliability (measure)
The probability that an entity cannot perform a required function under given conditions for a given time interval $[t_1, t_2]$.
Notes: 1. It is generally assumed that the entity is in a state to perform this required function at the beginning of the time interval.
2. The term 'unreliability' is also used to denote the unreliability performance quantified by this probability.
3. The unreliability is denoted by $\overline{R}(t)$.

Unreliability (performance)
The inability of an entity to perform a required function under given conditions in a given time interval.

Unsafety (measure)
The probability that an entity will cause critical or catastrophic events under given conditions.
Note: The term 'unsafety' is also used to designate unsafety performance.

Unsafety (performance)
The ability of an entity to cause critical or catastrophic events under given conditions.

Unscheduled maintenance
The maintenance carried out, not in accordance with an established time schedule, but after reception of an indication regarding the state of an item (IEC-50).

Up state
A state of an entity characterized by the fact that it can perform a required function.

Up time
The time interval during which an entity is in an up state (IEC-50).

Useful life
Under given conditions, the time interval beginning at a given instant of time and ending when the failure intensity becomes unacceptable or when the entity is considered unrepairable as a result of a fault (IEC-50).

Variance (of a random variable)
The mathematical expectation of the squared difference between a random variable and its mathematical expectation.

Vector of basic event states
A vector representing a combination of the states of all the basic events, each basic event being represented by a state.

Volatile fault
See intermittent fault.

Wear-out failure
A failure whose probability of occurrence increases with the passage of time, as a result of processes inherent in the entity (IEC-50).
Note: It is also called 'ageing failure'.

Wear-out failure period
That possible period during which the failure rate increases rapidly in comparison with the preceding period (IEC-271).

APPENDIX 2
AVAILABILITY OF A COMPONENT AND A SYSTEM ON STANDBY AND PERIODICALLY TESTED

A2.1 INTRODUCTION

In industry, large numbers of components are on standby and are periodically tested. These components are very important, particularly when they form part of safety systems for industrial installations. If this is the case, the components are in non-operation most of the time, but must be ready to start up and operate when an incident or accident occurs.

This appendix deals with the availability as a function of time and the mean availability of one of these components, and it is demonstrated that there is an optimum interval between tests. The availability of systems made up of these components is then examined. Reference [1] presents an early appraisal of these problems, an appraisal that was followed by numerous papers [2–7], notably by Signoret [2, 7].

A2.2 AVAILABILITY OF A COMPONENT ON STANDBY AND PERIODICALLY TESTED

We evaluate this availability on the basis of the following hypotheses:

- the component is started up at regular intervals to test whether a fault occurs or has occurred;

- the component in non-operation is liable to experience failures that will only be detected when the test subsequent to the appearance of the failure is performed;

- the component is repaired after a failure has been detected during a test.

The characteristic parameters of the component are as follows:

λ_a: failure rate in non-operation (due to latent fault); also called standby failure rate;
γ_d: failure rate (upon demand) due to start-up;
μ: repair rate;
T: interval between tests.

The component is brought into service at time $t = 0$ and the first test is performed at time T. The component was thus in operating state at time $t = 0$. The time parameter is as shown in Figure A2.1.

Let us consider the interval $k + 1$: $t = \delta + kT$. The estimated instantaneous availability of the component (on standby, in non-operation and before being started up) is denoted by $A_k(\delta)$ at instant δ of time interval $[kT, (k + 1)T]$. Immediately before a given test (k), the component could be unavailable for two reasons:

- it is being repaired subsequent to an earlier failure;

- it is in failed state resulting from a failure that has occurred during the period of non-operation since the preceding test $(k - 1)$.

We define $\bar{A}_{k-1}(T)$ as the unavailability of the component just before the test. $\bar{A}_{k-1}(T) + \gamma_d$ is the probability that the component will not start up during test k. At time t, we can assess the instantaneous availability of the component as follows:

- the component was available at $\delta = 0$ and no failure has occurred between 0 and δ:

$$[1 - (\bar{A}_{k-1}(T) + \gamma_d)]e^{-\lambda_a \delta}$$

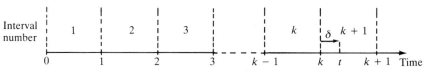

Figure A2.1

- the component was in failed state at $\delta = 0$, was repaired at $\delta' < \delta$ and has not failed in non-operation between δ' and δ:

$$(\bar{A}_{k-1}(T) + \gamma_d) \int_{\delta'=0}^{\delta} \mu e^{-\mu\delta'} e^{-\lambda_a(\delta-\delta')} d\delta'$$

$$= (\bar{A}_{k-1}(T) + \gamma_d) \frac{\mu}{\mu - \lambda_a} [e^{-\lambda_a \delta} - e^{-\mu\delta}] \quad \text{if} \quad \mu \neq \lambda_a$$

Therefore:

$$A_k(\delta) = \left[1 + (\bar{A}_{k-1}(T) + \gamma_d) \frac{\lambda_a}{\mu - \lambda_a}\right] e^{-\lambda_a \delta} - (\bar{A}_{k-1}(T) + \gamma_d) \frac{\mu}{\mu - \lambda_a} e^{-\mu\delta} \quad (A2.1)$$

If $\mu = \lambda_a$:

$$A_k(\delta) = [1 - (\bar{A}_{k-1}(T) + \gamma_d)(1 - \mu\delta)] e^{-\mu\delta} \quad (A2.2)$$

It should be pointed out that $A(\delta)$ is the estimated instantaneous availability of the component on standby, in non-operation and before being started up. The probability of the component starting up at instant δ is as follows:

$$A'_k(\delta) = A_k(\delta)(1 - \gamma_d)$$

Here we have assumed that the probability of a component failing to start up when it is brought into operation is the same as the probability of its failing to operate under test conditions, but this is clearly not always verified. Let us now consider $A_k(T)$: this is the probability that the component is in operating state just before test $k + 1$.
Therefore:

$$A_k(T) = \left[1 + (\bar{A}_{k-1}(T) + \gamma_d) \frac{\lambda_a}{\mu - \lambda_a}\right] e^{-\lambda_a T}$$

$$- (\bar{A}_{k-1}(T) + \gamma_d) \frac{\mu}{\mu - \lambda_a} e^{-\mu T} \quad (\mu \neq \lambda_a)$$

However

$$\bar{A}_{k-1}(T) = 1 - A_{k-1}(T)$$

$$A_k(T) = A_{k-1}(T) \left[\frac{\mu}{\mu - \lambda_a} e^{-\mu T} - \frac{\lambda_a}{\mu - \lambda_a} e^{-\lambda_a T}\right]$$

$$+ \gamma_d \left[\frac{\lambda_a}{\mu - \lambda_a} e^{-\lambda_a T} - \frac{\mu}{\mu - \lambda_a} e^{-\mu T}\right]$$

$$+ e^{-\lambda_a T} + \frac{\lambda_a}{\mu - \lambda_a} e^{-\lambda_a T} - \frac{\mu}{\mu - \lambda_a} e^{-\mu T} \quad (A2.3)$$

To simplify these equations, we can express them thus:

$$x = e^{-\lambda_a T} \qquad \alpha = \frac{\mu}{\mu - \lambda_a} \qquad (A2.4)$$

$$y = e^{-\mu T} \qquad \beta = x(1 - \alpha) + \alpha y \qquad (A2.5)$$

$$A_k(T) = \beta A_{k-1}(T) + \gamma_d[(\alpha - 1) - \alpha y] + x + (\alpha - 1)x - \alpha y$$

$$\boxed{A_k(T) = \beta A_{k-1}(T) + x - \beta \gamma_d - \beta}$$

This is expressed as follows:

$$A_k(T) + \frac{x - \beta \gamma_d - \beta}{\beta - 1} = \beta\left[A_{k-1}(T) + \frac{x - \beta \gamma_d - \beta}{\beta - 1}\right]$$

$A_k(T)$ is deduced by recursion:

$$\boxed{A_k(T) = \beta^k\left[A_0(T) + \frac{x - \beta \gamma_d - \beta}{\beta - 1}\right] - \frac{x - \beta \gamma_d - \beta}{\beta - 1} \qquad \text{if } \mu \neq \lambda_a}$$

$$(A2.6)$$

A similar calculation produces the following formula for $\mu = \lambda_a$:

$$\boxed{A_k(T) = \beta'^k\left[A_0(T) + \frac{x - \beta' \gamma_d - \beta'}{\beta' - 1}\right] - \frac{x - \beta' \gamma_d - \beta'}{\beta' - 1} \qquad \text{if } \mu = \lambda_a}$$

$$(A2.7)$$

where $\beta' = (1 - \mu T)x$.

What is the value of $A_0(t)$? Here, the component is assumed to be available at $t = 0$. Therefore:

$$A_0(T) = e^{-\lambda_a T}$$

On the basis of formulae (A2.1) and (A2.2), it is easy to express availability as a function of time, i.e.

$$kT \leq t \leq (k + 1)T \qquad (k \geq 1)$$

If $\mu \neq \lambda_a$

$$U_k(T) = \left[1 + (\bar{A}_{k-1}(T) + \gamma_d)\frac{\lambda_a}{\mu - \lambda_a}\right]e^{-\lambda_a k T} \qquad (A2.8)$$

$$V_k(T) = -(\bar{A}_{k-1}(T) + \gamma_d)\frac{\mu}{\mu - \lambda_a} e^{\mu k T} \qquad (A2.9)$$

The following formula is obtained:

$$A_k(t) = U_k(T)e^{-\lambda_a t} + V_k(T)e^{-\mu T} \qquad (A2.10)$$

If $\mu = \lambda_a$

$$U'_k(T) = [1 - (\bar{A}_{k-1}(T) + \gamma_d)(1 + \mu kT)]e^{-\mu kT} \qquad (A2.11)$$

$$V'_k(T) = (\bar{A}_{k-1}(T) + \gamma_d)e^{-\mu kT} \qquad (A2.12)$$

The following formula is produced:

$$\boxed{A_k(t) = U'_k(T)e^{-\mu t} + V'_k(T)\mu t e^{-\mu t}} \qquad (A2.13)$$

From this, it is easy to deduce:

- the unavailability of the component on standby and before start-up (except during testing);

- the unavailability of the component on standby and after start-up between two tests, which is denoted by $\gamma_k(t)$

$$\gamma_k(t) = \bar{A}_k(t) + \gamma_d \qquad (A2.14)$$

The curve representing $\bar{A}_k(t)$ is a sawtooth curve and is characteristic of this unavailability (Figure A2.2).

The shape of the curve is very different in the first time interval because the component is assumed to be available at $t = 0$. Failure during non-operation is thus the only possibility. In addition, during each test, a jump equal to γ_d represents the integration of the component's failure to start up when called upon to do so.

What are the minimum and maximum values of $A_k(T)$ or $\bar{A}_k(T)$ when a large number of tests are performed? It is easy to prove that $|\beta| < 1$ [5]; $A_k(T)$ will therefore tend towards a limit when a large number of tests are performed.

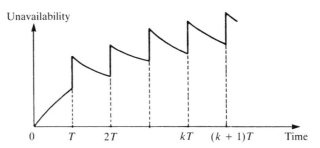

Figure A2.2 Unavailability as a function of time.

When $\mu \neq \lambda_a$

$$A_\infty(T) = \frac{x - \beta\gamma_d - \beta}{1 - \beta}$$

$$\bar{A}_\infty(T) = \frac{1 - x + \beta\gamma_d}{1 - \beta} \quad \text{(A2.15)}$$

When $\mu = \lambda_a$

$$A_\infty(T) = \frac{x - \beta'\gamma_d - \beta'}{1 - \beta'}$$

$$\bar{A}_\infty(T) = \frac{1 - x + \beta'\gamma_d}{1 - \beta'} \quad \text{(A2.16)}$$

It is interesting to develop an approximation of this expression in the case that is usually encountered: $\lambda_a \ll \mu$; $\lambda_a T \ll 1, \mu T \gg 1$. The conditions of validity will be discussed in Section 3. It is demonstrated that:

$$1 - \beta \simeq \frac{\mu}{\mu - \lambda_a}$$

$$\bar{A}_\infty(T) \simeq \lambda_a T \quad \text{(A2.17)}$$

This approximation is particularly useful, and allows us to deduce an approximation to unavailability at the moment the component is started up:

$$\gamma_\infty(T) \simeq \gamma_d + \lambda_a T \quad \text{(A2.18)}$$

By way of example, let us consider the characteristic parameters of dependability for a standby generator at a nuclear power station [5].

$$\lambda_a = 10^{-4}/\text{h}$$
$$\gamma_d = 1.5 \times 10^{-2}$$
$$\mu = 6 \times 10^{-3}/\text{h} \left(\frac{1}{\mu} \simeq 7 \text{ d}\right)$$
$$T = 240 \text{ h} = 10 \text{ d}$$

Reference [5] shows that the limit of unavailability is reached quickly with a good level of approximation at the 3rd test interval. Formula (A2.17) is also a good approximation.

A2.3 MEAN AVAILABILITY

It is interesting to calculate the mean availability of a component between two tests. Mean availability is defined thus:

$$A^m(t_1, t_2) = \frac{1}{t_2 - t_1} \int_{t_1}^{t_2} A(t) \, dt$$

By applying this formula, mean availability between tests k and $k+1$ is

$$A_k^m = \frac{1}{T}\left[\left[1 + (\bar{A}_{k-1}(T) + \gamma_d)\frac{\lambda_a}{\mu - \lambda_a}\right]\frac{1 - e^{-\lambda_a T}}{\lambda_a}\right.$$
$$\left. - (\bar{A}_{k-1}(T) + \gamma_d)\left(\frac{1 - e^{-\mu T}}{\mu - \lambda_a}\right)\right] \quad (A2.19)$$

The limit of this mean availability can be calculated when the number of tests increases. Using the notations introduced earlier, this limit is as follows:

$$A_\infty^m = \frac{1 - x}{\lambda_a T} + \frac{1 - x + \gamma_d}{1 - \beta} \frac{y - x}{(\mu - \lambda_a)T} \quad \text{if} \quad \mu \neq \lambda_a$$

$$A_\infty^m \simeq \frac{1 - y}{\mu T} - \frac{1 - x + \gamma_d}{1 - \beta'} y \quad \text{if} \quad \mu = \lambda_a$$

It is also interesting to give an approximation of this value when $\lambda_a \ll \mu$; $\lambda_a T \ll 1$, $\mu T \gg 1$:

$$A_\infty^m \simeq 1 - \frac{\lambda_a T}{2} - \frac{\lambda_a}{\mu} - \frac{\gamma_d}{\mu T}$$

or

$$\boxed{\bar{A}_\infty^m \simeq \frac{\lambda_a T}{2} + \frac{\lambda_a}{\mu} + \frac{\gamma_d}{\mu T}} \quad (A2.20)$$

The mean unavailability after start-up is thus

$$\boxed{\gamma_\infty^m \simeq \gamma_d + \frac{\lambda_a T}{2} + \frac{\lambda_a}{\mu} + \frac{\gamma_d}{\mu T}} \quad (A2.21)$$

Formula (A2.20) can be established directly using simple reasoning. When a component on standby and periodically tested is started up, it can be unavailable before start-up for several reasons:

- It has failed in non-operation since the last test: mean unavailability is $\lambda_a T/2$.

- It is being repaired subsequent to a failure observed during the latest test and caused by a failure in non-operation during the interval between preceding tests: the mean unavailability calculated on N intervals between tests is λ_a/μ.

- It is being repaired subsequent to a failure observed during the latest test and caused by a failure to start up: the mean unavailability calculated on N intervals between tests is

$$\frac{N\gamma_d}{NT\mu} = \frac{\gamma_d}{\mu T}$$

The formula for mean unavailability is

$$\bar{A}_\infty^m \simeq \frac{\lambda_a T}{2} + \frac{\lambda_a}{\mu} + \frac{\gamma_d}{\mu T}$$

Mean unavailability after start-up is therefore

$$\gamma_\infty^m \simeq \gamma_d + \frac{\lambda_a T}{2} + \frac{\lambda_a}{\mu} + \frac{\gamma_d}{\mu T}$$

If the test lasts for time θ, it can be the source of a failure during operation (λ: rate of failure during operation). It is easy to show that γ_∞^m can be expressed as follows:

$$\gamma_\infty^m \simeq \gamma_d + \frac{\lambda_a T}{2} + \frac{\lambda_a}{\mu} + \frac{\gamma_d}{\mu T} + \frac{\lambda \theta}{\mu T} \qquad (A2.22)$$

Let us re-examine the conditions of validity of this formula:

- $\lambda_a \ll \mu$. This condition is generally verified because failures in non-operation are infrequent and the corresponding repair time is short. This condition is equivalent to:

$$\text{MTTR} \ll \frac{1}{\lambda_a}.$$

- $\lambda_a T \ll 1$, $\mu T \gg 1$. This condition is equivalent to

$$\text{MTTR} \ll T \ll \frac{1}{\lambda_a}.$$

The interval between tests is generally greater than the repair time. If this is not the case, repairs generally will not have been completed before the next test and the test policy must be reviewed. In addition, the period between tests is generally less than $1/\lambda_a$. If this is not the case, a failure is practically certain to occur during a test and testing thus comes too late!

A2.4 OPTIMIZING INTERVALS BETWEEN TESTS

The interval between tests can be optimized with respect to the availability of a given component.

- If the interval between tests is too short, the number of failures on start-up and failures during testing will increase, and the probability of the system being under repair will thus also increase.

- If the interval between tests is too long, the probability of the component being in failed state because of a failure in non-operation will increase.

There is thus an optimum interval, which makes it possible to detect failures in non-operation with maximum effectiveness without requiring a prohibitive number of start-ups. The curve indicating the variation of mean unavailability as a function of the duration of the interval between tests is as shown in Figure A2.3.

By cancelling the derivative of γ_∞^m in relation to T, the optimum value is calculated as follows:

$$T_0 \simeq \sqrt{\left(\frac{2(\gamma_d + \lambda\theta)}{\lambda_a \mu}\right)} \quad (A2.23)$$

For example, with the values given above for a standby generator at a nuclear power station, $T_0 = 224$ h.

The following comments can be made on these results:

- It is not possible to obtain a mean unavailability of less than

$$\gamma_d + \lambda_a\left(T_0 + \frac{1}{\mu}\right)$$

by optimizing the interval between tests.

- For a given level of unavailability, there are two possible values for the interval between tests.

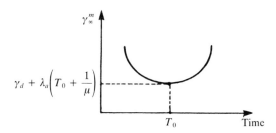

Figure A2.3 Mean unavailability as a function of the duration of intervals between tests.

- Unavailability varies only slightly around the optimum value; this makes it possible to choose an interval between tests without the unavailability varying significantly. In addition, uncertainties about dependability parameters must be taken into account. In particular, it is often difficult to determine the rate of failure in non-operation (λ_a) with sufficient precision.

A2.5 AVAILABILITY OF A SYSTEM ON STANDBY AND PERIODICALLY TESTED

In practice, sets of components are kept on standby and tested at different intervals to rationalize the testing policy. The expression for the estimated availability of a component on standby and periodically tested was given above in (A2.10) and (A2.13). The estimated availability of a system can be calculated by combining the estimated availability of each of its components according to the laws conventionally applied to event probability calculations. This approach is relatively simple only if it is assumed that availabilities are not interdependent. Examples of these calculations are given in references [5] and [6], using the example of standby generators at nuclear power stations. Computer programs also exist to calculate the availability of these systems (e.g. PHAMISS, see Chapter 20).

REFERENCES

[1] I. Bazovsky (1961) *Reliability Theory and Practice*, Space technology series, Prentice Hall, Englewood Cliffs, N.J.

[2] I. M. Jacobs and P. W. Marriott (1969) *Guideline for Determining Safe Test Interval and Repair Time for Engineered Safeguard*, APED, 5736.

[3] J. Halpern (1974) *A sequential testing procedure for a system's state identification*, IEEE Trans. Reliability, **R23**, October.

[4] S. C. Chay and M. Mazumdar (1975) *Determination of test intervals in certain repairable stand by protective systems*, IEEE Trans. Reliability, **R24** (3), August.

[5] J. P. Signoret (1976) *Disponibilité d'un système en attente périodiquement testé*, CEA-IPSN-DSN n° 113.

[6] J. P. Signoret (1976) *Optimisation de l'intervalle entre test et de la disponibilité de système en attente. Application à deux diesels de secours en parallèle*, IIIe Congrès National de Fiabilité. Perros-Guirec,

[7] J. P. Signoret (1976) *Disponibilité d'un système en attente périodiquement testé Calculs approchés. Optimisation*, CEA-IPSN-DSN No. 129.

APPENDIX 3
IMPORTANCE FACTORS

A3.1 INTRODUCTION

It is sometimes useful to attach a set of measures to each component of a system in order to assess its importance in the system as regards dependability measures. These measures will be called 'importance factors'.

The concept was introduced by Birnbaum [1] and developed by a number of other authors [2–4]; Lambert [5] has drawn up a detailed list and developed the first computer program (known as IMPORTANCE) to calculate these importance factors (see Section 20.7). References [6, 7] offer useful information on this subject.

By calculating the importance factors, various questions can be answered, such as:

- Which component should be improved first in order to increase the reliability of the system?

- Which minimal cut sets are the most important?

- If the system is in a failed state, which component should be repaired first?

A3.2 IMPORTANCE FACTORS

We will confine our study to the three most commonly used importance factors:

- the marginal importance factor;
- the critical importance factor;
- the diagnostic importance factor.

These importance factors can be calculated to establish the importance of a component or an event (e.g. a minimal cut set).

- **Marginal importance factor** This notion was introduced by Birnbaum [1] and is still called the Birnbaum importance factor. As an example, let us consider the availability of a system. The marginal importance factor of component i is the rate at which the availability of the system increases when the availability of the component i increases.

   Mathematically speaking, the marginal importance factor is the partial differential of the unavailability $\overline{A}(t)$ of the system as compared to the unavailability $\overline{a}_i(t)$ of component i:

$$F = \frac{\partial \overline{A}}{\partial \overline{a}_i}(t) \tag{A3.1}$$

It is shown [6] that this marginal importance factor can be interpreted as being the probability that the system will be in an operating state with i as the critical component, when i is operating.

- **Critical importance factor** This notion was introduced by Lambert [5]. It is the probability that the component i has caused the failure of the system, knowing that the system is in failed state.

   It should be pointed out that the component i is not necessarily the only component in failed state, but if this is the case, it is the most recent to have failed. It is shown that

$$F = \frac{\overline{a}_i(t)}{\overline{A}(t)} \frac{\partial \overline{A}}{\partial \overline{a}_i}(t) \tag{A3.2}$$

The critical importance factor of a minimal cut set can also be calculated. This notion was introduced by Fussell and Vesely [4] and is still called the Fussell–Vesely critical importance factor of a minimal cut set. It is the probability that the minimal cut set has caused the failure of the system, knowing that the system is in failed state. This importance factor is highly interesting; it also indicates the extent to which each minimal cut set contributes to the failure of the system.

   The calculation is simply

$$F = \frac{\overline{A}_i}{\overline{A}(t)} \tag{A3.3}$$

where $\overline{A}_i$ is the unavailability tied to the minimal cut set i.

- **Diagnostic importance factor** This notion was introduced by Vesely and has been used by Fussell [4]; it is still called the Vesely–Fussell importance factor. It is the probability that the component i has failed, knowing that the system is in failed state.

   This importance factor can be useful in diagnosing system failure causes, which explains its name. It is shown that

$$F = \frac{\overline{a}_i}{\overline{A}(t)} \overline{A}(t|\ \text{entity } i \text{ is in failed state}) \tag{A3.4}$$

REFERENCES

[1] Z. W. Birnbaum (1969) On the importance of different components and a multicomponent system, in *Multivariate Analysis*, II (ed. P. R. Krishnaiah), Academic Press, New York.

[2] R. E. Barlow and F. Proschan (1974) *Importance of Systems Components and Fault Tree Analysis*, Opeations research Center, Univ. of Calif., Berkeley, Rept. ORC 74-3.

[3] P. Chatterjee (1976) *Fault Tree Analysis: Reliability Theory and Systems Safety Analysis*, Operations Research Center, University of California, Berkeley, Rept. ORC 74-34.

[4] J. B. FUssell (1975) *How to hand-calculate system reliability characteristics,* IEEE Trans *Reliability*, **R-24** (3).

[5] H. E. Lambert (1975) *Fault Trees for Decision Making in System Analysis*, Ph.D. Thesis, Lawrence Livermore Laboratory.

[6] A. Pages and M. Gondran (1980) *Fiabilité des systèmes*, Collection de la Direction des Etudes et Recherches d'Electricité de France, Eyrolles.

[7] E. J. Henley and H. Kumamoto (1981) *Reliability Engineering and Risk Assessment*, Prentice Hall, Englewood Cliffs, N.J.

APPENDIX 4
ASSESSMENT OF UNCERTAINTIES

A4.1 INTRODUCTION

Dependability assessment generally takes into account various familiar parameters comprising elements of uncertainty on the one hand, and the analyst's judgments, which are sometimes subjective, on the other hand. The main causes of uncertainty in the analysis [1] are the following:

- *Uncertainties linked to the parameters*. For various reasons, the information on dependability available to us is uncertain: a small sample leading to a wide confidence interval, extrapolation of data from one installation to another, etc. Certain other parameters (delayed appearance of physical factors, time available after losing a system before undesirable effects ensue, etc.) connected with design or operation are also familiar but with elements of uncertainty;

- *Uncertainties connected with modelling*. These are due to the use of an approximate dependability model. It is particularly true in the modelling of failures with a common cause, human error or software bugs. Generally speaking, modelling can integrate all the relevant variables without assessing their relationship in sufficient detail;

- *Uncertainties connected with the non-exhaustive nature of the analysis*. The analyst cannot be totally sure that his modelling has taken all important factors, relevant failures and significant interactions into account.

Analysing uncertainties therefore consists in identifying all the uncertainties and their repercussions on the assessment. Usually, only the first source of uncertainty is taken into account; an attempt is then made to assess the uncertainty of the final result (a measure of dependability) caused by the parameter uncertainties. The analyst must nonetheless examine the other sources of uncertainty, even if they appear difficult to process.

Large-scale uncertainty propagation was first used in the initial probabilistic risk assessment (PRA) of the nuclear power stations in Surry and Peach Bottom [2] in the United States in the early 1970s; it is now widely used in the PRA of nuclear power stations [3].

Let us now look at the problems of processing and propagation of uncertainties.

A4.2 PROCESSING OF UNCERTAINTIES

We will restrict our observations to the processing of parameter uncertainties. Each parameter can be considered as a random variable characterized by its law of probability. The measures associated with uncertainty in knowledge of the parameter are usually the variable, the standard deviation and the confidence interval. Before proceeding further, let us come back to the two interpretations (objectivist and subjectivist) of probabilities (see Section 4.2) as they have consequences on the interpretation and processing of uncertainties.

Objectivist (or frequency) interpretation links probability to an observed frequency of an event; subjectivist (or Bayesian) interpretation defines probability as a degree of belief in a hypothesis.

Let us take the example of the confidence interval. From an objectivist viewpoint, one of the properties of the confidence interval of a parameter λ is the following: the probability that the true value of λ is within the confidence interval, is α (the level of confidence); it is calculated on the basis of frequency-type measures. From a subjectivist viewpoint, the confidence interval is based on a law of probability representing the analyst's degree of belief regarding the possible values of the parameter and reflecting his knowledge of the parameter. Thus, if his knowledge is represented by the probability density function the confidence interval $[I, S]$ is such that

$$\int_I^S f(\lambda) \, d\lambda = \alpha$$

The subjectivist approach to uncertainties generally facilitates the quantitative expression of uncertainties since it describes all the uncertainties in terms of laws of probability reflecting the knowledge—even partial—that the analyst possesses.

Let us examine the Bayesian approach which has occasionally been used by nuclear power stations [4]. Available generic data on dependability represent the variability of the components (their design, the conditions in which they are used, etc.) depending on the power station. They are considered as *a priori* data; knowledge of the operating results of a given nuclear power station enables *a posteriori* data characterized by a law of probability to be deduced, by applying the Bayes theorem (see Section 5.2). The *a posteriori* law of probability reflects the extent of the analyst's knowledge about the value of the parameter for this nuclear power station. Such laws of probability can therefore be used for a PRA of the power station and for calculating associated uncertainties.

We should note that, with an objectivist approach, dependability data can be processed as random variables owing to the variability of the data of different components and installations. The data are often related to similar components (or installations) but they are never exactly identical and operating conditions can vary from one installation to another. As for the Bayesian approach, a law of probability is associated with each data item; however, it will result from frequency-type measures.

In fact, the choice of approach depends on a number of factors such as the degree of data knowledge, the variability of data between the sample considered statistically and the sample to which the data will be applied, the possibility of obtaining *a priori* laws of probability, etc.

Which law of probability can we examine? Several laws of probability have been used in the PRA of nuclear power stations; however, the most widely used is the lognormal law.

Such a law is appropriate for the description of data which can vary by several factors. Thus, if X is a given random variable, X will generally be situated between X_0/f and $X_0 \times f$, where X_0 is a reference value and f a factor. For example, a data item estimated at 10^{-6} can vary between 10^{-7} ($10^{-6}/10$) and 10^{-5} ($10^{-6} \times 10$). If it is assumed that the limits between which X is fairly likely to vary are the confidence level limits (e.g. 90%), the parameters of the lognormal law can be easily deduced (see Section 4.3).

A4.3 PROPAGATION OF UNCERTAINTIES

Two methods are generally used to determine the uncertainty of the final result based on the knowledge of input data uncertainties: Monte Carlo type simulation and the moments method.

Let us consider the output variable z which is a function of the input variables x_i ($i = 1, \ldots, n$):

$$z = f(x_1, x_2, \ldots, x_i, \ldots, x_n) \tag{A4.1}$$

A4.3.1 Monte Carlo-type simulation

The data are compared to random variables whose laws of probability are known. One solution consists of carrying out a Monte Carlo-type simulation organized in the following manner:

- The value of all the input variables are drawn at random;

- the value of the function f is calculated using these particular values, using an analytical program (e.g., programs that calculate cause trees or consequence trees).

The previous procedure is repeated several times; an approximation of the law of probability of the output variable z is thus obtained.

Many computer programs that calculate cause trees use a simulation of this kind (see Chapter 20). The main limit to this technique is cost; some cases require a great many repetitions in order to gain sufficient knowledge of the law of probability of the output variable.

A4.3.2 Moments method

With this method, the first moments of the variables x must be known; the first moments of the output variable z [5,6] are then deduced. This method can be used with the cause tree method (CTM). Let us approach the OR and AND logic gates in this way:

- **OR logic gate** In this case

$$z = \sum_{i=1}^{N} x_i \qquad (A4.2)$$

The mean and the variance of z are

$$E[z] = \sum_{i=1}^{N} E[x_i] \qquad (A4.3)$$

$$V[z] = \sigma^2[z] = \sum_{i=1}^{N} \sigma^2[x_i] + \sum_{i=1}^{N-1} \sum_{j=i+1}^{N} \text{cov}(x_i, x_j) \qquad (A4.4)$$

where $\text{cov}(x_i, x_j)$ is the covariance of the two random variables:

$$\text{cov}(x_i, x_j) = E[(x_i - E[x_i])(x_j - E[x_j])]$$

The covariance is zero when the two random variables are independent.

- **AND logic gate** In this case

$$z = \prod_{i=1}^{N} x_i \qquad (A4.5)$$

Let us assume that the variables x_i are independent:

$$E[z] = \prod_{i=1}^{N} E[x_i] \qquad (A4.6)$$

$$V[z] = \sigma^2[z] = \prod_{i=1}^{N}(\sigma^2[x_i] + E[x_i]^2) - \prod_{i=1}^{N} E[x_i]^2 \qquad (A4.7)$$

These formulae make it possible to calculate the mean and the variance of the law of probability of an undesirable event modelled by the CTM, when the basic events are assumed to be independent. This also applies when the variable z is written in the form of quantities and products of independent random variables. When the variables are not independent, covariance terms have to be taken into account.

REFERENCES

[1] W. E. Vesely and D. M. Rasmuson (1984) Uncertainties in nuclear probabilistic risk analysis, *Risk Analysis*, **4** (4).
[2] *Reactor safety study* (1975) *An Assessment of Accident Risks in US Commercial Nuclear Power Plant*, WASH 1400 (NUREG 74/014), U.S. Nuclear Regulatory Commission.
[3] *PRA Procedures Guide* (1983) *A Guide to the Peformance of Probabilistic Risk Assessments for Nuclear Power Plants*, NUREG/CR-2300.
[4] G. Apostolakis (1978) Probability and risk assessment: the subjectivist viewpoint and some suggestions, *Nuclear Safety*, **19** (3).
[5] G. Apostolakis and Y. T. Lee (1977) Methods for the estimation of confidence bounds for the top-event unavailablity of fault trees, *Nuclear Engineering and Design*, **41**, 411–19.
[6] J. D. Murchland and G. G. Weber (1972) A moments method for the calculation of a confidence interval for the failure probability of a system. *Proceedings of the 1972 Annual Reliability and Maintainability Symposium*, Institute of Electrical and Electronics Engineers, 505–77.

APPENDIX 5
OPERATING RULES FOR CASES WHERE A SAFETY SYSTEM IS OBSERVED TO BE UNAVAILABLE

A5.1 INTRODUCTION

Let us consider the case of an industrial system that is potentially dangerous to the environment and that is equipped with a number of elementary safety/security systems. While the system is in operation, a safety system is observed to be partially unavailable (a redundant system with a failure, for example) or totally unavailable. What should the operator do?

- Leaving the system in operation increases the risk of a serious accident. The operator could therefore be tempted to stop the system immediately or within a short period of time.

- Stopping the system has a number of consequences, particularly from an economic viewpoint. Shutting down a nuclear facility, for example, disrupts the electricity distribution network, reduces electricity production and places an increased strain on equipment.

The system operator is thus faced with a dilemma. The operator wishes to retain a certain flexibility of operation and not be forced to stop the system every time a failure is observed, but he does not want to increase the risk of a serious accident to a potentially dangerous level.

The methods adopted to assess the risk of an accident can help the operator to make a decision. Much valuable work has been carried out on safety in the nuclear [1–3] and aeronautics [4] fields since the beginning of the 1970s and the methods developed are currently employed to varying degrees in France.

A5.2 METHOD

Relatively simple in principle, the method comprises two stages:

- *Assess the risk from the moment unavailability is observed.* For example, Figure A5.1 shows that the risk of a serious accident increases sharply if the operator continues to run the system (a nuclear reactor at full power for example) when a safety system is observed to be unavailable. The risk returns to a normal level when the system has been repaired. Note that the unbroken line represents the estimated risk in the case where the operator continues to run the system in the knowledge that a safety system is unavailable.

- *Limit the risk to a level deemed to be acceptable through the application of a risk criterion or objective.*

A risk can clearly be limited in a variety of manners. Figure A5.1 uses two types of threshold whose significance is given below:

- *Threshold S_1*. Indicates the maximum acceptable level of risk. If an assessment shows that the risk is greater than S_1, the system must be stopped.

- *Threshold S_2*. Indicates the risk of an accident during unavailability. Let us assume that the exact duration of the period of unavailability is known in principle. If an assessment shows that the risk exceeds S_2, the system must be stopped immediately.

The resulting rule of operation helps the operator to make a decision, i.e. to stop the system immediately, stop it under certain conditions or continue operation.

The risks encountered may be of an economic nature or related to human safety. A number of problems may arise in the application of the method

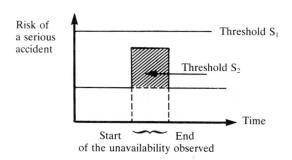

Figure A5.1 The risk of a serious accident.

defined to assess the level of risk. Let us look at three problems that are commonly encountered:

- The shutdown procedure entails a number of risks. The risk of continuing operation therefore must be measured against the risk of performing a shutdown. In certain cases, the wisest solution is to do nothing at all!

- It is important to ensure that the recommendations generated by the risk assessment are applied in a coherent manner. The assessment may indicate, for example, that in order to maintain the risk at an acceptable level, the system should be stopped within a specified period of time. In practice, however, when the specified period has elapsed, the operator may not see why he should stop the system if he has not finished repairs. He may argue that as no accidents have occurred, there is no risk in leaving the system running! At this point, the risk is no longer in check. An understanding of the principles of risk assessment must therefore be accompanied by a knowledge of the rules that the operator is likely to apply in practice.

- A certain period of time may elapse before the operator notices that the safety system is unavailable, i.e. in the case where the safety system is in standby and is only activated in the event of an accident. In this case, from the operator's viewpoint, it may appear that the best solution would be not to notice the unavailability of the system! In this way, a lower estimated risk would be calculated for the period subsequent to unavailability whilst the estimated risk for the period prior to unavailability would have no limit at all! However, in a global approach to risks, an assessment of system unavailability will consider the time taken to note the partial unavailability and the attendant risk will be covered by the global safety objective.

A5.3 EXAMPLES OF APPLICATION

A5.3.1 Aeronautical applications

Safety objectives have been applied in the French aeronautics sector for the Concorde and Airbus aircraft (see Section 19.5) since the end of the 1970s. Is it acceptable for an aircraft to take off if a safety system is known to be partially unavailable? A specific safety objective is described in reference [4]:

- A single combination of failures with catastrophic consequences and an hourly probability of between 10^{-9} and 10^{-7} is acceptable provided that all other combinations of failures have an hourly probability of less than 10^{-9}.

- A single combination of failures with critical consequences and an hourly probability of between 10^{-7} and 10^{-5} is acceptable provided that all other combinations of failures have an hourly probability of less than 10^{-7}.

An aircraft will not therefore take off when the hourly probability is outside the objective specified. When an aircraft is allowed to fly with a partial unavailability, the number of flights will be limited by the application of safety objectives.

A5.3.2 Nuclear applications

The work initially carried out in this field prompted the French electricity board (EDF) [5–8] to undertake further studies with a view to determining procedures to be respected in the event of the partial unavailability of a safety system in a nuclear facility. The result was a probabilistic approach that has been systematically applied in all types of French nuclear facilities (pressurized water and fast neutron) since 1982. We shall briefly describe the main characteristics of this approach [9].

First, it should be remembered that safety systems are always redundant, i.e. a function is reproduced at least twice. The term 'partial unavailability' will refer to the loss of one function on the redundant system, the other remaining available. If the whole safety system becomes unavailable, the facility will be shut down immediately.

We shall now specify the rule to be applied if a safety system is observed to be partially unavailable. The operator must accurately assess the time (τ) required to repair the faulty component. Then, he will use the written procedure concerning component unavailability, which specifies an authorized period (t).

- If $\tau \leq t$, the nuclear facility can continue to operate at full power. If, however, the operator realizes during repair that he will require more time than he originally thought, then the facility must be shut down.

- If $\tau \geq t$, the nuclear facility must be shut down immediately and switched to 'safe shutdown conditions'. The risk of a serious accident in this state, taking into account the partial unavailability of the system, is minimal.

We shall now look at the problem of determining the authorized period t. EDF applies a probabilistic criterion in agreement with the safety authorities.

> 'The additional probability of a serious accident (leading to core meltdown) occurring during the authorized period whilst the nuclear facility is operating at full power and taking into account the partial unavailability of the safety system, must not exceed 10^{-7}.'

A number of comments can usefully be made concerning this criterion:

- The additional probability of a serious accident is the difference between the level of probability calculated on the basis of partial unavailability and the level of probability calculated on the assumption that no partial unavailability has been observed. The hatched area on Figure A5.1 shows the additional probability of a serious accident.

- The threshold of 10^{-7} is considered to be sufficiently low for the number of partial unavailabilities not to be taken into account. In reality, only a few partial unavailabilities are observed per year in a nuclear facility. The risk is therefore comparable to the level specified in the safety objectives (see Section 19.5).

We shall now look at the manner in which the authorized period is calculated. Note that if the nuclear facility continues to operate at full power, the component must be repaired and put back into service at the end of the authorized period at the latest. The increase in the additional probability of a serious accident will be calculated taking into account a partial unavailability of a duration t. The calculation refers to the identification of accident sequences Sq_i, involving the unavailable component c_i.

- Let $P\,[Sq_i, t|c_i$ unavailable during $t]$ be the probability of occurrence of the accident sequences Sq_i during the period t, c_i being unavailable during the period t.

- Let $P\,[Sq_i, t]$ be the probability of occurrence of the accident sequences Sq_i during the period t.

The criterion is written

$$P\,[Sq_i, t|c_i \text{ unavailable during } t] - P[Sq_i, t] < 10^{-7} \qquad (A5.1)$$

The second probability is generally far more remote than the first. The authorized periods calculated vary from a few days to a few weeks.

The above method, which concerns spurious partial unavailabilities, has been extended to cover scheduled partial unavailabilities. In this case, the additional probability of a serious accident must not exceed 10^{-7} per year, taking into account all the partial unavailabilities planned on a particular component during that year.

REFERENCES

[1] B. Bohm, J. Blombach and W. Rosenhauer (1974) *Etudes d'indisponibilité de circuits de refroidissement de secours en vue de déterminer les stratégies de réparation et d'inspection* (Traduction française). *Atomwirstschaft*, **19**(7).

[2] I. M. Jacob and P. W. Marriott (1975) *Guidelines for Determining Safe Test Intervals and Repair Time for Engineering Safeguards*, General Electric, San José, California.

[3] D. P. Wagner, W. E. Vasely and L. A. Minton (1985) *Risk-based Evaluation of Technical Specifications*, EPRI NP-4317, Interim Report.

[4] C. Lievens (1976) *La séurité des systèmes*, Cepadues Editions, Toulouse.

[5] J. F. Greppo, M. Boursier, A. Carnino and A. Blin (1975) *Détermination par une méthode probabiliste d'une règle d'exploitation relative aux sources d'alimentation électrique 6.6 kV des tranches PWR type 900 MW*, EDF-SPT-D57.7381-02-CEA-DSN No. 56.

[6] B. Gachot, L. Ratti and P. François (1978) *Proposition de règle de décision en cas d'indisponibilité partielle d'un système important pour la sûreté pendant le fonctionnement en puissance d'une tranche PWR*, EDF-E-SE/FO 78-15A.

[7] C. Le Floch and A. Villemeur (1979) *Consignes d'exploitation en cas d'indisponibilité partielle de systèmes de sûreté: formulation générale et proposition d'une méthode de résolution*, EDF-HT/13/49/79.

[8] C. Le Floch and A. Villemeur (1981) *Probabilistic Assessment of Technical Specifications for PWR Safety Related Systems*, Third National Reliability Conference, Birmingham, UK.

[9] B. Magnon, A. Villemeur and C. Ancelin (1985) *Residual Heat Removal in the Creys-Malville Power Plant: From Reliability Analyses of Systems to Allowable Operating Times*, International ANS/ENS Topical Meeting on Probabilistic Safety Methods and Applications. San Francisco, USA.

APPENDIX 6
SEMI-MARKOVIAN PROCESS: MUT, MTTR AND MTTF

In this appendix, we show the MUT, MTTR and MTTF calculation methods of semi-Markovian processes [1]; specific methods, enabling simple written calculations to be done, are also given [1].

A6.1 MUT, MTTR AND MTTF CALCULATION

Formulae (14.76) enable the MUT (mean up time) and the MTTR (mean time to repair) to be calculated. As an example, let us develop the MUT calculation. Let $V^o(\infty)$ be the asymptotic exit intensity of all operating states:

$$V^o(\infty) = \frac{\sum\limits_{j=1}^{l} P_j(\infty)}{\text{MUT}} = \sum_{j=1}^{l} V_j(\infty)\left[\sum_{k=l+1}^{P} b_{jk}\right] \quad (A6.1)$$

$$\text{MUT} = \frac{\sum\limits_{j=1}^{l} \lim\limits_{s \to 0}[I - s\bar{F}(s)]_{ij}^{-1} s d_j}{\sum\limits_{j=1}^{l} \lim\limits_{s \to 0}[I - s\bar{F}(s)]_{ij}^{-1} s b_j^P} \quad \text{with} \quad b_j^P = \sum_{k=l+1}^{P} b_{jk}$$

$$\text{MUT} = \frac{\begin{vmatrix} 1 & -b_{12} & \cdots & -b_{1,p-1} & d_1 \\ -b_{21} & 1 & & & \vdots \\ \vdots & & & & \\ -b_{l1} & \cdots & \cdots & -b_{l,p-1} & d_l \\ \vdots & & & & 0 \\ -b_{p1} & \cdots & \cdots & -b_{p,p-1} & 0 \\ 1 & -b_{12} & \cdots & -b_{1,p-1} & b_1^P \\ -b_{21} & 1 & & & \\ \vdots & & & & \\ -b_{l1} & \cdots & \cdots & -b_{l,p-1} & b_l^P \\ \vdots & & & & 0 \\ -b_{p1} & \cdots & \cdots & -b_{p,p-1} & 0 \end{vmatrix}} \quad \text{(A6.2)}$$

More concisely,

$$\text{MUT} = \frac{\sum_{j=1}^{l}[\text{cofactor } I - B]_{j,i} d_j}{\sum_{j=1}^{l}[\text{cofactor } I - B]_{j,i} b_j^P}$$

Similarly, we obtain

$$\text{MTTR} = \frac{\sum_{j=l+1}^{P}[\text{cofactor } I - B]_{j,i} d_j}{\sum_{j=l+1}^{P}[\text{cofactor } I - B]_{j,i} b_j^L} \quad \text{with } b_j^L = \sum_{k=1}^{l} b_{jk} \quad \text{(A6.3)}$$

It can be seen that:

$$\text{MTBF} = \text{MUT} + \text{MTTR} = \frac{1}{V^\circ(\infty)} \quad \text{(A6.4)}$$

Let us now examine the MTTF in more detail (see Section 14.9.2). Having applied a method similar to the one shown in Section 14.9.3, entry and exit intensities are calculated:

$$\overline{W}_{ij}(s) = \sum_{k=1}^{l}[I - s\overline{F}'_l(s)]_{ik}^{-1} s\overline{F}_{kj}(s) \qquad i = 1, 2, \ldots, l; \forall j$$

$$\overline{V}_{ij}(s) = [I - s\overline{F}'_l(s)]_{ij}^{-1} s\overline{F}_j(s) \qquad i, j = 1, 2, \ldots, l$$

We note that

$$\begin{aligned} \overline{W}_{ij}(s) &= [I - s\overline{F}'_l(s)]_{ij}^{-1} & i \neq j \\ \overline{W}_{ii}(s) &= [I - s\overline{F}'_l(s)]_{ii}^{-1} - 1 \end{aligned} \quad \text{(A6.5)}$$

We may easily deduce the asymptotic entry numbers in all states and the asymptotic exit numbers for non-absorbing states, which are:

$$N_{ij}^{0}(\infty) = \lim_{s \to 0} s \overline{N}_{ij}(s) = \lim_{s \to 0} \overline{V}_{ij}(s) = [I - B'_l]_{ij}^{-1} \qquad (A6.6)$$

whence

$$\boxed{\text{MTTF}_i = \sum_{j=1}^{l} N_{ij}^{0}(\infty) d_j} \qquad (A6.7)$$

Thus, the MTTF is equal to the sum of the asymptotic durations of the operating states.

Moreover, the MTTF_i is also the mean duration of operating states when the initial state i is returned to after a failed state. Let us thus transform this semi-Markovian process into another semi-Markovian process in which the former absorbing states now have transitions (e.g. constant time) towards the initial state: it is easily shown that the MTTF_i is equal to the MUT of this process (using the formulae (14.63) and (A6.2)).

The MTTF_i is thus equal to the MUT of the transformed semi-Markovian process for which transitions now exist between the former absorbing states and the initial state i.

These considerations enable us to explain the relationship between failure rate and pseudo-rate of failure in a Markovian process (see Section 14.2). In fact, the pseudo-rate of failure is the conditional exit intensity of all operating states (see Section 4.5.5). The relationship (A6.1) shows that

$$\lim_{t \to \infty} \widetilde{\Lambda}(t) = \frac{1}{\text{MUT}}$$

Now let $\widetilde{\Lambda}(t)$ be the pseudo-rate of failure of the process arrived at by transforming the Markovian process using the above method. Thus

$$\lim_{t \to \infty} \Lambda_i(t) = \frac{1}{\text{MTTF}_i} = \frac{1}{\text{MUT}} = \lim_{t \to \infty} \widetilde{\Lambda}(t) \qquad (A6.8)$$

$\widetilde{\Lambda}(t)$ is therefore a good approximation of $\Lambda(t)$ for high values of t and is associated with the states graph in which the initial state is directly returned to from failed states. If the $\widetilde{\Lambda}(\infty)$ of the availability states graph is calculated (*a priori* different from the previous graph), the conditions (14.28) are obviously necessary for $\widetilde{\Lambda}(\infty)$ to remain a fair approximation of $\Lambda(\infty)$.

A6.2 ASYMPTOTIC FREQUENCY METHOD

For a process without an absorbing state, the fr_j are determined by the following equations:

$$fr_j = \sum_{k=1}^{P} fr_k b_{kj} \qquad \sum_{j=1}^{P} fr_j d_j = 1 \qquad (A6.9)$$

The other measures can be deduced from this:

$$P_j(\infty) = fr_j d_j \quad \text{MUT} = \frac{\sum_{j=1}^{l} fr_j d_j}{\sum_{j=1}^{l} fr_j b_j^P} \quad \text{MTTR} = \frac{\sum_{j=l+1}^{P} fr_j d_j}{\sum_{j=l+1}^{P} fr_j b_j^L} \quad (A6.10)$$

Equations (A6.9) can prove easy to use for certain processes (see Appendix A6.4). This is generally the case for reliable systems for which there exists a very strong probability—whatever the state—of returning to the initial state (e.g. state 1). We thus obtain: $fr_1 d_1 \simeq 1$.

The asymptotic frequency of encountering the other states is deduced from this. It is assumed that the dominant contributions to the frequency fr_j come from less degraded states (with fewer components in failed state):

$$fr_j \simeq fr_1 \left[\sum_{\substack{\text{Reduced} \\ \text{sequences} \\ \text{from state 1} \\ \text{to state } j}} \prod \begin{array}{c} \text{Asymptotic transition probabilities} \\ \text{(associated with the sequences)} \end{array} \right] \quad (A6.11)$$

Among the reduced state sequences, we will always consider those corresponding to the most direct paths from state 1 to state j (no repair transition). The formula developed in Section 14.3.5 is thus generalized for semi-Markovian processes using this formula.

A6.3 ASYMPTOTIC DURATION METHOD

The asymptotic entry and exit numbers of the states of a semi-Markovian process possessing absorbing states are determined by the following equations:

$$N_{ij}^e(\infty) = \sum_{k=1}^{l} N_{ik}^0(\infty) b_{kj}, \qquad \sum_{j=l+1}^{P} N_{ij}^e(\infty) = 1$$

$$j \neq i, j = 1, 2, \ldots, l \qquad\qquad N_{ij}^e(\infty) = N_{ij}^0(\infty)$$

$$j = l+1, \ldots p \qquad\qquad\qquad N_{ij}^0(\infty) = 0 \quad (A6.12)$$

$$j = i \qquad\qquad\qquad\qquad N_{ii}^0(\infty) - N_{ii}^e(\infty) = 1$$

MTTF approximations are obtained when we consider only those operating states with a significant asymptotic duration. Thus, if the system is reliable, the asymptotic transition probabilities being great, there exists—for each operating state—a very strong probability of returning to the initial state (e.g. state 1):

$$\text{MTTF}_1 \simeq N_{11}^0(\infty) d_1$$

Let us assume the existence of a single absorbing state (state $l+1$):

$$N_{11}^0(\infty)\left[\sum_{\substack{\text{Reduced}\\\text{sequences}\\\text{from state 1}\\\text{to state }l+1}} \prod \begin{array}{l}\text{Asymptotic transition probabilities}\\\text{(associated with the sequences)}\end{array}\right] \simeq 1 \quad (A6.13)$$

Among the reduced state sequences, we will always consider those corresponding to the most direct paths from state 1 to state $l+1$.

MTTF$_1$ approximations, then asymptotic failure rate are deduced from this. The approximate formula developed in Section 14.4.3 for a Markovian process is thus generalized for semi-Markovian processes using this formula.

A6.4 EXAMPLES

A considerable number of the examples of Markovian or semi-Markovian processes examined in this book could be processed according to the above methods. We will develop two in this Section:

- **The Markovian process shown in Figure 14.2** (see Section 14.3.3 and 14.3.5). Assuming that $\lambda_i \ll \mu_j \ \forall \ i, j$

$$fr_1 \frac{1}{\lambda_1 + \lambda_2} \simeq 1; \quad fr_2 \simeq fr_1 \frac{\lambda_1}{\lambda_1 + \lambda_2}; \quad fr_3 \simeq fr_1 \frac{\lambda_2}{\lambda_1 + \lambda_2};$$

$$fr_4 \simeq \lambda_1 \lambda_2 \left(\frac{1}{\mu_1} + \frac{1}{\mu_2}\right)$$

whence

$$\bar{A}(\infty) = P_4(\infty) = fr_4 d_4 \simeq \frac{\lambda_1 \lambda_2}{\mu_1 + \mu_2}\left[\frac{1}{\mu_1} + \frac{1}{\mu_2}\right]$$

- **The semi-Markovian process shown in Figure 18.9** (see Section 18.6). The following equations are obtained:

$$N_{1,i+1}^0(\infty) = N_{1,i}^0(\infty) b_{i,i+1} \quad \left(b_{i,i+1} \text{ will be expressed as } b_i = \frac{\gamma_i}{\lambda_i + \gamma_i}\right)$$

$$N_{11}^0(\infty) - N_{1,n}^0(\infty) b_n = 1$$

whence

$$N_{11}^0(\infty) = \left[1 - \prod_{i=1}^n b_i\right]^{-1}$$

$$\text{MTTF} = \sum_{j=1}^n \frac{1}{\gamma_j} \frac{\prod_{j=1}^j b_i}{1 - \prod_{i=1}^n b_i}$$

If $\gamma_i \gg \lambda_i$,

$$\text{MTTF} \simeq \frac{\sum_{i=1}^{n} \dfrac{1}{\gamma_i}}{\sum_{i=1}^{n} \dfrac{\lambda_i}{\gamma_i}}$$

REFERENCE

[1] A. Villemeur (1987) *Evaluation de la fiabilité, disponibilité et maintenabilité de systèmes réparables: la Méthode de l'Espace des Etats* EDF-DER HT/50/3.

APPENDIX 7

INTERNATIONAL STANDARDS ON DEPENDABILITY

In this appendix, the international standards prepared by the International Electrotechnical Commission (IEC) are listed; usually national rules corresponding to these international standards exist on the same subjects.

IEC Publications

Reference	Date	Title
IEC 271	1974	List of basic terms, definitions and related mathematics for reliability
IEC 271 A	1978	First supplement
IEC 271 B	1983	Second supplement
IEC 271 C	1985	Third supplement
IEC 272	1968	Preliminary reliability considerations
IEC 300	1984	Reliability and maintainability management
300-3-1	1991	Analysis techniques for dependability—guide on methodology
IEC 319	1978	Presentation of reliability data on electronic components (or parts)
IEC 362	1971	Guide for the collection of reliability, availability, and maintainability data from field performance of electronic items
IEC 409	1981	Guide for the inclusion of reliability clauses into specifications for components (or parts) for electronic equipment
IEC 410	1973	Sampling plans and procedures for inspection by attributes
IEC 419	1973	Guide for the inclusion of lot-by-lot and periodic inspection procedures on specifications for electronic components (or parts)
IEC 605		Equipment reliability testing
605-1	1978	Part 1: General requirements
	1982	Part 1: Amendment No. 1
605-3-1	1986	Part 3: Preferred test conditions. Indoor portable equipment—low degree of simulation

IEC Publications

Reference	Date	Title
605-3-2	1986	Part 3: Preferred test conditions. Equipment for stationary use in weather protected locations. High degree of simulation
605-4	1986	Part 4: Procedure for determining point estimates and confidence limits from equipment reliability determination tests
	1989	Part 4: Amendment No. 1
605-5	1982	Part 5: Compliance test plans for success ratio
605-6	1986	Part 6: Test for the validity of a constant failure rate assumption
	1989	Part 6: Amendment No. 1
605-7	1987	Part 7: Compliance test plans for failure rate and mean time between failures assuming constant failure rate
	1990	Part 7: Amendment No. 1
IEC 706		Guide on maintainability of equipment
706-1	1982	Part 1: Sections One, Two and Three—Introduction, requirements and maintainability programme
706-2	1990	Part 2: Section Five—Maintainability studies during the design phase
706-3	1987	Part 3: Sections Six and Seven—Verification and collection, analysis and presentation of data
IEC 812	1985	Analysis techniques for system reliability—procedure for failure mode and effects analysis (FMEA)
IEC 863	1986	Presentation of reliability, maintainability and availability predictions
IEC 1014	1989	Programmes for reliability growth
IEC 1025	1990	Fault tree analysis (FTA)
IEC 1078	1991	Reliability block diagram method
IEC 50 (191)	1991	International vocabulary, Chapter 191: Dependability and quality of service

INDEX

Accelerated test 336, 687
Acceptable 235, 336, 687
Accident 3, 5, 6, 8, 10, 100, 106, 148, 202, 234, 336, 687
   scenario 234, 336, 371, 687
   sequence 234, 336, 687
Active component 72, 336, 687
Active corrective maintenance time 336, 687
Active maintenance time 336, 687
Active preventive maintenance time 336, 687
Active redundancy 129, 337, 688
Administrative delay (for corrective maintenance) 337, 688
Ageing failure 337, 688
Aircraft 8, 102, 149, 202, 370, 405, 531, 733
Artificial intelligence 593
Asymptotic availability 65, 281, 337, 688
Asymptotic failure rate 185, 293, 337, 688
Asymptotic reduced transition probability matrix 325
Asymptotic repair rate 337, 688
Asymptotic transition probability 299
   matrix 324
Asymptotic unavailability 65, 337, 688
Automaintainability 337, 688
Availability (measure) 18, 55, 63, 143, 190, 278, 307, 320, 337, 491, 688, 715
Availability (performance) 18, 337, 491, 688, 715

Basic event 157, 338, 689
Bathtub curve 24
Bayes' theorem 46, 80
Bayesian estimation 80, 479, 729
Behaviour 407
Binomial distribution 49

Block diagram 125
Boolean algebra 42, 249
Bug 338, 454, 689
Burn-in 338, 689

Capability 338, 689
Cascade failures 338, 373, 374, 375, 689
Cataleptic failure 23, 25, 338, 689
Catastrophic event 23, 25, 338, 689
Catastrophic failure 23, 25, 338, 689
Cause 26, 147
Cause–consequence diagram 264, 338, 689
   method 264, 339, 608, 610
Cause tree 147, 149, 339, 551, 587, 690
   method 147, 149, 249, 339, 388, 606, 690
Chi-square distribution 53, 91
Coherent structure function 180, 339, 690
Coherent system 173, 180, 339, 690
Command failure 26, 162, 339, 375, 690
Common cause failures 261, 339, 367, 564, 690
Common mode failures 339, 373, 374, 690
Complete failure 23, 340, 691
Complexity measures 471
Compliance test 340, 691
Component 19, 71, 340, 691
Conditional events 45
Conditional failure intensity 58, 741
Confidence interval 77, 340, 691
Consequences (of an event) 25, 234, 340, 691
Consequence tree 234, 331, 340, 691
   method 234, 340, 390, 560, 608, 691
Constant failure rate period 24, 340, 691
Convolution 68
Corrective maintenance 340, 691
   time 340, 691

Covariance 340, 691
Critical component 341, 692
Critical event 25, 341, 692
Criticality analysis 119, 341, 692
Cumulative distribution function 48, 341, 692
Cut set 169, 341, 692

Data bank 71, 85, 437
Decision table 341, 692
  method 341, 583, 692
Deductive approach 34, 209, 237, 341, 610, 692
Defect 30, 341, 692
Deferred maintenance 341, 692
Degradation failure 23, 341, 692
Dependability
  assessment 17, 141, 342, 457, 497, 549, 579, 605, 613, 621, 693
  assurance 342, 620, 623, 693
  audit 342, 620, 693
  control 342, 619, 693
  improvement 342, 693
  management 342, 619, 693
  model 37, 342, 615, 693
  plan 342, 693
  program 342, 623, 693
  surveillance 342, 693
  target 529, 624
  verification 343, 486, 617, 694
Dependent failures 258, 343, 367, 694
Determinate fault 343, 694
Determination test 343, 694
Down time 343, 694
Durability 19, 343, 694
Duration 283, 343, 694

Early failure 24, 343, 694
  period 24, 343, 694
Effect 106, 111, 218, 223, 343, 694
Effectiveness (performance) 343, 694
Eigenvalues 279
Endurance test 344, 695
Entity 344, 695
Erlang distribution 53
Error factor 51, 92, 93, 94
Error mode 344, 695
Estimated 344, 695
  value 344, 695
Event tree 234
Exclusive events 42
Expectation (of a random variable) 48, 344, 695
Expert judgment 83, 437, 593

Expert systems 593
Exponential distribution 50
External gathered fault 205
Extrapolated 345, 696
  value 345, 696

Failed state 275, 345, 696
Fail safe 345, 696
Failure 22, 345, 696
  analysis 345, 392, 696
  cause 26, 345, 696
  density 57, 131, 345, 696
  intensity 58, 187, 323, 345, 696
    acceleration factor 345, 696
  mechanism 345, 696
  mode 28, 108, 111, 203, 345, 465, 696
Failure modes and effects analysis (FMEA) 106, 202, 346, 388, 605, 697
Failure modes, effects and criticality analysis (FMECA) 107, 119, 346, 697
Failure rate 57, 61, 65, 74, 85, 187, 276, 346, 476, 697
  acceleration factor 346, 697
  upon demand 75
Failure simulation 589
Fatal accident rate 508
Fault 27, 346, 465, 697
  analysis 346, 697
  correction 346, 697
  coverage 346, 697
  mode 346, 697
  tolerance 347, 495, 698
  tree 148
Faulty 347, 698
Field data 71, 347, 698
Field test 71, 347, 698
Free time 347, 698
Frequency 44, 283, 328, 347, 698, 741
Functional mode 347, 698

Gamma distribution 52
Gate 152
Gathered fault combination method 111, 202, 243, 347, 390, 607, 698
Gaussian distribution 50
Generic event 236, 347, 391, 698
Global gathered fault 111, 202, 243, 347, 698
Gradual failure 23, 347, 698

Hazard 100, 348, 376, 699
Hazards and operability studies 120
Hazop 120

Human
  availability 348, 421, 699
  error 348, 375, 383, 413, 418, 699
  factors 348, 405, 699
  failure 348, 419, 699
    mode 348, 420, 699
  fault 348, 420, 699
  incapability 348, 419, 699
  maintainability 348, 421, 699
  operator 348, 405, 699
  reliability 348, 405, 699
    assessment 348, 405, 421, 699

Idle time 348, 699
Immediate, necessary and sufficient causes 159
Implicant 169, 348, 700
Importance factors 39, 725
Incapability mode 348, 699
Incident 349, 700
Inclusion–exclusion principle 177
Inductive approach 34, 209, 243, 349, 580, 610, 700
Initiating event 235, 238, 244, 246
Instant of time 349, 700
Intermediate event 161
Intermittent fault 27, 349, 700
Internal gathered fault 203, 349, 700
Irrecoverability 349, 700

Kinetic tree theory 185

Laboratory test 349, 700
Laplace transform 67, 131, 280, 293, 298, 323
Latent fault 349, 700
Logistic delay 349, 700
Lognormal distribution 51, 83

Maintainability (measure) 18, 55, 305, 307, 320, 349, 465, 491, 495, 701
Maintainability (performance) 18, 350, 465, 491, 495, 701
Maintenance 350, 701
  support performance 350, 701
  time 350, 701
Major event 25, 350, 701
Markov graph 274
Maximum likelihood estimator 477
MDT 55, 350, 701
Mean (of a random variable) 48, 350, 701
Mean availability 350, 701, 721
Mean unavailability 350, 701
Measure 36, 351, 702

Mechanics 446
Mental activity 408
Minimal cut set 137, 169, 286, 301, 351, 702
Minimal cut state 286, 301, 351, 702
Minimal failed state 304, 351, 702
Minimal operating state 295, 351, 702
Minimal sequence 249
Minimal successful path 138, 351, 702
Minimal tie set 137, 351, 702
Minor event 25, 351, 702
Monte Carlo simulation 455, 565, 566, 730
MTBF 55, 285, 351, 702, 740
MTTF 55, 61, 293, 326, 351, 476, 480, 483, 489, 508, 703, 739
MTTR 55, 285, 351, 703, 739
MUT 55, 285, 295, 302, 703, 739

Non-coherent system 173, 180, 352, 703
Non-markovian graph 275, 322
Non-operating state 352, 703
Non-relevant failure 352, 703
Non-repaired entity 352, 611, 703
Non-required time 352, 703
Non-sequential system 352, 703
Normal distribution 50
Nuclear plant 10, 234, 371, 406, 456, 514, 536, 733
Number of failures 58, 185, 326
Number of passages 326, 742
Number of repairs 59, 326

Observed value 352, 703
Operation 352, 703
Operating state 275, 352, 703
Operating time 352, 703
Operational value 352, 703

Parallel system 128, 315
Part 20, 145
Part count method 149
Partial failure 23, 352, 704
Permanent fault 353, 704
Persistent fault 353, 704
Phased mission system 189, 262, 612
Poincaré's theorem 45
Poisson distribution 49
Posterior distribution 80
Precursor event 353, 704
Predicted 33, 353, 704
  value 33, 353, 704
Prediction 33, 353, 704
Preliminary hazard analysis 101, 353, 704

Preliminary hazard and risk analysis 101, 353, 704
Preventive maintenance 353, 704
   time 353, 705
Primary failure 26, 162, 354, 375, 705
Prime implicant 169, 173, 354, 705
Prior distribution 80
Probabilistic risk assessment 10, 234, 354, 515, 523, 705
Probabilistic safety assessment 10, 354, 515, 705
Probability 44
   density function 48, 354, 705

Qualitative analysis 35, 551, 554
Quality 17, 354, 705
   assurance 355, 469, 706
Quantitative analysis 36, 177, 230, 256, 427, 553, 554

Random process 42, 53, 355, 706
Random variable 47, 355, 706
Rasmussen report 10, 234, 516, 706
Recoverability 355, 706
Reduced cause tree 169, 355, 706
Reduced consequence tree 236, 247, 355, 706
Reduced transition probability matrix 325
Reduction 247, 355, 706
Redundancy 127, 355, 368, 380, 707
Relevant failure 355, 707
Reliability (measure) 15, 17, 54, 63, 127, 187, 191, 291, 307, 320, 356, 447, 465, 470, 487, 707
Reliability (performance) 15, 356, 447, 465, 470, 487, 707
Repair 356, 707
Repair density 55, 66, 68, 356, 611, 707
Repair intensity 59, 66, 68
Repair rate 57, 65, 304, 707
Repair time 21, 22, 356, 708
Repairable entity 63, 141, 183, 278, 291, 305, 307, 611
Repaired entity 356, 707
Repairmen 356, 707
Required function 21, 22, 356, 708
Required time 356, 708
Risk 100, 356, 505, 708, 734
   acceptance 508, 511

Safety (measure) 18, 357, 504, 708, 733
Safety (performance) 18, 357, 504, 708, 733
Safety factor 447

Scenario (of events) 234, 357, 708
Scheduled maintenance 357, 708
Screening test 357, 708
Secondary failure 26, 162, 357, 375, 708
Semi-markovian graph 323, 491, 739
Sequence (of events) 234, 357, 708
Sequential system 258, 261, 357, 612, 708
Series system 127, 307
Servability performance 357, 708
Service 357, 709
   accessibility performance 19, 358, 709
   retainability performance 19, 358, 709
   support performance 19, 358, 709
Serviceability performance 19, 357, 709
Set (of events, of states) 40
Significant event 25, 358, 709
Software 461, 549
   model 462, 474
   testing 467
Solid fault 358, 709
Standard 745
   deviation 48, 358, 709
Standby component 184, 715
Standby failure rate 716
Standby redundancy 358, 709
Standby state 358, 709
Standby time 358, 709
State graph 275, 358, 709, 739
State sequence 289, 295, 358, 710
State space method 144, 261, 275, 358, 489, 562, 608, 710, 739
State variable 149
Steady-state 359, 710
   availability 359, 710
   unavailability 359, 710
Step stress test 359, 710
Stochastic process 53
Strength 447
Stress 415, 447
   analysis 359, 710
   model 85, 359, 710
Structure function 139, 179, 359, 710
Success diagram 125, 135, 359, 710
   method 125, 359, 606, 711
Success path 126, 360, 711
Sudden failure 23
System 19, 21, 360, 711
   analysis 32, 360, 711

Task 413
Technical and functional analysis 35
Test 72, 360, 711, 715, 723
   data 72, 360, 711
Three Mile Island 11, 371, 513

INDEX

Tie set 126, 137, 360, 711
Time acceleration factor 360, 711
Time interval 360, 711
Total probability theorem 46, 135
Transient fault 360, 712
Transition density 323, 360, 712
Transition probability 323, 361, 712
Transition probability matrix 323
Transition rate 276, 361, 712
Transition rate matrix 279, 292
Truth table method 141, 197, 221, 361, 607, 712,

Unacceptable 235, 361, 712
Unavailability (measure) 18, 65, 361, 712, 715
Unavailability (performance) 18, 361, 712, 715, 733
Uncertainties 37, 565, 728
Unconditional failure intensity 58, 361, 712
Unconditional repair intensity 59, 361, 713

Undesirable event 153, 362, 713
Undetected fault time 362, 713
Unmaintainability (measure) 18, 304, 362, 713
Unmaintainability (performance) 18, 362, 713
Unreliability (measure) 17, 362, 713
Unreliability (performance) 17, 362, 713
Unsafety (measure) 18, 362, 713
Unsafety (performance) 18, 362, 714
Unscheduled maintenance 363, 714
Up state 363, 714
Up time 363, 714
Useful life 24, 363, 714

Variance 48, 363, 714
Vector of basic event states 363, 714
Volatile fault 363, 714

Wear-out failure 25, 363, 714
Wear-out failure period 25, 363, 714
Weibull distribution 53